Advances in Fusion and Processing of Glass III

For information on ordering titles published by The American Ceramic Society, or to request a publications catalog, please contact our Customer Service Department at: 614-794-5890 (phone), 614-794-5892 (fax), info@ceramics.org (e-mail), or write to Customer Service Department, P.O. Box 6136, Westerville, OH 43086-6136, USA.

Visit our on-line book catalog at <www.ceramics.org>.

Volume 141

Advances in Fusion and Processing of Glass III

Proceedings of the 7th International Conference on Advances in Fusion and Processing of Glass, July 27–31, 2003 in Rochester, New York.

Edited by

James R. Varner
Alfred University

Thomas P. Seward III
Alfred University

Helmut A. Schaeffer
Research Association of the German Glass Industry
(HVG)

Published by
The American Ceramic Society
735 Ceramic Place
Westerville, Ohio 43081
www.ceramics.org

Proceedings of the 7th International Conference on Advances in Fusion and Processing of Glass, July 27–31, 2003 in Rochester, New York.

COVER PHOTO: Inline batch redox sensor positioned in the charging area of a container glass furnace is courtesy of O.S. Verheijen and O.M.G.C. Op den Camp and appears as figure 6 in their paper "Advanced Operation Support System for Redox Control," which begins on page 421.

For information on ordering titles published by The American Ceramic Society, or to request a publications catalog, please call 614-794-5890.

4 3 2 1–07 06 05 04

ISSN 1042-1122
ISBN 1-57498-156-0

In Memoriam

With great sadness we note that Dr. Shen Lin Chang, of the Argonne National Laboratory, and Dr. Robert E. Moore, of the University of Missouri-Rolla, died within a few months preceding the conference. Dr. Chang had been an invited speaker and Dr. Moore a co-author on two of the papers and a research site director for the Center for Glass Research, one of the conference sponsoring organizations. While this was a sad loss for the conference, it paled in comparison to the loss felt by the entire glass and ceramic technical community and the families of these highly liked and respected men. We offer our condolences to all who have been associated with them.

Contents

Advances in the Glass Melting Process

Characterization of Glass Melts/Glass Melt Properties

Materials for Glassmaking

Advances in Glass Forming

Polyvalent Elements and Redox Behavior

Effects of Composition and Forming on Structure and Properties

Emissions, Recycling and Other Environmental Issues

Computer Modeling and Process Control

Secondary Processing

Preface

Glass continues to be a material of great scientific and technological interest. The economic pressures on the glass industry, the emphasis on global markets, and the worldwide attention to energy and environmental conservation that emerged in the late nineteen eighties when this conference series originated have not lessened, but rather increased. So has the need for an international forum where scientists, engineers, technologists and manufacturers can discuss and seek new solutions to the challenges of glass manufacturing, particularly as they pertain to melting and forming. Hence, the seventh in a series of international conferences "Advances in Fusion and Processing of Glass" was held in Rochester, NY, July 27 - 31, 2003.

Session topics included Advances in the Glass Melting Process; Characterization of Glass Melts/Melt Properties; Materials for Glassmaking; Advances in Glass Forming; Advances in Glass Surface Engineering; Polyvalent Elements and Redox Behavior; Effects of Composition and Forming on Structure and Properties; Emissions, Recycling and other Environmental Issues; Computer Modeling and Process Control; Secondary Processing; and Non-Traditional Processing. This proceedings volume contains a compilation of 47 of the papers and posters presented.

Dr. L. David Pye, Professor of Glass Science and founding director of the Center for Glass Research inaugurated this series of international conferences on glass in 1988 at Alfred University, Alfred, NY. With the strong support of the Research Association of the German Glass Industry (HVG) and the German Society of Glass Technology (DGG), the original conference has developed into a series alternating biennially or triennially between the United States (North America) and Germany. This conference, organized and co-chaired by Dr. James R. Varner and Dr. Thomas P. Seward, professors in the School of Ceramic Engineering and Materials Science at the New York State College of Ceramics at Alfred University (AU), and Dr. Helmut Schaeffer, director of the Research Association of the German Glass Industry (HVG) and the German Society of Glass Technology (DGG), attracted more than 140 participants from fourteen countries.

The conference concluded with a one-day workshop sponsored by the Glass Manufacturing Industry Council (GMIC), the NSF Industry/University Center for Glass Research (CGR) and the US Department of Energy Industrial Technologies Program (DOE-IPT) on "Evolutionary and Revolutionary Strategies for Keeping Glass Viable through the 21st Century." The workshop was chaired by Michael Greenman (GMIC Executive Director) and Thomas Seward (CGR Director, AU) and included a panel discussion facilitated by Dr. James Seebold (Chevron-Texaco - retired). Topics identified for further attention include improved fining and conditioning of glass, better process sensors, funding needed for research by the glass industry, mechanisms for achieving that funding, improvements in international communication and collaboration, and environmental and regulatory opportunities. While these discussions were extremely dynamic and thought provoking, the nature of the material was beyond the scope of what could be included in this volume.

Past conference dates and locations were: 1988, June 14-17, Alfred, NY, USA; 1990, October 22-25, Düsseldorf, Germany; 1992, June 10-12, New Orleans, LA, USA; 1995, May 22-24, Würzburg, Germany; 1997, July 27-31, Toronto, Canada; 2000, May 29-31, Ulm, Germany; 2003, July 27-31, Rochester, NY, USA; In Rochester, Dr. Schaeffer announced that the eighth conference in the series would be held in Germany in 2006.

James R. Varner

Thomas P. Seward III

Alfred, NY

—

Helmut A. Schaeffer

Offenbach, Germany

Acknowlegements

The conference organizers and proceedings editors would like to thank the organizations that helped sponsor the events: The New York State College of Ceramics, the School of Ceramic Engineering and Materials Science and the NSF Industry-University Center for Glass Research at Alfred University, the Glass Manufacturing Industry Council, and the U.S. Department of Energy Industrial Technologies Program, in cooperation with the Research Association of the German Glass Industry (HVG) and the German Society of Glass Technology (DGG). Thank you also to Dr. Oleg A. Prokhorenko of the Laboratory of Glass Properties, International, located in St. Petersburg, Russia, and Dearborn, Michigan, for helping sponsor social events related to the conference.

On July 1, 2003, Alfred University reorganized its engineering degree programs into a single School of Engineering. Because the School of Ceramic Engineering and Materials Science was deeply involved throughout the planning of this conference, we chose to acknowledge its sponsorship here, even though it technically no longer existed at the time of the conference.

The conference organizers would also like to thank the international team of session chairs who maintained the smooth flow of presentations and discussions: Mark Allendorf, Sandia National Laboratories, USA; Ian Pilkington, Pilkington plc, UK; Henry Schreiber, Virginia Military Institute, USA; Alain DeMeringo, St. Gobain Recherche, France; George Pecoraro, PPG Industries (retired), USA; Reinhard Conradt, RWTH Aachen Institut für Gesteinshüttenkunde, Germany; Alexis Clare, Alfred University, USA; Oleg Prokhorenko, Laboratory of Glass Properties, Russia and USA; Warren Wolf, WW Wolf Services (Owens Corning, retired), USA; and Hayo Müller-Simon, Hüttentechnische Vereinigung der Deutschen Glasindustrie, Germany.

The editors especially thank our Conference Coordinator Mrs. Marlene Wightman and her assistant Mrs. Barbara Timbrook, without whom little would have been accomplished. Their long hours and tireless effort ensured the success of the conference and their diligent correspondence with authors and organization of manuscripts facilitated a timely publication of the proceedings.

James R. Varner

Thomas P. Seward III

Alfred, NY

Helmut A. Schaeffer

Offenbach, Germany

Advances in the Glass Melting Process

ANALYSIS OF ADVANCED AND FAST FINING PROCESSES FOR GLASS MELTS

Ruud G.C. Beerkens,
Eindhoven University of Technology
TNO Glass Group
Rondom 1, P.O. Box 595
Eindhoven 5600 AN, The Netherlands
beerkens@tpd.tno.nl

ABSTRACT

The removal of dissolved gases and gas bubbles from viscous glass melts is one of the main process steps decisive for the glass quality and for the required residence time of the glass melt in glass furnaces. Models, describing the transfer of gas species from the melt phase to bubbles or reverse, can be applied to study the effect of several process parameters on the removal of gas bubbles from glass melts in industrial processes. Such parameters are: (1) the height of the tank in the fining section; (2) the temperature level; (3) oxidation state of the glass; (4) concentrations of fining agents added to the batch; (5) water content of the melt, which is closely related to water vapor pressure in furnace atmosphere; (6) application of sub-atmospheric furnace pressures; (7) pre-conditioning of the melt by other gases. Examples of the effect of oxidation state of the melt, preconditioning of the melt by helium gas, reduced furnace pressure and amounts of fining agents, or a combination of two fining agents on the removal of small seeds (fine bubbles) will be presented. Results of some fining tests are presented, showing the effect of additions of polyvalent ions to sulfate refined melts and the effect of pre-conditioning of the glass melt with highly soluble gas species on the efficiency of seed removal.

INTRODUCTION

The conventional way of fining is based on the addition of a certain amount of a compound or a combination of compounds, which start to decompose after exceeding a certain fining-onset temperature of the melt. At this temperature the total pressure of the gases dissolved in the melt exceeds the internal bubble pressure, which is slightly above 1 bar (1-1.5 bar). The bubbles will steadily grow.

Fining Process

During fining, liberated dissociation gases from fining agents will diffuse into existing bubbles and will also aid the stripping of other dissolved gases from the melt [1]. Decomposition of sulfate in glass melts, forming the fining gases oxygen and sulfur oxide, is the basis for the most well-known and probably most important fining process in the glass industry. A small seed or bubble hardly ascends in the highly viscous melt and the seed will not grow or shrink when an equilibrium exists between the partial vapor pressures of the gases in the bubble and the same types of gases dissolved in the molten glass.

As soon as the fining gases, such as oxygen and sulfur dioxide, dilute the gases originally present in the bubble, like nitrogen or carbon dioxide, the equilibrium between these gases dissolved in the melt and present in the bubble will be disturbed. The partial pressures of the original gases in the bubbles will decrease, and this leads to diffusion of the gases from the melt into the growing bubbles. During primary fining, bubbles grow, and consequently the bubble ascension rates in the melt increase. Because of the dilution of the gas in the bubbles by fining gases, a large part of the gas species, dissolved in the molten glass, will simultaneously be stripped from the melt.

In the past, many mathematical models have been developed for the quantitative description of (a) bubble growth, (b) bubble ascension, and (c) diffusion of gases into bubbles during the fining process at high temperature or from the bubbles into the melt during the refining process in glass melts. Detailed models for the transport of dissolved gases in the glass melt have been presented, solving the differential equations, describing the diffusion processes [2-9] of dissolved gas species in the melt surrounding a bubble. Other models, that describe the transfer of gases by using relations based on semi-empirically derived mass transfer coefficients [10-17], have been applied in the past to describe bubble growth or shrinkage in glass melts. However, only a few models take into account the effect of chemical reactions that produce gas in the melt. This is essential, because fining reactions in the melt release gases, which will determine bubble growth rates. The equilibrium constant of the decomposition reaction of the fining agent (for instance, sodium sulfate, Sb_2O_5, CeO_2, or As_2O_5) in the melt is strongly temperature dependent.

A detailed fining model will be presented. This model derives the gas production from fining agents and polyvalent-ion oxidation-reduction reactions dependent on temperature, oxidation state, and contents of fining agents. The model also calculates the diffusion of these fining gases and diffusion of other relevant gas species (Ar, H_2O, He, SO_2, O_2, N_2, CO_2) in the melt, from or into the gas bubbles. The model can estimate or predict the change of the size and composition of gas bubbles, depending on temperature, concentrations of dissolved gases in the melt, glass melt depth, concentration of fining agent(s), furnace pressure and oxidation state of the melt.

Almost all the industrial melting processes are performed at atmospheric pressure with hardly any physical aids for fining. The model presented in this study is also used to investigate the effect of physical means to enhance conventional fining processes in glass furnaces. Among these methods are sub-atmospheric fining [18-20], gas injection enhanced fining, application of a fining bank (fining with smaller tank depth) [21], and acoustic waves [22-29].

The effect of the application of these methods on the fining process will be described, and results of model calculations will show the potential of some of these fining enhancement methods. The model will be explained in the next section.

Some results of experiments on a laboratory scale will be presented to show the effect of advanced fining methods on the bubble removal efficiency.

Gas-injection-enhanced fining, sub-atmospheric fining, and application of a fining bank are promising ways to improve the removal rate of bubbles. The bubble simulation models will help the technologist to find the optimum process settings, or they can be used to optimize the concentrations of fining agents in the batch. Fining agents can be rather expensive or may lead to gaseous emissions into the environment (e.g., SO_2 during sulfate fining). Large additions of fining agents may also cause undesired foaming on top of the melt in the fining zone of the glass melt tank. Therefore, it is important to optimize the concentration level of these raw materials in industrial glass making.

Fining model

After the fusion or dissolution of most batch components, the glass melt contains dissolved gases and gas bubbles in sizes varying between about 20 micrometers up to several millimeters [30, 31]. The concentrations of dissolved gases in the fresh melt are still relatively high. The most important gases in the melt or in the gas bubbles are nitrogen, CO_2, water vapor, oxygen, SO_2, argon, and sometimes NO or CO. The bubbles originating from the batch melting reactions generally show high CO_2 contents.

The bubbles can be removed by two different processes--bubble ascension to the glass melt surface plus bubble collapse at the surface of the melt, or complete re-absorption of the gases by the melt. The first mechanism is called primary fining. During primary fining at temperatures exceeding the fining onset temperature, the fining gases diffuse steadily into the existing bubbles. The total equilibrium pressure of all dissolved gases in the melt exceeds the internal bubble pressure. The bubbles will grow continuously, and the ascension rate will increase progressively.

The rising velocity 'v' of a bubble or change in distance from the surface of the melt is given by the formula derived from Stokes' law:

$$v = dh/dt = - c \cdot \rho \cdot g \cdot R^2 / \eta \qquad (1)$$

The bubble ascension is proportional to R^2 and $1/\eta$. The value of c is in the range of 2/9 to 1/3, depending on the rigidity of the interface between the bubble –and the glass melt [32].

There is an equilibrium between the vapor pressure in the gas phase (p_i) and the concentration in the melt phase at the interface between the bubble and the melt (C_{ii}), for each gas species i:

$$p_i = C_{ii}/ L_i \qquad (2)$$

For gases dissociating in the melt, p_i may be not linearly proportional to C_{ii} and can be derived from: $p_i = C^n_{ii}/L_i$ (n is number of ions in the melt formed from one gas molecule,; for example, for H_2O dissolving in the melt, two OH^- ions are formed).

The gas species can dissolve chemically by reacting with glass melt components, or physically by occupying open sites in the structure of the melt without chemical bonding.

In general, the chemical solubility values of gases in glass melts are orders of magnitude higher than the physical solubility levels of a gas. However, the chemical solubility is much more temperature dependent. For most gas species, the maximum solubility of a glass melt can be determined as a function of temperature. The glass melt can be saturated by a gas species (at defined partial pressure) at a certain temperature. After the saturation procedure, the gases are completely stripped from the melt by helium bubbling or imposing a sub-atmospheric pressure. The quantity of dissolved gas stripped from a glass sample can be determined by means of mass spectrometry or gas chromatography [33, 34].

A special gas is the so-called fining gas. This gas dissolves chemically at low temperatures. Examples are SO_2 gas and oxygen. A fining agent dissolved in the molten glass releases these gases at increasing temperatures. For instance, sulfates produce SO_2 and oxygen gas. Goldman [35] investigated the sulfur reactions and sulfur solubility in glass melts as a function of the oxidation state of the melt. Other polyvalent ions present in the glass melt in the most oxidized state will be reduced by temperature increments, accompanied by the release of oxygen gas; for instance, $Sb^{5+} + O^{2-} \rightarrow Sb^{3+} + 1/2\,O_2$.

We need information on the reaction equilibrium constants of reduction-oxidation reactions and fining reactions, such as for the decomposition of sulfates [36], to determine the concentrations or partial equilibrium pressures of released fining gases in the melt. Löh et al. [37] showed the possibility of following the partial oxygen pressure in equilibrium with physically dissolved oxygen in a glass melt during the fining process by using an electrochemical oxygen sensor. Information on the oxygen activity or pressure can be used to study the decomposition of sulfates in the molten glass.

Since bubble ascension depends on the bubble size ($v \sim R^2$), the bubble growth strongly accelerates the primary fining process. Higher temperatures will lower the glass melt viscosity, and they will increase the amount of fining gases released by a fining agent. Thus, at increasing temperatures, the bubble ascension rate (see equation 1) will increase.

The total bubble pressure p_t is given by:

$$p_t = \Sigma p_i = p_o + \rho \cdot g \cdot H + 2\sigma/R \qquad (3)$$

The bubble pressure decreases as the bubble grows (R increases) and as the bubble ascends in the melt (H decreases).

The equilibrium pressure of all the fining gases dissolved in the melt taken together, may exceed the value of p_t, above a certain temperature level. The bubble will continuously grow, because equilibrium between the gases dissolved in the melt and in the bubble cannot be reached. The total equilibrium pressure of the dissolved gases remains above p_t. At this temperature level, the removal of gases becomes very efficient. This temperature is called the fining onset temperature T_{onset}.

When the concentration of a dissolved gas species is not in equilibrium with this gas entrapped in a bubble, an exchange of this gas (gas transport) will take place between the gas bubble and the melt.

Gas formation by oxidation-reduction reactions

It is essential to find the concentrations of the dissolved gases produced by chemical reactions in the melt. In this model, local chemical equilibrium conditions are assumed for the reduction-oxidation reactions and fining reactions. Therefore, it is likely that the oxygen pressure of the melt locally is in equilibrium with the different valency states of the polyvalent ions, for instance, the ferric-ferrous iron ratio ($[Fe^{3+}]/[Fe^{2+}]$) present in the melt. This is a reasonable assumption, since the diffusion rates are much slower than the reaction kinetics. The chemical equilibrium constants of these redox reactions depend on the glass composition and temperature. To determine the amount of gases released by these equilibrium reactions after a temperature change, we need to know the initial concentrations or vapor pressures of the gases (such as SO_2 and O_2) in the melt and the equilibrium constants as a function of temperature.

After calculation of the concentration of dissolved gases in the melt, the diffusion for each gas species into a bubble can be calculated from mass transfer relations, presented later in this paper. The increase of the quantity of gas in the bubble can be determined for each gas type. The gas volume and bubble pressure (equation 2) in the bubble can be calculated from the total number of gas molecules in the bubble and from the temperature. The bubble ascension rate in the melt can then be determined by equation 1.

Determination of the concentration of fining gases in the melt

The fining agent reacts in the melt during temperature changes. For most fining agents such a reaction concerns a reduction from a highly oxidized to a more reduced state. A redox equilibrium reaction of a polyvalent ion species M_j (like Fe, As, Sb, Ce, or Mn) can be expressed [38] as:

$$M_j^{nj+kj+} + kj/2\ O^{2-} \Leftrightarrow M_j^{nj+} + kj/4\ O_2 \tag{4}$$

Assuming a constant oxygen ion activity of the melt, we can define K_j:

$$K_j = [M_j^{nj+}] \cdot [O_2]^{kj/4}/[M_j^{nj+kj+}] = [M_1^{nj+}] \cdot L^P_{O2}{}^{kj/4} \cdot pO_2^{kj/4}/[M_j^{nj+kj+}] \tag{5a}$$

$$\text{or } K_{pj} = K_j / L^P_{O2}{}^{kj/4} = [M_j^{nj+}] \cdot pO_2^{j/4} / [M_j^{nj+kj+}] \tag{5b}$$

For instance, for Sb^{5+} and Sb^{3+}: $K_{pSb} = [Sb^{3+}] \cdot pO_2^{1/2}/[Sb^{5+}]$

Data for the physical solubility of oxygen in the glass melt, L^P_{O2}, should be available as a function of temperature, see annex 1, to determine K_{pj} from K_j.
For sulfate decomposition we can write:

$$SO_4^{2-} \Leftrightarrow SO_3^{2-} + 1/2\ O_2 + O^{2-} \tag{6}$$

For constant glass composition and oxygen ion activity:

$$K_s = [SO_3^{2-}] \cdot [O_2]^{1/2} / [SO_4^{2-}] \tag{7}$$

$[SO_3^{2-}]$ = chemically dissolved SO_2 (as sulfite, in which sulfur is present as S^{4+})

$$\text{or } K_{ps} = pSO_2 \cdot pO_2^{1/2} / [SO_4^{2-}] \tag{8}$$

K_s/K_{ps} is $L^c_{SO2} \cdot L^P_{O2}{}^{1/2}$. L^c_{SO2} is the chemical solubility of SO_2 (as sulfite) in the glass:

$$\text{At equilibrium, } pSO_2 = [SO_3^{2-}] / L^c_{SO2}$$

Setting the partial equilibrium pressure of the oxygen in the melt or physically dissolved oxygen concentration $[O_2]_{T_0}$ at a certain value at a starting temperature T_0, the concentrations of the polyvalent ions in the melt at T_0 can be calculated from known values of $K_{pj}(T_0)$ and $K_{ps}(T_0)$ and from the added total concentrations of the polyvalent elements. For instance, in oxidized melts (no sulfide formation):

$[SO_4^{2-}] + [SO_3^{2-}] = C_s$ (total concentration of S in mole/m^3 melt)
$[M_j^{nj+}] + [M_j^{nj+kj+}] = C_j$ (total concentration of element j in mole/m^3 melt)

The physically dissolved oxygen concentration $[O_2]$ can be determined from the oxygen pressure and physical solubility of oxygen.

An increase in temperature will lead to a change of $K_{pj}(T)$ and $K_{ps}(T)$. In case of no direct evaporation loss of polyvalent elements from the melt before fining, the values for C_s and C_j are constant.

For each element we have at least one equilibrium constant equation, one element balance equation, and two states. For elements with three oxidation states, at least two reaction equilibrium constants have to be determined. Thus, for each temperature change or during temperature changes, the concentrations of the polyvalent ions in their different valency states, plus the concentrations of sulfite, sulfate, oxygen gas and sulfur dioxide gas in the melt, can be determined from the known element balances and equilibrium constants [39].

The difference between the concentrations of the oxidized ions at temperature T compared to T_o is the same as the difference between the concentrations of the reduced ions at T_o compared to T. This difference determines the extra concentration of oxygen formed $\Delta[O_2]$ in the melt, due to the temperature increase. The difference in sulfate concentration leads to a change in the concentration of dissolved SO_2 or sulfite in the melt.

$$1/2\Delta[SO_3^{2-}] + \Sigma_j\, k_j/4 \cdot \Delta[M^{nj+}] = \Delta[O_2]$$
$$[O_2]_T = [O_2]_{To} + [O_2] \tag{9}$$

and (neglecting the very small amount of physically dissolved SO_2):

$$-\Delta[SO_4^{2-}] = \Delta[SO_3^{2-}] = \Delta(L_{SO2} \cdot pSO_2)$$
$$[SO_3^{2-}]_T = [SO_3^{2-}]_{To} + \Delta[SO_3^{2-}] \tag{10}$$

Important for the bubble growth model are the parameters $[O_2]_T$ and the chemically dissolved sulfur dioxide $[SO_3^{2-}]_T$. These concentrations of dissolved fining gases will drive the fining process and bubble growth. After a temperature increase, these concentrations can be determined using the method described above using the reaction equilibrium constant data (annex 1); the element mass balances; and, for example, solubility data given in annex 1 for a float-glass melt.

THE MULTI-COMPONENT DIFFUSION PROCESS

The bubble may grow or shrink by multi-gas-species diffusion. This is described by the following relation for the change of the total moles in a bubble (ideal gas):

$$\frac{d}{dt}\left[(4\pi R^3 \cdot p_t)/(3R_g \cdot T)\right] = 4\pi R^2 \cdot \Sigma_i\, \frac{Sh_i}{2} \cdot D_i \cdot (C_{si} - C_{ii}) \cdot \left(\frac{1}{R} + \frac{1}{\sqrt{\pi \cdot \dfrac{Sh_i}{2} \cdot D_i \cdot t}}\right) \tag{11}$$

The Sherwood number (Sh_i) for mass transfer takes into account the effect of convection of the molten glass (relative to the ascending gas bubble) on the mass transfer rate of gas into or from the bubble. This convection is caused by the flow pattern of the glass melt (there are velocity gradients in the melt) itself and the ascension of the bubble relative to the surrounding melt, due to buoyancy. The difference between the average flow velocity of the bubble and the surrounding melt determines the value of Sh_i. The value of Sh_i is determined by glass melt flow relative to the bubble, which influences the concentration profiles of the dissolved gases in the melt.

For most gases, if $C_{si} > C_{ii}$, bubble growth generally will take place; but, for most gases, when $C_{si} < C_{ii}$, bubble shrinkage (re-absorption) will occur.

At high temperatures, the fining gas solubility in the melt decreases (the value of C_{si} for the physically dissolved fining gas i will increase), and the bubbles will grow. During cooling, the solubility of the fining gases increases again, and the value of C_{si} for the fining gases may become much smaller than C_{ii}. The re-absorption of the fining gases or other gases, which can dissolve in the melt at decreasing temperatures, is called the refining process. Both the primary fining process at high temperatures as well as the bubble dissolution process during controlled cooling of the melt can be simulated with the same model.

The calculation of gas exchange between the bubbles and the melt and bubble ascension rate by equation 11 requires initial conditions: Initial bubble composition and size; position of bubble in the melt; initial concentration level of dissolved gases in the melt; concentration of fining agent, and the oxidation state (the oxidation state determines the decomposition equilibrium of the fining reaction) or oxygen equilibrium pressure at a defined / fixed temperature T_o.

Very important data required for the modeling of the bubble behavior in glass melts are the diffusion coefficient (D_i) values, the gas solubility (L_i) values, and fining reaction equilibrium constants. These values are often strongly temperature and glass composition dependent. For the gases nitrogen, helium, and oxygen, the diffusion coefficient is measured from the mass transfer of physically dissolved gases. The diffusion coefficient for dissolved SO_2 in the soda-lime-silica melt is estimated from [16]. For CO_2 and H_2O, the chemical solubility and diffusion of the chemically dissolved species is determined. Unfortunately, for most glass types, accurate experimentally derived values of these properties are missing. Estimated values from literature and our own measurements are given in annex 1 for soda-lime-silica compositions, more specifically, a typical commercial float-glass composition.

Important is the derivation of the Sherwood number in equation 11. Here, the Sherwood number for a bubble with diameter 2R is the ratio between the diameter of the bubble and the Nernst concentration boundary layer thickness, δ_{Ni}:

$$2R/Sh_i = \delta_{Ni} = (C_{si}-C_{ii})/(\delta C_i/\delta r)_{r=R} \qquad (12)$$

The Sherwood number can be applied to describe mass transfer processes at quasi-stationary conditions. However, since in the initial stages (low value for

$$\sqrt{\pi \cdot \frac{Sh_i}{2} \cdot D_i \cdot t}$$

time t) the concentration profiles of dissolved gases in the melt surrounding the bubble are considered not to be stationary, the last term with the parameter t is introduced in equation 11.

The Nernst boundary layer thickness or the value of $(\delta C_i/\delta r)_{r=R}$, depends on three phenomena: (1) The time (a concentration profile will be established during the diffusion process; the expression given above takes this effect into account); (2) the flow of the melt relative to the bubble, due to the ascension of the bubble in the melt and to the velocity gradients in the melt caused by convection processes on a macro-scale; and (3) chemical reactions in the boundary layer as described by Beerkens and de Waal [40]. Chemical reactions, such as redox reactions in the melt surrounding the bubble, will decrease the boundary layer thickness and increase the value of the Sherwood number for the reacting gas species.

The Sherwood number depends on the velocity of the melt relative to the bubble. This velocity v is given for laminar flow surrounding a spherical subject [40]:

$$Sh_i = 1 + (1+Re\cdot Sc_i)^{1/3} = 1+ (1+ 2\cdot v\cdot R/D_{c,i})^{1/3} \qquad (13)$$

Here Re is the Reynolds number related to the bubble diameter and velocity v of the melt relative to the bubble, and Sc_i is the Schmidt number for gas species i dissolved in the melt. Beerkens and de Waal [41] showed that, for reactive gases, instead of the diffusion coefficient of a gas species i in the melt, the value of $D_{c,i}$ should be substituted in the relation for the boundary layer thickness or Sherwood relation, as given in equation 13. Generally, for reacting species $D_{c,i} \ll D_i$, because chemical equilibrium reactions in the boundary layer will counteract the change (flattening) of the concentration profiles (and the concentration profile becomes a profile belonging to a lower diffusion coefficient) by diffusion, and consequently the value of $(\delta C_i/\delta r)_{r=R}$ is much higher with chemical reactions of gas i in the melt than without chemical reactions, thus the boundary layer thickness will decrease, and the Sherwood number will increase. The value of $D_{c,i}$ for oxygen can be estimated from the concentration of polyvalent elements in the melt and the redox reaction equilibrium constants [41]. Here, it is assumed that the boundary layer thickness for SO_2 dissolved in the melt is equal to the boundary layer thickness of oxygen, since the concentration of SO_2 and O_2 in sulfate-fined melts are coupled by the sulfate decomposition reaction [36]. For gases which will not be involved in redox or fining reactions, the diffusion coefficients should be substituted for $D_{c,i}$ in equation 13. In equation 11, D_i is the diffusion coefficient of gas i in the melt.

The velocity v, the average velocity of the melt relative to the bubble, in glass melt tanks is often determined by the macro-flow conditions. Its value depends on the bubble radius and the velocity gradients in the melt and the bubble ascension, due to buoyancy. In the case of no convection other than the buoyancy-driven flow of the bubble, v can be determined by equation 1.

Solution Procedure

For a pre-described temperature course, T(t), of a melt with dissolved gases such as nitrogen, water, sulfur dioxide, oxygen, carbon dioxide, and noble gases, plus addition of polyvalent elements and a fining agent to the batch, the concentration of fining gases can be calculated at each small time step dt using equations 5, 7, 9, and 10. The initial oxygen concentration in the melt, which characterizes and sets the oxidation state of the melt, should be known. For instance, the partial oxygen pressure at a temperature T_0 will determine the oxygen concentration $[O_2]$ in the melt with these fining agents at each prevailing temperature (Equation 9) after further heating or cooling of the melt. The oxygen equilibrium pressure or concentration of physically dissolved oxygen at a reference temperature characterizes the oxidation state of this melt with the prevailing composition. However, different glass compositions with the same oxygen pressure at the same temperature may show a different oxidation-reduction behavior, depending on the concentrations of polyvalent elements in the melt.

From a combination of equations 1, 3, 11, and 13, we finally can derive for the bubble size, assuming no temperature dependency of the surface tension:

$$dR/dt = (F_1 + F_2 \cdot dT/dt + F_3)/A \tag{14}$$

$$F_1 = \Sigma_i \frac{Sh_i}{2} \cdot D_i \cdot (C_{si} - C_{ii}) \cdot \left(\frac{1}{R} + \frac{1}{\sqrt{\pi \cdot \frac{Sh_i}{2} \cdot D_i \cdot t}} \right)$$

For: $t \gg 2R^2/(\pi \cdot Sh_i \cdot D_i)$ it follows: $F_1 = \Sigma_i \frac{Sh_i}{2R} \cdot D_i \cdot (C_{si} - C_{ii})$

$$F_2 = (p_o \cdot R + \rho \cdot g \cdot h \cdot R + 2\sigma) /(3R_g \cdot T^2)$$

$$F_3 = 2\rho^2 \cdot g^2 \cdot R^3 / (27 R_g \cdot T \cdot \mu)$$

$$A = (p_o + \rho \cdot g \cdot h + 4\sigma/3R) / (R_g \cdot T)$$

For each time step dt, or in discrete form Δt, the Sherwood number and mass transfer coefficient per species i can be determined and the mass transfer per gas species can be calculated from the partial pressures in the bubble and concentration values of dissolved gases in the melt. The terms A, F_1, F_2, and F_3 can be calculated for each time t, and equation 14 gives the bubble growth or $\Delta R \approx \Delta t \cdot dR/dt$. Per time step, the increase or decrease of the amount of gases per species can be derived, and the change in bubble composition is calculated. More sophisticated numerical methods use the values for the concentrations and bubble radius at $t+1/2 \Delta t$, instead of the values at time t, to estimate R at $t+\Delta t$, using an iterative procedure.

MODELING RESULTS AND ANALYSIS OF FINING PROCESSES USING THE AID OF PHYSICAL PROCESSES

The model has been applied to analyze and study the effect of different additives to the melt, like fining agents, other polyvalent elements, or reducing agents and to investigate the impact of sub-atmospheric pressure, helium bubbling, or the water-vapor content of the melt on the removal of small seeds (primary fining process) from a glass melt. In this study, the base case refers to a float-glass melt using sodium sulfate as a fining agent. The glass is molten in a natural–gas,air-fired furnace, which typically brings about 40-50 moles water/m³ melt in the glass [42].

Level of Addition of Fining Agent: Sodium Sulfate

An increase in the concentration of sulfate in the melt will lower the fining onset temperature. The total partial pressure of the fining gases, $pO_2 + pSO_2$, will become higher at an increased value of $[SO_4^{2-}]$ at the same temperature (K_{ps} is constant, see equation 8); or in other words, when increasing the sulfate content in the melt, the fining gases reach a total pressure above 1 bar at a lower temperature. For a very oxidized float glass, the total fining gas pressure is given by Figure 1 as a function of temperature for three sulfate concentration levels in a dry melt and in a melt with 40 mol water/m³.

The bubble growth rate and the bubble ascension in a melt have both been predicted using the previously described model. A typical depth of a glass melt bath is on the order of 1 meter. It is the task of the primary fining process to remove small nitrogen or CO_2 seeds, typically on the order of $R_o=0.025$ up to 0.15 mm.

Figure 2 shows the bubble growth rate for three sulfate concentration levels at a temperature of 1723 K of a CO_2-seed with an initial radius of 0.1 mm starting at the bottom of a tank with 1-meter depth. The oxidation state is

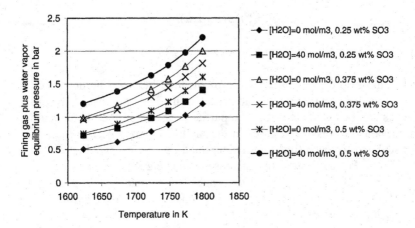

Figure 1. Pressure of fining gases (including water vapor) in float-glass melt, dependent on the total sulfur concentration (given in weight % SO₃) for two different water concentration levels.

characterized by an oxygen equilibrium pressure of the melt at 1473 K. This oxygen pressure is set at 100 Pascal in this demonstration case. The effect of the amount of sulfate in the melt on bubble growth, as expected, is very distinct. Especially at low sulfate levels, the extra addition of fining agent is very effective.

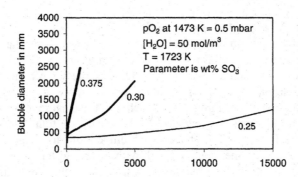

Figure 2. Bubble growth during primary fining in float-glass melt for three sulfate concentration levels at 1723 K and at mildly reducing conditions.

Oxidation State of the Melt

Addition of reducing components, such as coke or organic material, to the batch will lower the oxidation state of the melt. The pO_2 value (oxygen partial pressure in equilibrium with dissolved oxygen in the glass melt) will decrease. Referring to reaction equation 6, this means that this reaction shifts to the right side; and, although the pO_2 value will be low, the pSO_2 value will become higher. Since the sulfate decomposition reaction forms 2 SO_2 molecules per one O_2 molecule, the effect of reducing agents at a higher pSO_2 value is larger than the effect at a lower pO_2 level. This means that, for the same sulfate concentration in the melt, the fining starts at a lower temperature when adding reducing components. At the same temperature and with the same sulfate content, increasingly reducing conditions will increase the pressure of the fining gases. This will enhance the release of fining gases, and this may improve the fining process.

Figure 3 shows the effect the oxidation state of the float-glass melt on the growth of a bubble. Clearly, the re-absorption of oxygen by reducing components in the melt will shift the sulfate decomposition reaction to the gas-producing side and will increase the partial pressure of SO_2 in the melt, leading to an increased growth at lower oxidation states.

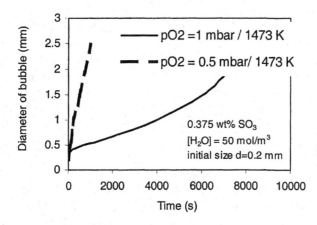

Figure 3. Bubble growth in float-glass melt for two different glass-melt oxidation states characterized by pO_2 (mbar) at 1473 K.

Figure 4 shows the calculated time required for the fining process in order to bring the small seed from the bottom of the melt to the surface at different temperatures as a function of the oxidation state of the float-glass melt.

Indeed, in industrial practice sometimes coke or other reducing agents are added in small quantities to improve the fining process. However, a larger addition will lead to a higher concentration of ferrous iron (Fe^{2+}) in the glass, causing a greenish color. Very large coke additions will even lead to sulfide formation, and the sulfate level will decrease rapidly; the glass may become amber colored.

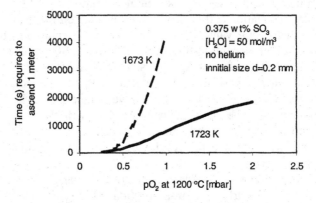

Figure 4. Effect of glass-melt oxidation state on the removal time for a small bubble during primary fining of a float-glass melt in a 1-meter-deep tank (ascension from bottom to surface).

Addition of Other Polyvalent ions

The level of decomposition of sulfate can be influenced by the addition of other polyvalent compounds or elements. In the case that the standard reaction enthalpy change (ΔH^o) of the redox reaction of such an additional redox couple is much lower than for the sulfate decomposition reaction, the oxygen liberated from the sulfate, after increasing the glass-melt temperature, is absorbed by the redox couple with the lower enthalpy change. The extra amount of oxygen liberated by the sulfate is partly re-absorbed by the other redox couple. Thus, reaction 6 will shift to the right side, forming sulfite (SO_3^{2-}).

The addition of iron oxides in the melt will hardly influence the release of oxygen and sulfur dioxide from the sulfate during the sulfate fining stage, according to the modeling study. The standard enthalpy change for the production of 1 mole of oxygen from the reduction of ferric iron (206000 Joules per 0.5 mole O_2) is not much lower compared to the standard enthalpy change for the reduction of sulfate (271000 Joule per 0.5 mole oxygen). However, when adding cerium oxides to the melt, we observe a larger impact (60000 Joules per 0.5 mole O_2 production) on the SO_2 release from sulfate in the melt. The sulfate decomposition is enhanced by the cerium-oxide redox reactions. This means, that at increasing temperatures, the presence of cerium oxide in the melt will stimulate sulfite and SO_2 formation during primary fining.

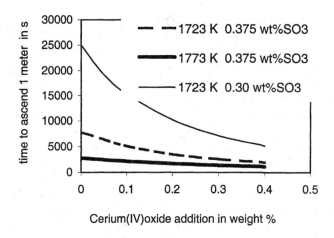

Figure 5. The effect of cerium-oxide addition to a sulfate-containing float-glass melt on the required primary fining time at two different temperatures and sulfur (sulfate) concentration levels.

This can offer the possibility of lowering the residence time in the fining zone or lowering the temperature during fining.

Figure 5 shows the effect of cerium-oxide additions to the batch on the required primary fining time for a float-glass melting tank with a depth of 1 meter for two different temperatures and sulfate levels. According to the model, the fining onset temperature will decrease considerably when cerium oxide, even in low quantities, is used in the batch. In oxidized glasses, the addition of cerium oxide will improve the sodium sulfate fining process without having a non-desired effect, as caused by reducing conditions, such as the formation of ferrous iron (Fe^{2+}) with a change of the glass color. Addition of coke will also decrease the fining onset temperature, but this can lead to a greenish color of the final glass.

Müller-Simon [39, 43] described the effect of iron oxides in the melt. Gaseous sulfur species, such as SO_2, react with ferric iron during cooling of oxidized melts, forming sulfates; and, in amber glass melts, SO_2 or S_2 can react with ferrous iron, forming the Fe^{3+}-S^{2-} chromophor. In green soda-lime-silica glass melts, SO_2 bubbles are not completely absorbed during cooling for oxygen equilibrium pressures $<< 2 \ 10^{-4}$ bar, because of the limited reaction of Fe^{3+} with SO_2, forming sulfate. Thus, the presence of iron in the melt improves the re-absorption of sulfur-gas-containing bubbles. Müller-Simon [39] showed that, at oxidizing melting conditions ($pO_2 > 5.10^{-4}$ bar at 1200 °C), an increased concentration of chromium oxide in the melt also leads to improved absorption of SO_2 bubbles in the melt during cooling.

Water Content in the Glass Melt

Laimböck [42] showed that the water content of the melt is strongly dependent on the furnace atmosphere during the melting-in of the raw materials and in the initial phases of the fining process, often involving foaming. A glass molten in an electric furnace often contains less than 20-30 moles water per m^3. Typical concentrations of water found in glasses prepared in an air-fired furnace are 350-400 mg per kg glass; this is about 40-50 mole H_2O/m^3. Water shows a high solubility and diffusion coefficient and therefore a high water permeability. A high water concentration in the melt means that water will contribute strongly to the multi-component gas diffusion process. The bubble will not only grow due to the release of fining gases, but also due to diffusion of other gases, especially by water diffusion into the bubble. The fining gases dilute the water vapor in the bubble and vice versa. This will support the diffusion rates of the other gases as well, since the dilution processes keep the concentrations of the gases in the bubble (and the C_{ii} values in equation 11) relatively low, and this will increase the driving force for diffusion of all gases into the bubbles. An increase in the water concentration in the melt will lower also the fining onset temperature. The water in the melt gives a contribution to the vapor pressure in the bubble, and the pSO_2 plus pO_2 partial pressures together need not exceed 1 bar to onset the fining in a water-rich melt. For instance, in a melt with about 70 mole water $/m^3$, the fining already starts at $pO_2+pSO_2 > 0.65$-0.70 bar.

Figure 6 shows the removal time of a bubble during fining at three different water concentration levels. When melting in an oxygen-gas-fired furnace, the water concentration in the glass can exceed a level of 70 mole/m^3. We clearly see from the modeling results the effect of the water content of the melt on the fining process.

This means that there is a potential for reduction of the amount of fining agent to the batch after conversion of an air-fired furnace into an all-oxygen-fired glass furnace. Without changing the batch recipe, more gas volume will be produced by water-rich melts. This may lead to strong foam formation [40], as indeed has been observed in the glass industry. Therefore, batch adaptations may be necessary when changing to oxygen firing.

Sub-Atmospheric Fining

In the past, it has been proposed that a lower pressure will enhance the primary fining process by creating larger bubbles and faster bubble growth rates. Patents [18] show the design of a sub-atmospheric fining process for industrial glass melting. Bihuniak and co-authors [44] reported about the application on an industrial tank, operating at sub-atmospheric pressure, in order to remove gases and gas bubbles from the glass melt. Laboratory studies [20] indeed show the strongly increased bubble sizes when reducing the pressure of a melt during fining. Sub-atmospheric fining requires modifications in the melting equipment and the use of refractory materials, which hardly show out-gassing or porosity. The combination of sub-atmospheric fining and a shallow fining tank with foam barriers seems to give a promising way to eliminate bubbles and foam from the melt.

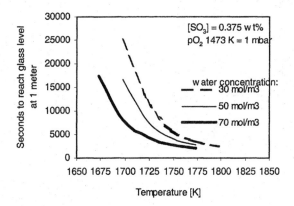

Figure 6. Ascension of bubbles influenced by the concentration of dissolved water in the sulfate-fined float-glass melt.

In the model, the pressure above the melt can be changed, and the impact of the furnace pressure on bubble growth can be predicted. A melt with for instance CO_2-rich bubbles with a radius of 0.1 mm entering a space under lower pressure will encounter: (1) Increased bubble sizes due to the lower bubble pressure--the same quantity of gas will require a large volume; (2) because of the low pressures in the bubbles, the mass transfer of the gases in the melt into the low-pressure bubbles will increase, and this will lead to additional bubble growth; (3) the increased bubble growth and gas release will enhance the removal of all gases (stripping) from the melt, lowering the sensitivity for re-boil.

Figure 7 shows the effect of the low furnace pressure (0.3 bar instead of 1 bar) on the time required to remove a bubble from the melt at a depth of 1 meter. It is possible to decrease the sulfate content in the batch considerably and still obtain a faster fining process. The sub-atmospheric fining can take place in a vertical shaft through which the melt, transferred for instance by the suction forces caused by the low pressure of the melt, can be poured through a spout from a tank at ambient pressure into an evacuated chamber.

Often the addition of sulfates or other fining agents has to be reduced considerably in order to avoid excessive foaming in the low-pressure chamber.

Preconditioning of a Melt by a Gas with High Solubility and High Diffusion Coefficient

Bubbling of the glass melt with a certain gas type, prior to the primary fining process will condition the melt in such a way that: (1) Some of the gases originally dissolved in the melt are stripped during this bubbling process; (2) a new gas species will be dissolving in the glass melt; (3) the bubbling may influence some chemical reactions, such as gas-releasing reactions in the melt, for instance by changing the oxygen equilibrium pressures (this may lead to redox reactions).

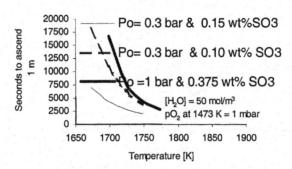

Figure 7. The effect of a reduction of ambient pressure from 1 bar to 0.3 bar on bubble removal time from a float-glass melt contained in a 1-meter-deep tank.

In order to transfer a sufficient gas volume from the bubbles into the melt during this bubbling process, the bubbles should not be too large. Large bubbles would require a large gas volume to achieve a total gas-exchanging surface area, required for sufficient mass transfer of gas into the melt (surface area divided by volume is larger for smaller bubbles). Hence, large bubbles will escape from the melt within much shorter times than smaller bubbles. On the other hand, the bubbles should not be too small, since this may lead to an extra number of very small bubbles in the melt. Small bubbles may only partly dissolve, may not reach the glass surface, and can end up in the product with different gas contents compared to the initial bubble composition due to gas exchange during their trajectories in the furnace.

For a gas species with a high solubility and diffusion in the melt, the transfer from the bubbles to the melt can be very efficient. Especially for bubbles of a diameter of 1-3 cm, it is possible to almost saturate the melt (about 70-80 % of saturation) by intense bubbling. About 100 bubbles per kg melt are required to achieve this saturation level. A gas species which fulfils the requirement of high solubility and diffusion in a silicate melt is helium.

Although during the helium bubbling process about 25 % of the dissolved CO_2 will be stripped and about 50 % of the nitrogen, the other gas concentrations are not very much affected. In the present modeling studies, the stripping of gases during helium bubbling prior to fining has been taken into account but is not discussed in the framework of this paper [45].

Figure 8 shows the effect of bubble growth during fining in a melt preconditioned by helium to only 20 % saturation.

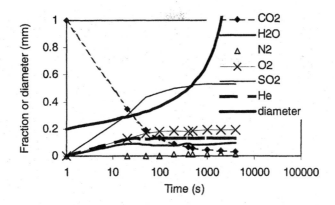

Figure 8. Bubble growth and bubble composition during primary fining in a float-glass melt bubbled with helium prior to fining (20 % helium saturation).

Figure 9 shows the bubble growth and bubble composition for a melt with up to 75 % helium saturation level. Helium contributes significantly to the bubble composition and will have an important impact on the bubble size. For instance, according to the modeling studies, at fining temperatures of 1725 K, the sulfate concentration can be decreased from 0.375 weight % down to 0.24 weight % when bubbling the melt with helium, up to 40 % of the helium saturation level, and still obtaining at least the same fining efficiency.

The saturation by helium of a melt with a high water concentration is more difficult since water in the melt will diffuse rapidly in the helium bubbles and dilute the helium in the bubbles. The bubble growth will decrease the bubble residence time in the melt. The transfer of water from the melt into the helium bubbles dilutes the helium, and the driving force for helium diffusion from the bubble into the melt will decrease. This all will lead to a lower mass transfer of helium from the bubbles into the melt, and limits the saturation of the melt by helium. Thus bubbling with helium is more effective in an air-fired furnace compared to the situation of an oxygen-fired furnace.

Figure 10 shows the impact of the helium concentration in the melt (before fining) on the time required to remove a CO_2 bubble from a depth of 1 meter and with initial radius of 0.1 mm. The stripping of CO_2 and nitrogen by the helium bubbling process has been taken into account.

The application of helium bubbling prior to primary fining will offer the possibility to lower the fining temperature or to decrease the amount of fining agent, or to melt sulfate-containing flint- or float-glass melts at more oxidizing conditions.

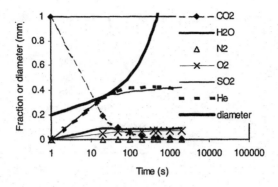

Figure 9. Bubble growth and bubble composition during primary fining in a float-glass melt intensively bubbled with helium prior to fining (75 % helium saturation).

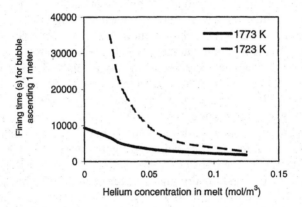

Figure 10. Effect of helium concentration in glass melt on required fining time of a float-glass melt in a 1-meter-deep tank.

Acoustic or Ultrasonic Fining

The application of sonic and, more specifically, ultrasonic waves imposed on glass melts has been investigated already more than 60 years ago by Krüger [22] using a sound field with a frequency of 16 kHz. At 1350 °C the application of these ultrasonic waves enhanced the removal of bubbles considerably. Other investigators also found improved fining behavior (bubble removal), especially when combining ultrasonic waves with a stationary magnetic field [23].

The application of transverse-traveling sound waves by a transducer leads to alternating compression and dilatation of the liquid volume elements. This can have a series of effects on the bubbles in this fluid.

In a melt with a certain depth, a transducer delivering sound waves with constant frequency, preferably between 10 and 20 kHz, can generate a standing wave over the height between the bottom of the vessel or tank and the surface, as explained by Wondergem-de Best [46].

From the surface down to the bottom, the standing wave shows points with both maximum compression and dilatation. The liquid particles will reach maximum displacement. These points are the so-called displacement antinodes. At these points the pressures reach constant values. The displacement antinodes coincide with pressure nodes. The distance between two adjacent pressure nodes is $\frac{1}{2}\lambda$. Between these pressure nodes, pressure antinodes and displacement nodes (no displacement) at a distance $\frac{1}{4}\lambda$ from the displacement antinodes can be observed, see figure 11. At the displacement nodes (no displacement of the liquid) the pressure varies widely from minimum to maximum pressure levels--pressure antinodes.

Due to the presence of an ultrasonic standing wave in a melt, the bubble size and diffusion rates of gases from the melt to the bubbles can be influenced, as well as the movement of the bubbles in the melt.

In the standing wave, bubbles with a size below a certain radius R_0 will experience an average force in the direction of pressure antinodes where the fluid displacement in the wave is minimum. These small bubbles can follow the frequency of the imposed wave. The bubble resonance frequency (proportional to $1/R_0$ and proportional to the square root of the pressure of the melt at the bubble position: $p^{1/2}$) is higher than the applied frequency.

At its maximum pressure value, the pressure acts on a relatively small (contracted) bubble, and the bubble experiences a force to the pressure node. Then, the bubble will be forced to move slowly to the pressure node. However, at the pressure minimum value, the low pressure acts on a larger (expanding) bubble, and the bubble is forced to the pressure minimum (at the nearest pressure antinode). This force to the pressure antinode is larger than the force of the smaller bubble at the pressure maximum towards the pressure node. Thus, the net result is an average movement of the small bubble in the direction of the pressure antinode.

Larger bubbles, which cannot follow the frequency of the applied acoustic field, because of the low resonance frequency of large bubbles, are forced towards pressure nodes. A large bubble (contraction and expansion cannot follow the pressure fluctuations) will experience an average force (integrating the force over the time, which depends on the pressure gradient) in the direction of the pressure node. At frequencies of 10 kHz or higher in a glass melt, most bubbles with a diameter larger than 1 mm will agglomerate in the pressure nodes [46].

Thus, standing waves can be used to enhance the agglomeration of bubbles in the pressure nodes and antinodes. Agglomeration of bubbles can lead to bubble growth (merging of bubbles), and this will increase the bubble ascension, for instance after switching off the acoustic field.

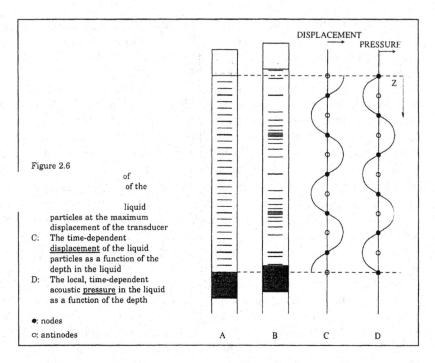

Figure 2.6

of
of the

liquid
particles at the maximum
displacement of the transducer
C: The time-dependent
    displacement of the liquid
    particles as a function of the
    depth in the liquid
D: The local, time-dependent
    acoustic pressure in the liquid
    as a function of the depth

●: nodes
○: antinodes

A   B   C   D

DISPLACEMENT
PRESSURE
Z

Figure 11. Schematic presentation of nodes and antinodes in a standing wave imposed on a fluid [46].

Attraction between bubbles due to Bjerknes force

Oscillating bubbles in a sound field will encounter forces from each other. The velocity of the melt between bubbles caused by the oscillation of the bubbles will be higher than the velocity at the other side of the bubble. This will lead to a lower pressure between the bubbles than in the surrounding liquid, and this results in an attractive force between the bubbles. This is called the Bjerknes force [24].

Beside these agglomeration effects, the acoustic waves will introduce pressure differences in the bubble and convection in radial and tangential directions of the melt surrounding the bubbles, leading to rectified diffusion.

Rectified diffusion

The periodically contracting and expanding bubble in a standing wave leads to pressure variations in the bubble. During bubble contraction, the bubble size decreases and the internal gas pressure will increase. There is an extra driving force for diffusion of gas from the contracting bubble into the melt. On the other hand, during expansion, the bubble pressure decreases, and gases will diffuse more rapidly into the low-pressure expanding bubble. Since the extra diffusion rate in the direction of the bubble is maximum when the size of the bubble is maximum, the extra net diffusion of gases into the bubble is larger than the diffusion during contraction (smaller bubbles) into the melt. This phenomenon is called rectified diffusion. In glass melts, the effect of rectified diffusion seems to be very small, because strong contraction and strong expansion would require very strong pressure variations in the sound wave, and such variations can hardly be realized in practice in highly viscous melts.

Shell effect

The melt surrounding the bubble will have a boundary layer through which gas diffusion takes place. The concentration profile of the dissolved gas species in the radial direction is pressed together during the expansion of the bubble (the thickness of the Nernst boundary layer will decrease). The concentration gradient of a dissolved gas in the melt at the bubble interface will become steeper during the expansion stage. This will enhance the diffusion process of gases from the melt into the low-pressure bubble. During the contraction stage, the boundary layer thickness will increase, and the diffusion of gases from the high-pressure bubble into the melt will be reduced due to the less-steep gas concentration gradient in the melt. This effect will lead to a net bubble-growth effect and will aid the effect of the rectified diffusion.

Acoustic streaming

A pulsating bubble in a sound wave or sound field will cause crispations or vibrating waves on the surface of the bubble. These crispations renew the melt at the surface continuously and keep the gas concentration gradients in the surrounding melt at a very high level. During the primary fining process, this can lead to a strong increase of the diffusion of gas species in the melt, which increases the bubble growth during primary fining [26, 27].

Spinosa and Ensminger [28] showed the application of an acoustic field on the removal of gas bubbles from glass melts.

EXPERIMENTS

It is not within the scope of this paper to present a detailed experimental study on the bubble-removal efficiency of different fining processes or fining enhancement methods. However, a few tests have been performed and are discussed here to demonstrate that the expected effect of helium bubbling and the addition of cerium oxide, supporting the fining process in sulfate-containing melts, as shown by the modeling studies, can be observed in reality.

Addition of Polyvalent Ions to Sulfate_Containing Batches

Raw-material batches, all containing the following raw materials (total 143 grams batch for 122.3 grams of glass) have been prepared: Pure silica sand (88.1 grams), soda (28.5 grams), limestone (19.3 grams), magnesium oxide (5 grams), sodium sulfate (0.9 grams (gives 0.415 wt % SO_3 in the glass melt)), and aluminum oxide (1 gram).

The raw materials are pure compounds, resulting in a glass with < 0.01 weight-% iron oxides, unless iron oxides have deliberately been added. Additions of some polyvalent compounds other than sulfate have been supplied to the base batch composition: Batch I (no additive), Batch II (0.03 grams Fe_2O_3), Batch III (0.10 gram Fe_2O_3), and Batch IV (0.3 gram CeO_2 (gives 0.24 wt % $CewO2$ in the glass)).

To all these batches carbon has been added in such way that the redox number according to the Simpson method [47] is the same for all batches. These batches form 122.3 grams of flint glass. The oxidation state is taken to be the same for all these batches, because this oxidation state of the melt will considerably influence the bubble growth rates.

Melting tests have been performed with these base batches in which the melt is isothermally treated in platinum crucibles in an electric furnace (air atmosphere) for 1.5 hours at 1675 K.

After this melting and a fining stage, the glasses have been annealed, and polished glass slabs have been prepared from the obtained samples. The conditions have been chosen in such way that only partly degassed melts (fining not complete) are obtained. By this method, one is able to observe differences in bubble-removal efficiency.

The photograph in Figure 12 shows the three glass samples obtained from the batches given above. Clearly, the glass prepared from the cerium-oxide-doped batch shows the lowest seed number. The glass prepared with the 0.3 grams CeO_2 addition (0.014 mole Ce/kg melt) shows a much better fining quality than the glass without cerium oxide or the glass with iron oxide (0.01 mole Fe/kg melt). These iron-oxide concentration levels are common in float glasses, for instance. The seed count in the prepared cerium-oxide-containing glass is considerably lower. This is in agreement with the trends presented in figure 5.

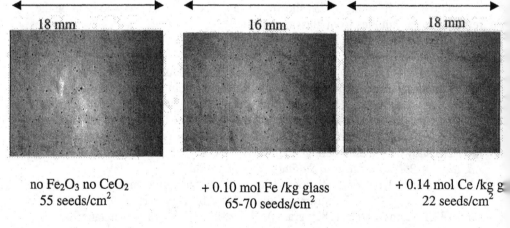

18 mm   16 mm   18 mm

no Fe_2O_3 no CeO_2
55 seeds/cm^2

+ 0.10 mol Fe /kg glass
65-70 seeds/cm^2

+ 0.14 mol Ce /kg g
22 seeds/cm^2

Figure 12. Seeds in float-glass samples containing sulfate, with addition of iron oxides or cerium oxides. Primary fining during 90 minutes at 1675 K and after fining, slow cooling with 1 K/min. Largest bubble diameter in glass without Fe_2O_3 and CeO_2: d = 220 μm. Largest bubble in cerium-oxide-containing glass: d = 120 μm.

Melting Experiments with Helium Bubbling

Glass samples from soda-lime-silica melts intensively bubbled with helium prior to primary fining have been compared to glass pieces from glass melts not bubbled, or bubbled with other gases.

The batch compositions and main test conditions for these experiments are given in table I.

The experimental series with helium bubbling at 1575 K prior to fining at 1700 or 1725 K, but without a secondary fining stage, showed still a large number of bubbles. These bubbles contained more than 40-60 volume % helium, and the bubble pressures appeared to be lower than normal. The bubble compositions have been analyzed by mass spectrometry; the bubbles still contained a large quantity of helium. These tests showed that slow cooling is essential for the secondary fining process; i.e., the glass melt plus seeds should not be quenched too fast. After slow cooling from 1700 K down to about 1475 K, the seeds are much smaller and contain hardly any helium.

The best results have been found for the case of a sulfate-fined melt where helium bubbling has been applied. After primary fining, the melt has been cooled down slowly. Only a few fine seeds with a diameter smaller than 150 micrometer have been found. Without helium bubbling, or instead of helium only oxygen bubbling, the obtained glass sample contained much larger bubbles. The CO_2 concentration of the bubbles found in the glass not exposed to helium bubbling is very high (up to > 90 vol.%). This indicates that the stripping of CO_2 from the melt is not very efficient. The smaller bubbles are found in the glass conditioned with helium prior to fining, and these bubbles showed much smaller CO_2 contents.

A batch without sulfate delivered a glass with a very high number of seeds and bubbles, even when bubbling with helium. The bubbles are rather large. A combination of helium bubbling and 0.25-0.5 weight % sodium sulfate in the batch plus a secondary fining stage (slow cooling from 1700 K down to 1475 K) shows the best results. Table II shows some summarized overall results.

CONCLUSIONS & DISCUSSION

The application of the fining model presented here offers the possibility for evaluating potential methods for the enhancement of glass-melt fining processes; i.e., removal of dissolved gases and gas bubbles. This model describes the production of gases from fining agents and reduction-oxidation reactions in the melt, and it predicts the diffusion of all present gases from the melt to a bubble or vice versa. The model predicts a strong improvement in fining quality when adding small amounts of cerium oxide to oxidized float-glass-type melts. Also, helium bubbling prior to fining will enhance bubble removal in the primary fining stages. These predictions are in agreement with the trends found by laboratory experiments. The fining model also predicts a much faster bubble growth and fining at sub-atmospheric pressure, which is also found in a very pronounced way on laboratory and industrial scales [18, 20, 44].

Another application of the model not shown here is the derivation of the bubble-shrinkage kinetics and bubble composition changes during the cooling process. This offers the possibility of finding the origin of analyzed bubbles found in rejected glass products.

Other physical methods, such as application of acoustic standing waves in the melt, may support the required mass-transfer processes by rectified diffusion, acoustic streaming, the shell effect, and the Bjerknes effect [24]; and these standing acoustic waves can lead to agglomeration of small bubbles. The current model described in this paper, however, is not capable at this stage to predict bubble growth and mass transfer of gases to bubbles in melts exposed to acoustic waves.

ACKNOWLEDGEMENT

The author appreciates the contribution of Dr. Sho Kobayashi from Praxair Inc. on the studies of helium-supported fining processes. The experimental support provided by Arjen Steiner and Olga Zwaneveld from TNO and Martien Hendriks and Ad Verbeek, both from L.G. Philips Displays, is also gratefully acklowledged for its contribution to this paper.

NOMENCLATURE

C_{gi} = concentration of dissolved gas i in melt (mole/m^3)

C_{ii} = concentration of gas i in the melt in equilibrium with the vapor pressure of gas i in the bubble (mole/m^3)

$= L_i \cdot p_{bi}$,

C_{si} = concentration of the gas i in the glass melt (mole/m^3)

D_i = diffusion coefficient of gas i in the molten glass (m^2/s)

$D_{c,i}$ = coefficient determining concentration profile of reactive gases in the melt

[41] (m^2/s)

c = factor with a value between 2/9 and 1/3. In the case where the bubble surface can be considered to be rigid, then: c = 2/9 (Stokes), but for a completely mobile bubble surface: c = 1/3 (Hadamard-Rybczynski).

g = acceleration of gravity (9.8 m/s^2),

H = height of melt above the bubble (m)

kj = number of electrons involved in the reduction process of a polyvalent ion

K_j = reaction equilibrium constant (mole/m^3)$^{kj/4}$

K_{pj} = reaction equilibrium constant based on oxygen pressure (Pa$^{kj/4}$)

K_s = reaction equilibrium constant of sulfate decomposition (mole/m^3)$^{1/2}$

K_{ps} = reaction equilibrium constant of sulfate decomposition based on gas pressures in (Pa$^{3/2}$/(mole/m^3))

L_i = Henry solubility of gas i in the melt at 1 Pascal vapor pressure (mole/m^3·Pa)

[$[M_j^{nj+}]$ = concentration of reduced-form polyvalent element M_j (mole/m^3)

$[M_j^{nj+kj+}]$ = concentration of oxidized-form polyvalent element M_j (mole/m^3)

nj = valency of element j in most-reduced state

$[O_2]$ = oxygen concentration (physically dissolved) (mole/m^3)

p_o = atmospheric pressure (Pa)

p_i = partial vapor pressure of gas i within the bubble (Pa)

p_{bi} = partial pressure of gas i in the bubble (Pa)

p_t = total pressure in the bubble (Pa)

R = bubble radius (m).

Re = Reynolds number = $v·2R·\rho/\eta$

R_g = gas constant = 8.31432 (J/(mole·K))

Sc_i = Schmidt number for component i in melt = $\eta/(\rho· D_{c,i})$

Sh_i = Sherwood number for component i in melt

$[SO_2]$ = sulfur dioxide concentration (physically dissolved) (mole/m^3)

$[SO_4^{2-}]$ = sulfate concentration (mole/m^3)

$[SO_3^{2-}]$ = sulfite concentration (mole/m^3)

t = time (s)

T = temperature in K

v = velocity of bubble (relative to the melt) (m/s),

SYMBOLS

ρ = density of the glass melt (kg/m^3),

η = viscosity of the glass melt (Pa·s).

σ = surface tension of glass melt (N/m)

δ_{ni} = thickness of Nernst boundary layer component i in the surrounding bubble (m)

ANNEX 1
PROPERTIES OF SODA-LIME-SILICA (FLOAT-GLASS TYPE) GLASS MELT AND PROPERTIES OF GASES IN THESE MELTS

SOLUBILITIES OF GASES (L_i in mole/(m^3·Pa))

$L_{N2} = 1.1 ·10^{-5}$ ·EXP (-6633/T), physical solubility	lit. 48
$L_{CO2} = 5.6 ·10^{-7}$ ·EXP (3120/T)	lit. 49, 50
$L_{He} = 3.0 ·10^{-6}$ ·EXP (-1005/T), physical solubility	lit. 51, 52
$L_{O2} = 1.37 ·10^{-4}$ ·EXP (-6633/T), physical solubility	estimated

from 48, 52

$L_{H2O} = 0.68$ ·EXP (-613/T)	lit. 53
$L_{SO2} = 6.44·10^{-7}$·EXP(7860/T)	lit. 50

DIFFUSION COEFFICIENTS OF GASES (D_i in m^2/s)

$D_{N2} = 4.3 \cdot 10^{-5} \cdot EXP(-19364/T)$	lit. 54
$D_{CO2} = 1.92 \cdot 10^{-5} \cdot EXP(-21516/T)$	lit. 16
$D_{He} = 3.3 \cdot 10^{-7} \cdot EXP(-4931/T)$	lit. 55, 56
$D_{O2} = 4.2 \cdot 10^{-3} \cdot EXP(-26646/T)$	lit. 57
$D_{H2O} = 1.2 \cdot 10^{-5} \cdot EXP(-18320/T)$	lit. 58
$D_{SO2} = 4.45 \cdot 10^{-7} \cdot EXP(-15360/T)$	lit. 16

(estimated)

REACTION EQUILIBRIUM DATA, p in Pascal, concentration mole/m^3

$pSO_2 \cdot pO_2^{1/2}/[SO_4^{2-}] = 1.66 \cdot 10^{13} \cdot EXP(-271200/R_g \cdot T)$	lit. 36
$[Ce^{3+}] \cdot pO_2^{1/4}/[Ce^{4+}] = 650.7 \cdot EXP(-42000/R_g \cdot T)$	lit. 46
$[Fe^{2+}] \cdot pO_2^{1/4}/[Fe^{3+}] = 2788 \cdot EXP(-103000/R_g \cdot T)$	lit. 46

ESTIMATION OF CONCENTRATION OF GASES IN THE GLASS MELT (mole/m^3)

Nitrogen: 0.005 mole/m^3 (25-50 % lower concentration after helium bubbling)

Carbon dioxide: 0.05 mole/m^3 (10-25 % lower concentration after helium bubbling)

Water in case of air-natural-gas firing: 40 mole/m^3

Without preconditioning of melt with helium: 0 mole He/m^3 melt

Argon concentration* is not taken into account.

(* This information however is important in the diagnosis of bubble sources derived from the composition of a reject bubble)

REFERENCES

[1] R.G.C. Beerkens, "The role of gases in glass melting processes," *Glastech. Ber. Glass Sci. Technol.*, **68** (1995), no. 5, pp. 369-380

[2] M.C. Weinberg, P.I.K. Onorato and D.R. Uhlmann, "Behaviour of bubbles in glass melts: I, Dissolution of a stationary bubble containing single gas," *J. Am. Ceram. Soc.*, **63** (1980) nr. 3-4, pp. 175-180

[3] M.C. Weinberg, P.I.K. Onorato and D.R. Uhlmann, "Behaviour of bubbles in glass melts: II, Dissolution of a stationary bubble containing a diffusing and non diffusing gas," *J. Am. Ceram. Soc.*, **63** (1980) nr. 7-8, pp. 435-438

[4] P.I.K.Onorato, M.C. Weinberg and D.R. Uhlmann, "Behaviour of bubbles in glass melts: III, Dissolution and growth of a rising bubble containing a single gas," *J. Am. Ceram. Soc.*, **64** (1981) nr. 11, pp. 676-672

[5] M.C. Weinberg, "Dissolution of a stationary bubble in a glass melt with a reversible chemical reaction/rapid forward reaction rate constant," *J. Am. Ceram. Soc.*, **65** (1982) nr. 10, pp. 479-485

[6]M.C. Weinberg, "Behavior of a two component gas bubble in a glass melt with chemical reactions," *Proceedings XIII International Congress on Glass (1983) Glastech. Ber. Sonderband* **56K** (1983)

[7]M. Cable and J. R. Frade, "Theoretical analysis of the dissolution of multi-component gas bubbles," *Glastech. Ber.,* **60** (1987) nr. 11, pp. 335-362

[8]J. L. Duda and J. S. Vrentas. "Mathematical Analysis of Bubble Dissolution," *AIChE Journal,* **15** (1969) no. 3, pp. 351-356

[9]C. Parton and D. Dollimore. "The application of the contracting sphere equation to the behavior of oxygen bubbles in molten glass," *Thermochimica Acta,* **19** (1977), pp. 25-36

[10]J.J. Ramos: "Behaviour of Multicomponent Gas Bubbles in Glass Melts," *J. Am. Ceram. Soc.,* **69** (1987) nr. 2, pp. 149-154

[11]R.G.C. Beerkens: "Chemical equilibrium reactions as driving forces for growth of gas bubbles during fining," 2nd International Conference on Fusion and Processing of Glass, Düsseldorf (1990). *Glastech. Ber. Sonderband* **63K** (1990), pp. 222-242

[12]F. Krämer: "Mathematisches Modell der Veränderung von Glasblasen in Glasschmelzen," *Glastech. Ber.,* **52** (1979) nr. 2, pp. 43-50

[13]L. Nèmec: "The behaviour of bubbles in glass melts. Part 2. Bubble size controlled by diffusion," *Glass Technol.,* **21** (1980) nr. 3. pp. 139-144

[14]M. Mühlbauer and L. Nèmec: "Einfluß der Gase einer Glasschmelze auf das Verhalten von Gasblasen unter isothermer Bedingungen," *Glastech. Ber.,* **54** (1981), nr. 12, pp. 389-399

[15]L. Nèmec: "The behaviour of bubbles in glass melts. Part 1. Bubble size controlled by diffusion and chemical reaction," *Glass Technol.* **21** (1980) nr. 3. pp. 134-138

[16]L. Nèmec and M. Mühlbauer: "Verhalten von Glasblasen in der Schmelze bei konstanter Temperatur," *Glastech. Ber.,* **54** (1981) nr. 4, pp. 99-108

[17]S. Kawachi and Y. Kawase: "Evaluation of bubble removing performance in a TV glass furnace. P 1 Mathematical formulation," *Glastech. Ber. Glass Sci. Technol.,* **71** (1998), nr. 4, pp. 83-91

[18]H. Itoh, R. Kitamura, M. Sakai, K. Sekine and M. Okada: "Process for producing a glass for cathode ray tubes," European Patent Application Asahi Glass Co. Ltd. Tokyo 100-8405 (JP) Patent nr. EP-1245539 A1 (2002)

[19]R. Kitamura, H. Itoh, Y. Taskei and T. Kawaguchi: "Mathematical model of the bubble growth at reduced pressures," *Proc. Int. Congr. Glass,* **Vol. 2.** Extended Abstracts, Edinburgh, Scotland, 1.-6. July 2001, pp. 361-362

[20]W. Wintzer, S. Römhild and D. Ehrt: "Einfluss von Unterdruck auf die Läuterung von Glasschmelzen," Kurzreferate **76.** *Glastechnische Tagung der Deutsche Glastechnische Gesellschaft (2002),* 27.-29. Mai, Bad Soden am Taunus, pp. 99-102

[21]R. Ehrig, J. Wiegand and E. Neubauer: "Five years of operational experience with the SORG LoNOx® Melter," Glastech. *Ber. Glass Sci. Technol.,* **68** (1995) nr. 2, pp. 73

[22]F. Krüger: "Über die Entgasung von Glasschmelzen durch Schallwellen," *Glastech. Ber.*, **16** (1938) nr. 7, pp. 233-236

[23]C. Eden: "Ultraschall-Entgasung von Glasschmelzen im Hochfrequenz-induktionsofen," *Glastech. Ber.*, **25** (1952) nr. 3, pp. 83-86

[24]V.A. Shutilov.: "Fundamental Physics of Ultrasound. Chapter 5.3: Steady forces acting on suspended particles in an ultrasonic field," *Gordon & Breach Science Publishers*, London (1988) pp. 139-140.

[25]A. Eller, H.G. Flynn, H.G.: "Rectified diffusion during nonlinear pulsations of cavitation bubbles," *Journal of the Acoustic Society of America*, **37** (1965) nr. 3, pp. 493-503

[26]R.K. Gould: "Rectified diffusion in the presence of, and absence of, acoustic streaming," *Journal of the Acoustic Society of America*, **56** (1974) nr. 6, pp. 1740-1746

[27]M.A. Mironov: "Influence of microstreaming on the growth of gas bubbles due to rectified diffusion," *Soc. Phys. Acoust.*, **23** (1977), nr. 5, pp. 476-478

[28]E.D Spinosa and E.D Ensminger: "Sonic energy as a means to reduce energy consumption during glass melting," *Ceram. Engng. Sci. Proc.*, **7** (1986) nr. 3-4, pp. 410-425

[29]A. Eller: "Force on a bubble in a standing acoustic wave," *Journal of the Acoustic Society of America*, **43** (1968) nr. 1, pp. 170-171

[30]H.O. Mulfinger: "Gasanalytische Verfolgung des Läutervorganges im Tiegel und in Schmelzwanne," *Glastech. Ber.* **49** (1976) nr. 10 pp. 232-245

[31]L. Nèmec: "Refining in the Glass-melting Process," *J. Am. Ceram. Soc.* **60** (1977), nr. 9-10, pp. 436-440

[32]J. Gailhbaud and M. Zortea: "Recherches sur la coalescence des bulles dans un liquide visquex," *Rev. Gen. Therm.* **8** (1969), pp. 433-453

[33]J. Zluticky: "Bestimmung des im Glas gelösten Kohlen- und Schwefeldioxides über die gaschromatografischer Analyse," *Glastech. Ber.*, **47** (1974), nr. 10, pp. 232-238

[34]J. Zluticky, J. Stverak and L. Nèmec: "Bestimmung der im Glas gelöste Gase über die Hochtemperatur-Vakuum-Extraktion und die gaschromato-grafische Analyse," *Glastech. Ber.*, **45** (1972) nr. 9, pp. 406-409

[35]D.S. Goldman, "Redox and sulfur solubility in glass melts," *In Gas bubbles in glass. Technical Committee 14 of the International Commission on Glass. Inst. Nat. Verre Charleroi.* 1985, pp. 74-91

[36]R.G.C. Beerkens, "Sulfate Decomposition and Sodium Activity in Soda Lime Silica Glass Melts," *Submitted to J. Am. Ceram. Soc. 2003*

[37]I. Löh, T. Frey and H. A. Schaeffer, "Continuous determination of the oxidation state in a soda-lime-silica glass melt during refining," *J. Am. Ceram. Soc. Communications of the American Ceramic Society 12* (1981) nr. 12, pp. C-168-C-169

[38]C. Rüssel and E. Freude: "Voltammetric studies of the redox behaviour of various multivalent ions in soda-lime-silica glass melts," *Phys. Chem. Glasses*, **30** (1989) nr. 2. pp. 62-68

[39]H. Müller-Simon, "Oxygen balance in sulfur-containing glass melts," *Glastech. Ber. Glass Sci. Technol.* **71** (1998) nr. 6, pp. 157-165

[40]R.B. Bird, W.E. Stewart and E.N. Lightfoot: "Transport Phenomena," *John Wiley and Sons, (1960)* New York-Chichester, Brisbane, Toronto, Singapore, International Edition

[41]R.G.C. Beerkens and H. de Waal: "Mechanism of Oxygen Diffusion in Glass melts containing variable-valence ions," *J. Am. Ceram. Soc.*, **73** (1990), nr. 7., pp. 1857-61

[42]P. R. Laimböck: "Foaming of Glass Melts," *PhD thesis Eindhoven University of Technology (1998)*, ISBN 90-386-0518-8

[43]H. Müller-Simon: "On the interaction between oxygen, iron, sulfur in industrial glass melts," *Glastech. Ber. Glass Sci. Technol.* **67** (1994) nr. 11, pp. 297-303

[44]P. P. Bihuniak, R.L. Cerutti, R.L. Schwenninger, J.M. Matesa and L.J. Shelestak: "Manufacturing of high performance flat glass by continuous vacuum refining (abstract)," *Proceedings of the 4th International Conference on Advances in Fusion and Processing of Glass. Glastech. Ber. Glass Sci. Technol.*, **68C2** (1995) pp. 3

[45]R.G.C. Beerkens: "Enhanced fining of soda-lime glass melts by helium bubbling, phase 1," *TNO report HAM-RPT-00-190 for PRAXAIR Inc.*, 27. July 2000, The Netherlands

[46]A. Wondergem-de Best: "Redox behaviour and fining of molten glass," *PhD thesis Eindhoven University of Technology*, The Netherlands (1994)

[47]W. Simpson and D. D. Myers: "The redox number concept and its use by the glass technologist," *Glass Technol.*, **19** (1978) nr. 4, pp. 82-85

[48]H.O. Mulfinger, A. Dietzel and J.M.F. Navarro: "Physikalische Löslichkeit von Helium, Neon und Stickstoff in Glasschmelzen," *Glastech. Ber.*, **45** (1972) nr. 9, pp. 389-396

[49]C. Kröger and N. Goldmann: "Kohlendioxid Löslichkeit in Gläsern," *Glastech. Ber.*, **35** (1962) nr. 11, pp. 459-478

[50]R.G.C. Beerkens and M. van Kersbergen: "Redox reactions and Properties of Gases in Glass Melts," *Final Report NCNG-NOVEM.* July 1996 internal Report TNO HAM-RPT-9577

[51]H.O. Mulfinger and H. Scholze: "Löslichkeit und Diffusion von Helium in Glasschmelzen, 1. Löslichkeit," *Glastech. Ber.*, **35** (1962) nr. 11, pp. 466-478

[52]H. Scholze: "Gases in Glass," Proceedings VIII International Congress on Glass (1968) *Society of Glass Technology* Sheffield, pp. 69-83

[53]H. Franz and H. Scholze: "Die Löslichkeit von H_2O Dampf in Glasschmelzen verschiedener Basizität," *Glastech. Berichte*, **36** (1963) nr. 9, pp. 347-356

[54]G.H. Frischat, O. Bushmann and H. Meyer: "Diffusion von Stickstoff im Glasschmelzen," *Glastech. Ber.* **51** (1987), nr. 12, pp. 321-327

[55]H.O. Mulfinger and H. Scholze: "Löslichkeit und Diffusion von Helium in Glasschmelzen, 2. Diffusion," *Glastech. Ber.* **35** (1962) nr. 11, pp. 495-500

[56]G.H. Frischat and H.J. Oel: "Bestimmung des Diffusionskoeffizienten von Helium in Glasschmelzen aus der Abnahme einer Blase," *Glastech. Ber.* **38** (1965) nr. 4, pp. 156-166

[57]R.H. Doremus: "Diffusion of oxygen from contracting bubbles in molten glass," *J. Am. Ceram. Soc.* **43** (1960), pp. 655-661

[58]L. Nemec: "Diffusion-controlled dissolving water vapor bubbles in molten glass," *Glass Technol.* **10** (1969), pp. 176-181

Table I. Batch composition and test conditions for helium-enhanced fining

Batch	grams	
Sand	60.1	
Soda	19.15	
Limestone	19.52	
Sodium sulfate	0.25	
Cokes	1.05	
Temperature course		
Heating up	20 K/min	
1st temperature level for He-bubbling	1575 K	Bubbling during 60 minutes
bubbling helium	0 or 10 ml/min	
2nd level	1700 K	Fining during 75 minutes
cooling down to 1475 K	1 K/min	secondary fining
cooling down to 875 K	20 K/min	
annealing	2 K/ min	

Table II. Observations after fining of soda-lime-silica melt with/without helium bubbling

Conditions	seed level	D seed size in mm	% CO_2 in bubble
Sulfate fining plus helium bubbling and secondary fining	± 1 /cm^2	0.145	40-90
Sulfate fining, no bubbling and secondary fining	Roughly estimated on 5 / cm^2	0.25	95
No fining agent, helium bubbling <u>no secondary fining</u> (seeds of about 0.1-0.25 mm)	> 10 /cm^2	0.36	35-50 * > 50 % helium ** very low pressure
Oxygen bubbling, no secondary fining	> 10 /cm^2	Not analyzed	Not analyzed

THE GLASS MELTING PROCESS – TREATED AS A CYCLIC PROCESS OF AN IMPERFECT HEAT EXCHANGER

Reinhard Conradt
Aachen University
Institute of Mineral Engineering and
Department of Glass and Ceramic Composites
Mauerstrasse 5
52064 Aachen
Germany

ABSTRACT
The description of finite time heat transfer in heat engines in general, and in heat exchangers in specific, is concisely translated to the glass melting process: In this process, there are two principal heat fluxes. These are: the cumulative heat flux crossing the furnace lining, and the heat flux from the combustion space into the basin. There are two principal mass flows, i.e., through the combustion space, and through the basin, and one principal heat sink (i.e., the batch). The treatment provides a tool which allows to swiftly identify the extremals and optimum conditions of a process, i.e., of a given batch melted in a specific furnace, and to anticipate the effects of any parameter changes like: pull rate, batch composition, flame emissivity, ratio of heat recovery, etc. in a quantitative way.

INTRODUCTION
The conventional glass melting process is a continuously operated high-temperature process converting a batch of raw materials into a workable melt. During this process, heat is transferred from the hot stream of a high-temperature heat reservoir to a compartment (a chemical reactor) in which the batch-to-melt conversion is brought about. In the majority of cases, the heat required for the process is made available by the combustion of fossil fuel. The heat set free in the combustions space is transferred to the batch and the glass melt passing through the basin. At the working end of the basin, a mass flow of a single-phase glass melt (essentially free of bubbles and crystalline relicts) is continuously extracted at a constant temperature. This mass flow goes together with a continuous extraction of heat from the furnace. It thus takes the role of a "cold stream" in the

system. From this point of view, the entire set-up of a glass melting tank furnace may be considered as a heat exchanger system comprising a hot stream and a cold stream involving chemical turnover. It is the purpose of the present paper to point out the performance of such a system in general, and to translate the general principles to the special situation of glass melting tank furnaces. The treatment will allow to identify and quantitatively determine conditions for optimal heat exploitation for a given production situation. The basis of the treatment is a reliable quantification of the amounts of heat (per t of molten glass) involved in the process. This is accomplished by applying the 1^{rst} theorem of thermodynamics. The further elaboration is essentially based on the 2nd theorem.

1^{rst} THEOREM TREATMENT: HEAT AND MASS BALANCE

Figure 1 illustrates the heat balance of a typical glass melting tank furnace with an offgas heat exchanger system. The individual terms are explained in the figure itself. The heat balance represents a snap shot of energy utilization valid for a given batch composition, fuel composition, air (or oxygen) factor, and pull rate. If any of these quantities is changed, then the relative proportions of the quantities given in Figure 1 change accordingly. The balance given in Figure 1 is valid either for amounts of heat H per t of molten glass (kWh/t), or for heat fluxes q referred to the total melting area (kW/m^2). Heat amounts H are converted to heat fluxes q by multiplication with the pull rate given in t/(m^2·h).

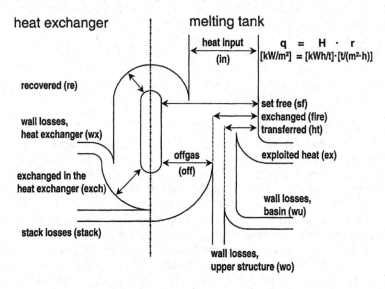

Figure 1.   Heat balance of a glass melting tank furnace with offgas heat recovery system; the balance is valid either for amounts of heat H per t of molten glass, or for heat fluxes q referred to the melting area; r = pull rate.

From Figure 1, five principal equations are derived describing the amounts of heat set free in the combustion space, exchanged with the furnace body, transferred to the basin, conveyed to the heat exchanger, and exchanged in the heat exchanger, respectively. They read, in the quoted sequence,

$$H_{sf} = H_{in} + H_{re} \tag{1}$$

$$H_{fire} = H_{sf} - H_{off} = H_{ex} + H_{wo} + H_{wu} \tag{2}$$

$$H_{ht} = H_{ex} + H_{wu} \tag{3}$$

$$H_{off} = H_{stack} + H_{re} + H_{wx} \tag{4}$$

$$H_{exch} = H_{off} - H_{stack} = H_{re} + H_{wx} \tag{5}$$

As pointed out earlier [1, 2], the quantities H_{in}, H_{off}, H_{re}, and H_{stack} related to the hot stream can be calculated via combustion calculations (as explained in many textbooks) in a fully quantitative way. As a prerequisite, the composition of the fuel (yielding its net calorific value H_{NCV}, oxygen demand, and offgas composition), the air (or oxygen) excess, and the temperature levels T_{off}, T_{re}, and T_{stack} need to be known. By this, all quantities in Equations (1), (4), and (5) are determined, which may be used for a continuous evaluation of furnace performance. The quantification of Equations (2) and (3) is more difficult. In view of the complicated geometry of a furnace, the decrease of refractory wall thickness during operation time, the radiation losses through peep holes and the hardly assessable convective flow around the furnace, a direct quantitative access to the quantities H_{wo} and H_{wu} is not very promising. Therefore, much effort has been devoted [3-9] to quantify the theoretical heat demand or exploited heat H_{ex}, either by experiment or by thermodynamic calculations. The exploited heat H_{ex} is composed of the so-called chemical heat demand $\Delta H°_{chem}$, which is equivalent to the standard heat of batch-to-melt conversion (at 25 °C, 1 bar),

$$\text{batch } (25 °C) \rightarrow \text{glass + batch gases } (25 °C) \tag{6}$$

and the heat $\Delta H_T(gl)$ physically stored in the glass melt (relative to 25 °C) at temperature T_{ex} at which the melt is extracted from the basin:

$$H_{ex} = (1 - y_C) \cdot \Delta H°_{chem} + \Delta H_T(gl) \tag{7}$$

Here, y_C denotes the weight fraction of cullet (referred to 1 t of molten glass), which does not contribute to $\Delta H°_{chem}$. The heat content of the batch gases $\Delta H_T(gas)$ at T_{off} is not comprised in H_{ex}, but rather treated as a contribution to the combustion space balance. $\Delta H°_{chem}$ is derived directly form the batch composition. This is demonstrated in Table 1 for a very simple model batch (no refiners; chemically pure raw materials; feldspar = albite). The batch composition unambiguously determines the nominal oxide composition of the resulting glass, the gas species, and the composition of the crystalline counterpart of the glass in terms of phases k. The standard heat of formation of the glass (1 t) from its oxides in thus found [10] as

$$H°(gl) = \sum_k n_k \cdot \left(H°_k + H_k^{vit} \right) \tag{8}$$

where n_k is the molar amount of phase k per 1 t of glass, $H°_k$ and H_k^{vit} is the standard enthalpy, and enthalpy of vitrification below T_g, respectively. For real raw materials, the standard enthalpy of each individual raw material i has to be assessed from its chemical composition in analogy to Equation (8), however, without using the term H_k^{vit}. Thus, the cumulative standard enthalpy of the batch H°(batch) is obtained. The contribution H°(gas) is determined in a straight-forward way from the batch gas composition. Finally, as verified by a comparison with experiments [3, 8],

$$\Delta H°_{chem} = H°(gl) + H°(gas) - H°(batch) \tag{9}$$

gives a surprisingly accurate account of the chemical heat demand. For the heat $\Delta H_T(gl)$ physically stored in the glass melt, oxide increment systems have been suggested [4-7] which work quite satisfactorily within narrow composition ranges. An approach by

$$\Delta H_T(gl) = H_T(melt) - H°(gl) \tag{10}$$

with

$$H_T(melt) = \sum_k n_k \cdot H_{T,k}(melt) \tag{11}$$

is more accurate and more versatile compositionally. In Equations (10) and (11), $H_T(melt)$ denotes the heat of the melt at temperature T_{ex}. $H_{T,k}(melt)$ is the heat of the individual phase k at T_{ex} taken in the liquid state. Figure 2 shows the quantities ΔH_T calculated for the batch, the glass (melt) and the gases from Table 1, together with the calculated c_P curve of the glass. This plot allows to determine the chemical heat demand, not only at 25 °C, but at any arbitrary temperature:

$$\Delta H_{chem} = \Delta H°_{chem} + \Delta H_T(gl) + \Delta H_T(gas) - \Delta H_T(batch) \tag{12}$$

With H_{ex} known, the cumulative wall losses $H_{wo} + H_{wu}$ in the heat balance can be calculated by Equation (2).

Table 1. Simple model batch composition (raw materials i), resulting glass composition (oxides j; CO_2), and corresponding crystalline reference system (phases k), given in terms of a mass balance; the resulting caloric quantities are (per t of molten glass):
$H°$(batch) = 4589.5 kWh, $H°$(gl) = 3942.3 kWh, $H°$(gas) = 503.0 kWh, $\Delta H°_{chem}$ = 144.2 kWh; ΔH_T(gl) = 404.1 kWh (25 to 1200 °C)

raw material i	kg/t	oxide j	kg/t		phase k	kg/t
sand	666.96	SiO_2	720.00		SiO_2	227.05
feldspar	77.15	Al_2O_3	15.00		$Na_2O·Al_2O_3·6SiO_2$	77.15
dolomite	182.98	MgO	40.00		$MgO·SiO_2$	99.62
limestone	34.54	CaO	75.00		$Na_2O·3CaO·6SiO_2$	263.34
soda ash	240.91	Na_2O	150.00		$Na_2O·2SiO_2$	332.84
		CO_2	202.54			
sum	1202.54		1202.54			1000.00

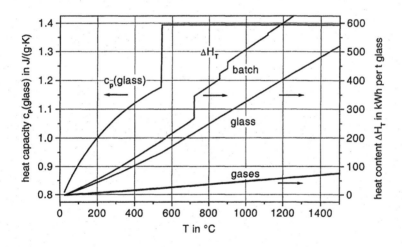

Figure 2. Caloric quantities of the batch given in Table 1

We may finally try to obtain an estimate of H_{ht} by evaluating the quasi-stationary radiation exchange between combustion space, crown, and basin. In a simplified way, this is accomplished as sketched in Figure 3 [11], where the radiation heat fluxes (emission and reflection) q_1 from the melt, q_2 to the melt, q_3 from the crown, and q_4 to the crown, respectively, are presented by their shares due to Planck radiation and reflection. This yields an equation system

$$q_1 = C_S \cdot \varepsilon_{glass} \cdot T_{glass}^4 + \left(1 - \varepsilon_{glass}\right) \cdot q_2$$

$$q_2 = C_S \cdot \varepsilon_{gas} \cdot T_{gas}^4 + \left(1 - \varepsilon_{gas}\right) \cdot q_3$$

$$q_3 = C_S \cdot \varepsilon_{wo} \cdot T_{wo}^4 + \left(1 - \varepsilon_{wo}\right) \cdot q_4 \qquad (13)$$

$$q_4 = C_S \cdot \varepsilon_{gas} \cdot T_{gas}^4 + \left(1 - \varepsilon_{gas}\right) \cdot q_1$$

which is readily resolved for the individual q_i, $i = 1, \ldots, 4$. C_S denotes the Stefan Boltzmann constant, and ε is the emissivity. Reasonable estimates for the ε values and temperatures of the glass, the gas and the crown (wo) need to be known by direct observation or by experience. The heat fluxes q_{ht} and q_{wo} are obtained by $q_{ht} = q_2 - q_1$ and $q_{wo} = q_4 - q_3$, respectively. Again, the general relation $q = H \cdot r$ between heat fluxes q (referred to the melting area), heat amounts H per t of molten glass, and the pull rate r is employed. Thus, for a given pull rate r, the heat quantities $H_{ht} = q_{ht}/r$ and $H_{wo} = q_{wo}/r$ are found. The result is a comprehensive determination of each individual heat quantity in the set of balance Equations (1) to (5).

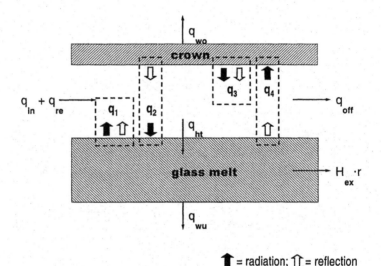

= radiation; = reflection

Figure 3. Radiative heat exchange across the combustion space, composed of shares of Planck radiation and reflection; q_1 = heat flux from the melt; q_2 = heat flux to the melt; q_3 = heat flux from the crown; q_4 = heat flux to the crown; the other indices have the meaning given in Figure 1; H_{ex} = exploited heat; r = pull rate

2nd THEOREM TREATMENT: THE IMPERFECT HEAT EXCHANGER

When resting with a 1st theorem treatment of a glass furnace, misleading conclusions may be drawn on the achievable limit of heat exploitation. This may trigger wrong decisions, not only related to technology, but also to economics and energy taxation policy. For example, the quantity of theoretical heat demand H_{ex} may be – and has been – misinterpreted as a lower limit of the actual heat demand H_{in} which can be approached by an appropriate technology. The model power plant sketched in Figure 4 after [12] illustrates the details of finite time heat transfer between a hot stream passing through a combustion space and a power generating, reversibly operating compartment.

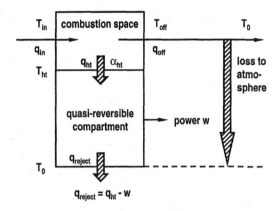

$$q_{reject} = q_{ht} - w$$

Figure 4. Heat fluxes in a model power station after [12]; T_0 = temperature of the environment; α_{ht} = internal heat transfer coefficient; q_{reject} flux of rejected heat; other indices have the same meaning as in Figure 1

In the simple model, no wall losses are taken into account. In the course of the heat exchange, the hot stream is cooled from T_{in} to T_{off} while transferring the heat flux q_{ht} to the lower compartment. The lower compartment generates power w and therefore rejects heat $q_{reject} = q_{ht} - w$. If the mass flows entering and leaving the combustion space are identical and have constant heat capacity, then the efficiency of heat transfer η_{ht} is given by

$$\eta_{ht} = \frac{q_{ht}}{q_{in}} = \frac{q_{reject} + w}{q_{in}} = \frac{T_{in} - T_{off}}{T_{in} - T_0} = 1 - \frac{T_{off} - T_0}{T_{in} - T_0} = 1 - \frac{\vartheta_{off}}{\vartheta_{in}} \qquad (14)$$

Note that the symbol T denotes an absolute temperatures, while ϑ denotes the temperature difference between T and the absolute temperature T_0 of the atmosphere. With respect to ϑ, η_{ht} takes the same form as the Carnot efficiency. The model in Figure 4 is readily transferred to the case of a glass melting tank furnace by replacing T_{in} by the adiabatic combustion temperature T_{ad} of the fuel. Replacing q_{off} by $q_{off} - q_{re}$ allows for heat recovery at an efficiency $\eta_{re} = q_{re}/q_{off}$, thereby reducing the temperature of the hot stream to T_{stack}. A glass furnace does, of course, not generate power. Nevertheless, it comprises a quasi-reversible cycle, presented as a generalized Carnot cycle in a S-T diagram as shown in Figure 5.

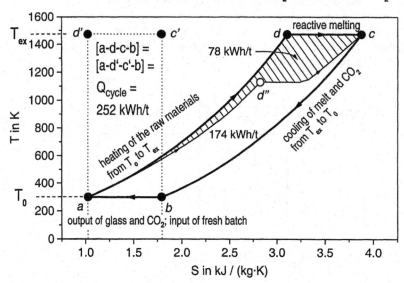

sand + limestone + soda ash → 74 SiO_2 +10 CaO + 16 Na_2O

Figure 5.   Generalized Carnot cycle of the batch-to-melt conversion process of a
simple model batch, operating between T_{ex} = 1473 K and T_0 = 298 K;
curve a → d illustrates fictive heating-up of the raw materials without
conversion; curve a → d″ → c presents a realistic path

The area enclosed by the cycle a-d-c-b is equivalent to a heat quantity (per t of
molten glass) denoted by Q_{cycle}. Like before, multiplication with the pull rate r
yields a corresponding heat flux. This heat flux Q_{cycle}·r is equivalent to the power
term w shown in Figure 4. Q_{cycle} does not explicitly appear in the balance in
Figure 1. It is directly related to the batch-to-melt conversion. Quantitative
information on Q_{cycle} and hence, on w, allows to evaluate the efficiency at
maximal batch-to-melt conversion η_w = w/q_{ht} by means of the Curzon-Ahlborn
expression [12]: η_w = 1 - $\sqrt{(T_0/T_{ad})}$. In combination with Equation (14), this yields
an estimate of the required energy input H_{in} = q_{in}/r from the quantity w and the
temperature levels T_0, T_{off}, and T_{ad}. The estimate is independent of the more
conservative approach followed below.

Finally, the offgas temperature T_{off} is derived from the adiabatic combustion
temperature T_{ad} and the temperature T_{ex} at which the melt is extracted from the
basin. In the glass melting process, both temperature levels T_{ad} and T_{ex} are fixed
while the offgas temperature is a result of heat loss to the basin and heat loss
through the crown. The related heat transfer is given in terms of so-called
numbers of transfer units [13] NTU_{ht} and NTU_{wo}, respectively. In general terms,
NTU denotes the dimensionless ratio of a heat transfer rate per K temperature
difference across a boundary and the heat capacity of a mass flow passing this
boundary, thereby delivering or receiving heat. Thus NTU → 0 denotes zero heat

transfer while NTU $\to \infty$ denotes heat transfer in perfect tune with the mass flow. In the absence of wall losses, the offgas temperature $\vartheta^{\circ}_{\text{off}}$ is given by

$$\frac{\vartheta^{\circ}_{\text{off}} - \vartheta_{\text{ex}}}{\vartheta_{\text{ad}} - \vartheta_{\text{ex}}} = \exp\left(-\text{NTU}_{\text{ht}}\right) \tag{15}$$

$\vartheta^{\circ}_{\text{off}}$ is decreased by wall loss through the crown to the actual value ϑ_{off} like

$$\frac{\vartheta_{\text{off}}}{\vartheta^{\circ}_{\text{off}}} = \exp\left(-\text{NTU}_{\text{wo}}\right) \tag{16}$$

Both Equations (15) and (16) yield an expression for the ratio ζ (r) of offgas and pull temperature:

$$\frac{\vartheta_{\text{off}}}{\vartheta_{\text{ex}}} = \zeta(\text{r}) = \left[1 + \left(\frac{\vartheta_{\text{ad}}}{\vartheta_{\text{ex}}} - 1\right) \cdot \exp(-\text{NTU}_{\text{ht}})\right] \cdot \exp(-\text{NTU}_{\text{wo}}) \tag{17}$$

The NTU, and hence ζ, depend on the pull rate r. All temperatures are given in terms of their difference ϑ to the environment. Finally, let us introduce the estimate $q_{\text{in}}/\vartheta_{\text{ad}} \approx q_{\text{off}}/\vartheta_{\text{off}}$ and combine Equations (1), (2), and (17). In doing this, $\text{NTU}_w = (\text{H}_{\text{wo}} + \text{H}_{\text{wu}})/\text{H}_{\text{ex}}$ is introduced as a cumulative wall loss number. This results in a furnace performance equation

$$\frac{\text{H}_{\text{ex}}}{\text{H}_{\text{in}}} = \frac{1 - \left(1 - \eta_{\text{re}}\right) \cdot \zeta \cdot \dfrac{\vartheta_{\text{ex}}}{\vartheta_{\text{ad}}}}{1 + \text{NTU}_{\text{w}}} \tag{18}$$

capturing the essential features of a glass melting tank furnace. It correctly reflects the dependence on the pull rate with a point of optimal performance (minimal $\text{H}_{\text{ex}}/\text{H}_{\text{in}}$) between an extremely wasteful low pull regime and a moderately wasteful but furnace tearing (high q_{wo} !) high pull regime. Let us use Equation (18) to estimate the maximum efficiency obtainable by the state-of-the-art melting technology. Let us therefore adopt $\zeta = 1$, $\vartheta_{\text{ex}} = 1200$ K, $\vartheta_{\text{ad}} = 2600$ K (air-fuel combustion), $\text{H}_{\text{ex}} = 430$ kWh/t (at a cullet fraction $y_C = 0.7$), and a cumulative wall loss number $\text{NTU}_w = 0.6$ (according to data from [2], this is an optimistic assumption). The efficiency of the heat exchanger system is restricted to $\eta_{\text{re}} < 0.75$ for physical reasons. This is because combustion offgases have higher heat capacities than air and offgas volumes exceed air volumes. Based on the values adopted above, Equation (18) yields an efficiency of $\text{H}_{\text{ex}}/\text{H}_{\text{in}} = 0.55$, hence, a heat consumption of $\text{H}_{\text{in}} = 770$ kWh per t of molten glass.

SUMMARY
    A generalized treatment of the batch-to-melt conversion and of the performance of glass melting tank furnaces by the 1^{rst} and 2nd theorem of

thermodynamics reveals, and helps to quantitatively determine, the conditions for optimal heat exploitation in a given production situation. These comprise an optimal pull rate with a maximum efficiency of heat exploitation located between an extremely wasteful low-pull regime and a furnace tearing high pull regime. Beyond this, a lower threshold for the process heat demand of glass melting tank furnace is assessed as approx. 750 kWh/t. Optimally tuned U-flame furnaces already reach 900 kWh/t [11].

REFERENCES

[1]W. Trier, Glasschmelzöfen – Konstruktion und Betriebsverhalten. *Springer Verlag*, Berlin 1984.

[2]R. Conradt, in: HVG-Fortbildungskurs: „Wärmetransportprozesse bei der Herstellung und Formgebung von Glas", *The German Society of Glass Technology*, Offenbach 2002.

[3]C. Kröger, *Glastech. Ber.* 26 (1953), 202-214.

[4]C. Kröger, W. Janetzko and G. Kreitlow, *Glastech. Ber.* 31 (1958), 221-228.

[5]H.E. Schwiete and G. Ziegler, *Glastech. Ber.* 28 (1955), 137-146.

[6]J. Moore and D.E. Sharp, *J. Am. Ceram. Soc.* 41 (1958), 461-463.

[7]O.D. Gudovich and V.I. Primenko, *Soc. J. Glass Phys. Chem.* 11 (1985), 206-211.

[8]C. Madivate, F. Müller and W. Wilsmann, *Glastech. Ber.* 69 (1996), 167-178.

[9]R. Conradt, *Glastech. Ber.* 63K (1990), 134-143.

[10]R. Conradt, Z. Metallkunde 92 (2001), 1158-1162.

[11]M. Lindig, in: HVG-Fortbildungskurs: „Wärmetransportprozesse bei der Herstellung und Formgebung von Glas", *The German Society of Glass Technology*, Offenbach 2002.

[12]P. Salamon, K.H. Hoffmann, S. Schubert, R.S. Berry and B. Andresen, *J. Non-Equilibrium Thermodynamics* 26 (2001), 73-83.

[13]VDI-Wärmeatlas, A8-A17, *Springer Verlag*, Berlin 1997.

ELECTROMAGNETIC INDUCTION HEATING IN MOLTEN GLASS AT 60 Hz WITH NO SUSCEPTORS

E. Carrillo, M.A. Barron and J. Gonzalez
Division de Ciencias Basicas e Ingenieria
Universidad Autonoma Metropolitana-Atzcapotzalco
Apdo. Postal 118-338
Mexico, D.F., 07050 Mexico

ABSTRACT

Theoretical eddy currents at 60 Hz in molten glass, which is contained in a toroidal vessel, are analyzed when a traveling wave is imposed at its surface. Preliminary findings show that poloidal poly-phased excitation sources promote the occurrence of induced fields in relatively small volumes. Besides, thermal effects and fluid movement are studied showing that high power densities are possible, which is typical of this electrode-less technology in metals. Given that molten glass is a very viscous fluid, electro-magnetically driven flows would be negligible.

INTRODUCTION

Induction heating, a mature technology in metals industry, uses no electrodes for passing electric current through conductive materials for ohmic release, since currents are induced according to Faraday's law. The two basics elements for induction devices are the excitation source, i.e., an inductor winding, and the electrically conducting material to be linked, which will be referred to from now on as the work-piece. The induced electromagnetic field is mainly located in a small region of the work-piece periphery; this region is known as the skin depth, which is related to both the electric properties of the work-piece and the frequency of the excitation source[1]:

$$\delta = \sqrt{\frac{2}{\sigma\mu\omega}} \tag{1}$$

where δ is the skin depth [m], ω is the electric angular frequency [s^{-1}], and μ and σ are the electric conductivity [S·m^{-1}] and the magnetic permeability of the work-piece [H·m^{-1}], respectively. As is well known, a key factor in designing induction devices consists in confining δ inside the work-piece to ensure a proper electromagnetic linkage. A way to heat molten glass through an inductive method at 60 Hz is by using a non-reacting material, known as a susceptor, having both high melting temperature and electrical conductivity. Since the susceptor acts as a work-piece, the Joule dissipation is mainly carried out in it; in this way, the molten glass becomes indirectly heated by the susceptor[2,3]. Here some inefficiency may arise, because the susceptor occupies useful volume otherwise available for glass. On the other hand, Gibbs[4] proposes a vertical arrangement of several molybdenum shanks located in the furnace bottom along the current path to enhance magnetic linkage for glass induction heating at 60 Hz. If a susceptor is undesirable, high frequency needs to be employed in order for the molten glass to become susceptible for induction. Nowadays, lowering working frequency seems appealing, because in this way the frequency converter and its associated costs are avoided. This could justify efforts to explore the possibility of using even 60 Hz in a glass induction furnace without having neither susceptor nor an extra-large furnace size; since, according to Equation (1), a cylindrical furnace would have a radius around 18 m. Matesa[1] et al. report the heating of molten glass at frequencies lower than 1 kHz using a device whose size is below that calculated from Equation (1). By placing an iron core in an open-channel induction furnace, Freidberg et al.[5] report the vitrification of nuclear hazardous wastes, employing frequencies up to 1 kHz within 1 m^3 of work-piece; however, at 60 Hz they had to increase the volume up to 10 m^3 to get a proper electric efficiency. On other hand, it is reported[6] that traveling fields produced by inductors having a poly-phased configuration lower skin depth[7]. Finally, it is important to highlight that the cylindrical geometry is the most common one for induction furnaces; however, the electromagnetic end effects are difficult to mathematically model.

PROPOSED DEVICE

To overcome the lack of electromagnetic linkage due to the poor electric conductivity of molten glass when it is confined in a small volume, an analysis of eddy currents at 60 Hz using a rotating excitation source is undertaken. In addition, to avoid dealing with electric end effects, a toroidal-shaped device with no susceptors is proposed; this geometry can be considered as a stretched out solenoid whose ends are joined. The corresponding coordinate system is shown in Figure 1. External inductor windings around the torus may have two directions which are co-axial to the torus magnetic axis:

a) A poloidal winding whose helical pitch follows the angular coordinate φ .
b) A toroidal winding whose helical pitch follows the angular coordinate θ.

In Figure 1 are shown the proposed external windings, which induce the eddy currents in the work-piece. The toroidal vessel is filled with molten glass, which in this case acts as the work-piece. The exit and return paths of coils resemble the polar pitch of an electrical machine; in this work it is supposed that this arrangement favors the induction process at 60 Hz. In order to evaluate the capability of the above windings to induce electromagnetic fields in the molten glass, one must first find those conditions where field distributions are analogue to that of molten metals. The analysis will be circumscribed to the conductive fluid; in this work, the gap between inductor and the work-piece, which acts both as an electric and a thermal insulator, will be assumed negligible; therefore, the traveling magnetic strength field at the torus surface (H$_s$) will be the same as the resulting wave superposition (H$_o$) from the poly-pashed arrangement.

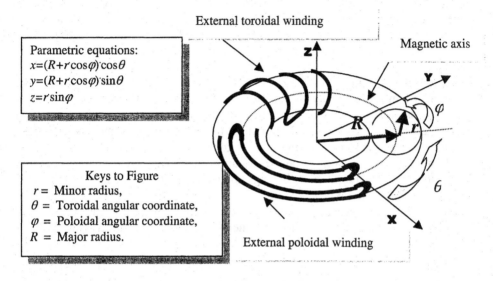

Figure 1. Toroidal coordinate system showing the windings.

BASE CASE

By combining the classical electromagnetic Maxwell equations, the expression describing the behavior of eddy currents is obtained, which is known as the electromagnetic diffusion equation[8]:

$$\nabla^2 \vec{H} + \sigma\mu\nabla\times\left(\vec{v}\times\vec{H}\right) = i\sigma\cdot\mu\cdot\omega\frac{\partial\vec{H}}{\partial t} \qquad (2)$$

Equation (2) is expressed in terms of the magnetic strength vector **H**. Included is the contribution of velocity v [ms^{-1}] of the conductive fluid to the process; i is equal to $\sqrt{-1}$. In the frequency domain, the dependence of **H** on time t is expressed in terms of harmonic variables for stationary fields. When the variable is a traveling wave, its functionality is represented by:

$$H_w(r,\theta,\varphi,t) \equiv \hat{H}_w(r)\exp\left[i(\omega t - \vec{k}_w \bullet \vec{r})\right] \qquad (3)$$

where the subscript w may be either θ or φ, since there is no way to wind coils along the minor radius. In addition, k_w is the wave propagation vector whose components are the wave numbers that feature the traveling behavior of the electromagnetic field. Those wave numbers may arise from the external arrangement of an excitation source characterized by a certain poly-phase configuration and polar pitch similar to a stator of an induction motor[9]. Substituting Equation (3) in (2) and simplifying, both poloidal and toroidal Helmholtz scalar ordinary differential equations are obtained, which depend on the minor radius; these equations can be solved using the Newton-shooting method[10]. Next, a base case is presented in order to find the winding configuration that yields in the work-piece a magnetic field distribution analogue to that of a metal. A working volume of 1.23 m^3 contained in a torus vessel with 0.25 m minor radius (r) and 1.0 m major radius (R) is considered, which is smaller than that of 10 m^3 reported by Freidberg et al.[5] at 60 Hz. Besides, an isothermal motionless molten glass with an electrical conductivity of 28 S·m^{-1} is assumed. A magnetic strength field H_w of 1000 A·m^{-1} at the outer surface of the torus, featured by a wave number k_w of 6, is imposed. From the above remarks, the real magnetic strength distributions were obtained and are shown in Figure 2. One can see that the most suitable distribution belongs to the poloidal magnetic strength field, where the induced field reacts against the original field at the magnetic axis, according to the Lenz law. This is the way in which induced electromagnetic fields are distributed in metals, and therefore from now the analysis will be focused on poloidal arrangements.

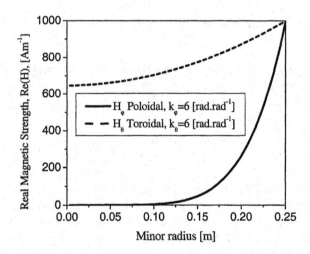

Figure 2. Poloidal and toroidal induced magnetic strength field distributions.

EDDY CURRENT EFFECTS

The interest in studying eddy currents is based on the assessment of two main effects, namely the ohmic heating and the Lorentz body force, both of them being the source terms for the thermal energy equation and the Navier-Stokes equations, respectively. The source terms are expressed in terms of the magnetic strength. The thermal energy equation featuring the scalar field temperature T is:

$$\rho C_p \left(\vec{v} \bullet \nabla T \right) = \nabla \bullet \left(k_T \nabla T \right) - \underbrace{\left(\frac{\nabla \times \vec{H}}{\sigma} - \vec{v} \times \mu_0 \vec{H} \right) \bullet \nabla \times \vec{H}}_{\text{Joule losses}} \tag{4}$$

where heat capacity (C_p) and thermal conductivity (k_T) will be considered as constants, namely 1256 J·K^{-1}kg^{-1} and 75 W·m^{-1}s^{-1}, respectively. The last right term of the above equation is the Ohm law for conductive fluids; here, the induced current density J [A·m^{-2}] is implicit in the Ampere law for low frequencies, which is $\nabla \times$ H = J. The toroidal induced density current (J_θ) is obtained from distributions of poloidal magnetic strength (H_φ) and its derivatives, being both density current and magnetic strength orthogonal to each other. On the other hand, the steady-state Navier-Stokes equations dealing with viscous conductive fluids such as molten glasses under laminar regime are, in vectorial form, as follows:

$$\rho \left(\vec{v} \bullet \nabla \right) \vec{v} = -\nabla P + \eta \nabla^2 \vec{v} + \rho \vec{g} + \underbrace{\left[\nabla \times \vec{H} \right] \times \mu_0 \vec{H}}_{\text{Lorentz body force}} \tag{5}$$

where P is pressure [Pa], g is gravity acceleration [m·s^{-2}]. The last right term in Equation (5) is the Lorentz body force, namely F= J$\times\mu_0$H, whose components are perpendicular each other, as states the Flemming rule, i.e., the right-hand rule[15]. As a result, the prevailing Lorentz force is aiming at the magnetic axis. In this way, if any fluid movement results, this would occur along the minor radial direction. The Lorentz body forces acting along angular coordinates are negligible, so flow equations in those directions will not be considered.

COUPLED GOVERNING EQUATIONS

Equations (2), (4), and (5) constitute a coupled system that yields a strong non-linear system. They were reduced to ordinary differential equations and solved by the Newton-shooting method. The remarks of Section 3 for the poloidal case are considered here. The steady-state Joule losses over the whole volume are 2196 kW, and the resulting temperature in the center of the magnetic axis becomes 1643 K, as is shown in Figure 3. This power matches the k_T{dT/dr}$_{surface}$ heat exchange through the $4\pi^2 R r_s$ torus outer surface, yielding 222 kWm^{-2} of heat flux density. Physical properties of molten glass as a function of temperature are shown in Table I. These properties were used in the governing equations. The operational heat flux density attainable is around 240 kWm^{-2} in high-pull fuel-fired furnaces, and 120 kWm^{-2} for high-melting-rate electric furnaces[12] with cold tops. These data can be compared with that obtained in the proposed device, which is 222 kWm^{-2}.

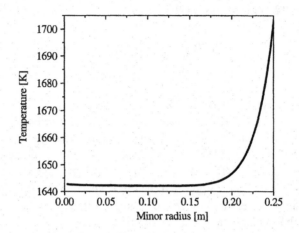

Figure 3. Temperature distribution for a 2196 kW powered torus featured by a poloidal winding.

Table I. Physical properties of soda-lime molten glass[11]

Physical property	Units	Temperature functionality
Density (ρ)	kg.m^{-3}	$2500 \cdot (1 - \beta \cdot T)$
Expansion coefficient (β)	K^{-1}	$5.3 \cdot 10^{-5}$
Dynamic viscosity (υ)	m^2.s^{-1}	$\exp\left(-14.29 + \frac{10525.07}{T-424}\right)$
Electric conductivity (σ)	S·m^{-1}	$563 \exp\left(-\frac{4890}{T}\right)$

The combined body forces acting on the molten glass, namely buoyancy and Lorenz forces, will have little effect on melt velocities, as is shown in Figure 4. This could be beneficial in terms of reducing refractory erosion, since the highest radial velocity and gradient velocity, see Figure 5, are located in the proximity of the confining walls. Since molten glass has a high electrical resistivity, a relatively low induced magnetic strength is required for massive heating. Nevertheless, the concomitant Lorentz forces would not be able to yield significant movements in the viscous fluid. This explains why the proposed device, employing a power of 2196 kW, cannot achieve the velocities of 0.010 ms^{-1} reported by Jutila[13] and Fekolin and Stupak[14] for the motion of molten glass by electromagnetic means.

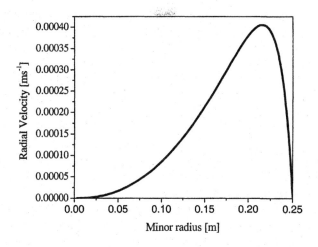

Figure 4. Velocity driven by buoyant and Lorenz forces.

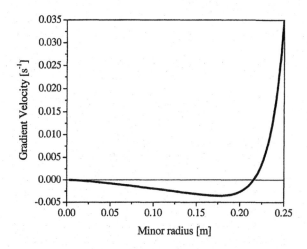

Figure 5. Gradient velocity driven by buoyant and Lorenz forces.

CONCLUSIONS

The feasibility of inducing eddy currents using 60 Hz frequency in molten glass contained in a toroidal vessel has been studied. This led to proposing a poly-phase poloidal winding as a suitable excitation source. Besides, it is found that a working volume of 1.23 m^3 in the toroidal vessel is enough to yield a proper electromagnetic linkage at 60 Hz, which is smaller than that of 10 m^3 reported by other authors[5]. In order to experimentally test our proposal, a prototype is being constructed in which an ionic liquid at room temperature with the same electrical conductivity as that of the molten glass will be employed instead of the melt. Currently, physical simulations are being carried out.

REFERENCES

[1]J. M. Matesa, K. Won, and H. Demarest, "Method and apparatus for inductively heating molten glass or the like (two stages: heating, radiant and induction)," U.S. Pat. No. 4 610 711, Sept. 9, 1986.

[2]J. Ferguson, "Electric furnace", U.S. Pat. No. 2, 101 675, Dec. 7, 1937.

[3]J. Ferguson. "Improvements in and relating to electric furnaces," U.K. Pat. No. 533 977, Feb 25, 1941.

[4]M. C. Gibbs, "The supply and applications of electricity to the glass industry," *Glass Technology,* **6** [4] 120-121 (1965).

[5]J. Freidberg, P. Shajii, K.W. Wenzel, and J.R. Lierzer, "Electrode-less melter for vitrification of nuclear waste," pp. 33-38 in *Scientific Basis for Nuclear Waste Management XX Materials Research Society Symposium,* Proceedings, v 465, Materials Research Society, Pittsburg, PA, USA, 1997.

[6]J. Szekely and C. Chang, "Turbulent electromagnetically driven flow in metals processing.Part 1:Formulation," *Iron and Steelmaking,* [3] 190-195 (1977).

[7]H. Iwata, K. Yamada, T. Fujita, and K. Hayashi. "Electromagnetic stirring of molten core in continuos casting of high carbon steel," *Transactions ISIJ,***16**, 374- 381 (1976).

[8]A. Krawczyk and J.A. Tegopoulos, pp. 5-7, 102-103 in *Numerical Modeling of Eddy Currents,* Clarendon Press, Oxford, 1993.

[9]F. Williams, J. Eastham, and L. Piggott, "Analysis and design of pole-change motors using phase-mixing techniques," Proc. IEEE, **111** [1] 80-94 (1964).

[10]A. Constantinides, in *Applied Numerical Methods with Personal Computers,* pp. 418-439, McGraw-Hill Company, 1987.

[11]A. Ungan, I.G. Choi, and R.U. Payli, "Numerical Study of Melting Conditions in a Deep Glass Melter", pp. 639 in *Advances in Fusion and Processing of Glass,* Edited by Arun K Varshneya,. Dennis F. Bickford and Peter Bihuniac, *Ceramic Transactions,* Vol. 29, 1993.

[12]W. Trier, in *Glass Furnaces- Design construction and operation,* pp.128,134,190, Society of Glass Technology, Sheffield,U.K., 1987.

[13]S. T. Jutila, "Magnetohydrodinamic behavior of isothermal electrically conducting liquid glass," *Glass Technology,* **11** [5] 135-138 (1970).

[14]V. N. Fekolin and F.A. Stupak, "The application of the magnetohydrodinamic effect to agitation of glass," Steklo i Keramika, [12]10-12 (1984).

[15]R. P. Feyman, R.B. Leighton, and M. Sams, in *The Feynman Lectures on Physics,* pp. 13-12, Addison-Wesley Publishing Company, Massachusetts, 1966.

FULL OXY CONVERSION OF A FLOAT FURNACE EQUIPPED WITH SEPARATED JETS ALGLASS FC BURNERS: FROM 0D MODEL TO 3D CHARACTERIZATION

Joumani Youssef, Imbernon Christian, Tsiava Remi and Leroux Bertrand
AIR LIQUIDE Claude-Delorme Research Center
1, chemin de la Porte des Loges
Les Loges en Josas – BP 126
78354 Jouy-en-Josas

ABSTRACT

To prepare the conversion of Float furnaces to full-oxy combustion, Air Liquide has developed adapted in-house numerical tools. The relevance of these codes has been demonstrated here by replacing the air burners in the TC21 Ford furnace by ALGLASS FC burners. First a 3D simulation has been carried out on the referenced air case. Results are checked to be in good agreement with experimental data. The strategy to find the oxy inlet power consists in using a 0D tool. The power to introduce is found to be 20.2 MW. That corresponds to a reduction in fuel consumption of 28%. Thanks to a dedicated 1D tool coupled with an optimization software, an optimized longitudinal power distribution has been found. Then, a 3D "Full-oxy" simulation has been achieved. The 3D temperature fields and heat fluxes are close to those obtained in the nominal "Air case." Requirements on glass temperature at exit and maximum crown temperature are fulfilled.

INTRODUCTION

Full oxy-firing has been accepted as a cost-effective process for melting many types of glass with the benefit of reducing emissions such as NO_x while improving the operational flexibility. For a variety of reasons, there has been only limited use up to now of full oxy-firing in float glass melting tanks [1]. As float glass tanks are typically quite large, and the glass product has extremely stringent quality demands, the potential economic consequences of a glass tank producing low-quality product has limited the use of full oxy-combustion technology. This occurs despite the fact that many glass makers observe an increase in the quality of glass product from full oxy-fired tanks and some important energy savings.

To help glassmakers, Air Liquide has developed an expertise in the conversion of glass furnaces [2-8]. We usually use numerical simulations that have become a useful way to predict heat transfer to the glass [9].

The purpose of this article is to illustrate the strategy usually chosen through the TC21 case [10]. The air burners are replaced by Air Liquide ALGLASS FC oxy-burners.

ALGLASS FC BURNERS

Air Liquide develops and sells a large variety of oxy-burners from 100 kW to 4 MW. For the application studied here, we have chosen the ALGLASS FC^{TM} burners, which develop low momentum, flat flame. Natural gas is injected through three injectors on the lower part of the block, while oxygen is injected on the higher part with a certain angle directed towards fuel flows. Such an arrangement avoids a rising flame (natural gas is lighter than O_2) and reduces foam formation, since the atmosphere near the glass bath is less oxidizing, and partial pressures of O_2 and H_2O are not so high as with other oxy burners.

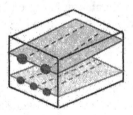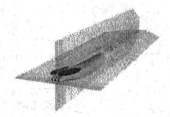

Fig. 1: geometry of the ALGLASS FC burners

Other advantages of the burners are as follows:
- Highly luminous flame,
- Flat flame maximizing the covering surface,
- Low NOx emission,
- Large operating range (between 30% and 150% of its nominal power).

Prior to industrial computations, flames obtained with ALGLASS FC^{TM} burners have been calibrated. The calibration of the models has been obtained by taking into account measurements performed in an Air Liquide pilot furnace [11]: local concentrations of species, temperature measurements, and heat-transfer profiles.

NUMERICAL TOOLS
Airlog® : A Global Heat and Mass Transfer Tool
Prior to any simulation, a heat and mass balance has to be achieved to assess both natural gas and oxygen consumptions. This is carried out by the AIRLOG software, a calculation tool developed by AL, which may be adapted to any kind of furnace. It solves the following energy equation :

$$Q_{fuel} = Q_{walls} + Q_{glass} + Q_{bath} + Q_{flue\ gas}$$

Q_{fuel} = total power of the burners,
Q_{walls} = heat losses through superstructures,
Q_{glass} = energy of the glass to exit at the desired temperature,
Q_{bath} = heat losses through the glass tank walls,
$Q_{flue\ gas}$ = heat of the flue gas.

The way to use it consists in choosing an unknown among the variables of the energy equation and by entering the information about the other variables. Generally, it is rather easy to measure fumes temperature and composition, hence one can have access to more cryptic data, such as heat losses or parasitic air inflow. An adapted graphical user interface enables the post processing of results.

GLASS1D : A Tool For the Glass Industry
The tool previously described is very relevant to have coarse data, but it does not give any temperature or heat flux profiles. Consequently, Air Liquide has developed a 1D software, GLASS1D, which does 1D mass balance, solves 1D energy equation and calculates 2D radiative heat transfer [12, 13].

This tool has been validated for several kinds of furnaces and shows interesting perspectives. It is indeed possible to:
- estimate quickly and globally the impact of an oxy-boosting,
- evaluate the influence of an operating parameter,
- optimize a design parameter,
- compute steady states for process control,

with really less CPU time than any full 3D simulation (about 30 min for GLASS1D against 3 days for a 3D simulation).

This tool is relevant to have a first estimation, but it does not replace a 3D simulation. Actually, the 1D tool does not give information on the molten glass behavior, nor on NO_x production nor on how the flame behaves. 3D simulation remains the last step for a complete furnace modeling.

ATHENA : a CFD Code For Furnaces

Once the basic configuration is known or defined thanks to the 0D and 1D tools, a 3D computation can be run. Therefore, an in-house space combustion simulator called ATHENA [8, 14] is used by coupling it to the glass tank model (GTM) from the TNO institute [15].

The coupling of the ATHENA code to the glass tank model (GTM) code from the TNO institute is made in the following way: assuming a constant temperature on the load at initial time, ATHENA is run and computes heat fluxes at the glass surface. This 2D distribution of heat fluxes is used by GTM as a boundary condition. GTM then computes a new temperature distribution at the surface. This procedure is repeated until convergence criteria are reached.

Parallel computing is possible with the ATHENA code even for radiative heat transfer. Moreover, dynamic memory allocation allows intensive use. A 4-millions-cells mesh takes four days on four processors to converge.

RESULTS

The use of the numerical tools previously presented is now illustrated through the oxy conversion of the TC21 Ford furnace.

1st Step: 3D/AIR Simulation

The first step consists in carrying out a 3D simulation in air-firing conditions. The total inlet power is 40.0 MW. As this case is well documented [5-7], let us focus on the main results obtained by our 3D simulation. The global heat and mass transfers are in agreement with our 0D tool and with published results [15]. Fig. 2 compares crown temperature and glass temperature profiles to experimental data.

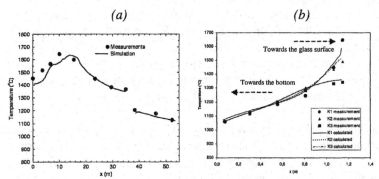

(a)   *(b)*

Fig. 2: Temperature profiles (a) along the crown centreline and (b) in the glass bath

Comparison between measurement and numerical results

The two sets of numerical data are in good agreement with experimental measurements. In the glass bath, the maximum discrepancy observed is 40K, but is on average about 15K. The temperature of exhaust gas is found to be 1630°C. The temperature of the glass at the exit is 1170°C, as found in [15]. 11 MW are transferred from the combustion chamber to the glass surface.

2nd Step: 0D oxy-conversion

The second step consists in determining the oxygen power to be injected in the furnace thanks to the 0D tool. The way to proceed consists in doing the 0D balance in the air case first. The software has an option that enables finding the oxy-fuel power for an oxy-conversion, by keeping constant the glass temperature at exit.

The only main difference concerns the flue gas temperature. According to our experience, we have supposed a temperature of 1450°C in the oxy case. It is 200K lower than in the air case.

A power of 20.2 MW is given by AIRLOG software. It corresponds to a volume flow rate of 1935 Nm^3/h for fuel and 4185 Nm^3/h for oxygen. The specific consumption of fuel is then 28% lower than in the air case.

However, it does not give any information about the longitudinal distribution of the total power. It is well known that in oxy furnaces the profiles of power density are not similar to those of the air case. To know what is the best power distribution, we have to use the GLASS1D software.

3rd Step: Air simulation with GLASS1D

First of all, it is necessary to perform an air simulation to identify the configuration to be carried out in the oxy case. This simulation will be used as a basis for the 1D oxy simulation. All input data are equal to the 3D ones. First, heat fluxes are checked to be in good agreement with CFD simulation. For example, the glass temperatures at exit are 1172°C for 1D and 1170°C for 3D.

Fig. 3a compares the crown temperature profile to the experimental data, and Fig. 3b does this at the glass surface. Note that there are no experimental data at the glass surface. We have therefore used the data provided by the sensors that are only 2 cm below the glass surface.

The 1D curves are close to the experimental data. Numerical profiles have the same shapes as those deduced from measurements, even if GLASS1D tends to overestimate the experimental data. In the glass bath, such a result is expected, since the real temperatures at the surface are certainly a little bit different than those measured 2 cm below.

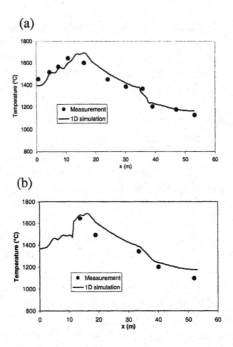

(a)

(b)

Fig. 3: Temperature profiles (a) along the crown centreline and (b) at glass surface
Comparison between 1D and measurements in the air configuration

4th Step: Oxy simulation with GLASS1D

Using all the data coming from the 1D simulation, it is possible to carry out full-oxy simulation. In this case, the oxy-burners have been chosen in staggered positions. Two outlets or chimneys are defined near the doghouse entry. To find the power distribution, we have coupled the 1D tool to FRONTIER, an optimization software. By giving:

- A constraint: the total power
- Input variables: the minimum and maximum power allowed for each burner
- A target: the glass temperature at exit

the optimization software randomly performs several computations, using the gradient method, in order to find the best power distribution with respect to the target. The obtained one is shown in Fig. 4.

Advances in Fusion and Processing of Glass III

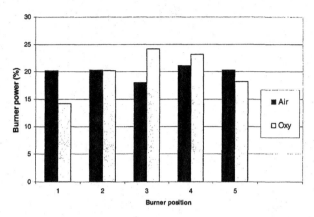

Fig. 4: Power distribution

As seen in Fig. 4, the power distribution is more homogeneous in the air case than in the oxy configuration. In the latter, there is a maximum in the middle of the melting zone. This program may be used to test other influencing parameters such as the pull rate, a change in refractory, or a change in furnace geometry.

5th Step: 3D/Oxy simulation

Once the power distribution is defined, it is possible to perform a 3D simulation. The oxy-burners are ALGLASS FC burners. All injectors are finely meshed; no premixed combustion is assumed. Results of this 3D simulation are compared with air-firing results (see Fig . 5 for 3D temperature profiles in oxy and air configurations).

Fig. 5 shows that the full-oxy conversion respects the requirements existing in air configuration; i.e., the maximum crown temperature is only 10 K greater than in the air case. The glass temperature profiles obtained by oxy-simulation are very close to air simulation and, therefore, to experimental data. The temperature of the glass at the exit is 1180°C. It is only 10 K greater than those obtained by the air simulation.

These results therefore enable one to validate the choice of the total power inlet and the corresponding power distribution.

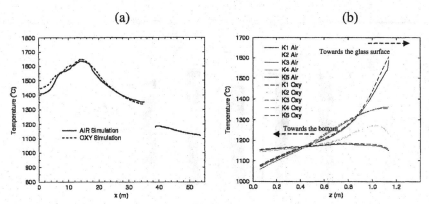

(a)   (b)

Fig. 5 : Temperature profiles (a) along the crown centreline and
(b) in the glass bath. Comparison between 3D/Air and 3D/Oxy results.

CONCLUSIONS

For a long time, Air Liquide has prepared the future conversion of float
furnaces to oxy-firing conditions. To help glassmakers to convert their furnaces,
Air Liquide has developed a large range of numerical tools dedicated to the glass
industry.

A 0D software can be used to carry out a heat and mass balance. Once an air-
fired furnace is calculated, a specific option is available to find the total inlet
power needed for a full-oxy conversion.

A 1D program has also been developed to compute the physical behaviour of
a glass furnace: radiative heat transfer, heat losses, 1D mass & heat balance and
melting process of glass. Coupled with an optimisation software, this program has
been used to find the best power distribution when converting a float furnace to
oxy-fired one.

A 3D modelling package composed by ATHENA, Air Liquide software
coupled with GTM, the TNO code, is also available.

Here, these three tools have been combined to virtually convert the Ford
Nashville TC21 furnace to oxy-firing conditions. It has been shown that using the
proposed dedicated tools is a good way to convert a furnace. The oxy-burner
power corresponds to a specific consumption of fuel 28% lower than those of the
air case. All other parameters (glass temperatures profiles, crown temperatures,
batch length...) are similar to air-case ones.

The door is now open for further improvement and optimisation of the oxy-firing design.

ACKNOWLEDGEMENTS
The authors want to thank M. Till, Air Liquide Modeling Expert and J-F. Simon, Air Liquide Glass Expert for their precious help.

REFERENCES
[1] W.G. Pike, J.J. Shea and L.R. McCoy,"The Evolution of float furnaces," *Glass*, **Feb** 65-69 (1992).
[2] D. Jouvaud, J-F. L'Huissier and B. Genies, "Glass melting with Pure oxygen Combustion : Modeling of Convective and Radiative Heat Transfers," *Cer. Eng. and Sci. Proc.*, **9** [3] 221-231 (1988).
[3] R. Ruiz, S. Wayman, B. Jurcik, L. Philippe and J-Y. Iatrides, "Oxy-fuel Furnace design Considerations," *Cer. Eng. and Sci. Proc.*, **6** [2] 179-189 (1995).
[4] F. Ammouri, C. Champinot, W. Bechara, E. Djavdan , M. Till and B. Marié, "Influence of oxy-firing on radiation transfer to the glass melt in an industrial furnace: importance of spectral radiation model," *Glass Science & technology*, **70** [7] 201-206 (1997).
[5] D. Shamp, O. Marin, C. Champinot, B. Jurcik and R. Grosman, "Oxy-fuel furnace design optimization using coupled combustion/glass bath numerical simulation," *Proceedings of the 59th Annual Conference on Glass Problems*, Columbus, OH, October 27-28 (1998).
[6] C. Schnepper, O. Marin, C. Champinot and J-F. Simon, "A modeling study comparing an air- and an oxy-fuel fired float glass melting tank," *Proceedings of the XVIII International Congress on Glass*, San Francisco, July 5-10 (1998).
[7] C. Schnepper, B. Jurcik, C. Champinot and J-F. Simon, "Coupled combustion space-glass bath modeling of a float glass melting tank using full oxy-combustion," *Proceedings of the Advances in Fusion and Processing of Glass*, Toronto, July (1997).
[8] M. Till, O. Marin, O. Louédin and B. Labégorre, "Numerical Simulation of Industrial Processes," *American/Japanese Flame Research Commitees, International Symposium*, Kauai, Hawaï, September 9-12 (2001).
[9] M.K. Choudhary, "Mathematical Modeling of Transport Phenomena in Glass Furnaces: an overview of status and needs," *Verre*, **6** [6] 14-19 (2000).
[10] C. Champinot, "TC21 Round Robin Discussion" *IV International Seminar on Mathematical Simulation in Glass Melting*, Vsetin, Czeck republic (1997).
[11] B. Leroux, J-F. Simon, P. Duperray, R. Tsiava and R. Soula, "Ultra low Nox oxy-combustion system for glass furnaces with adjustable flame length and heat transfer profile," *Glass Odissey*, Montpellier, France (2002).
[12] F.C. Lockwood and N.G. Shah, "A new radiation solution method for incorporation in general combustion prediction procedures," *18th Symposium on Combustion*, 1405-1414 (1980).

[13] A. Soufiani and E. Davdjan, "A comparison between Weighted Sum of Gray gases and statistical narrow band models for combustion applications," *Combustion and Flame*, **97** 240-250 (1994).

[14] C. Champinot, M. Till, C. Haung-Naudin and A. Klug, "Gekoppelte Berechnungen von glas und atmosphaere," *Glas Ingenieur*, **Feb** (1996).

[15] P. Boerstoel, A.M. Lankhorst and E. Muysenberg, "Complete simulation of the glass tank and combustion chamber of the former Ford Nashville float furnace including melter, refiner and working end," *IV International Seminar on Mathematical Simulation in Glass Melting*, Vsetin, Czeck republic (1997).

A METHOD FOR MAKING ARSENIC OXIDE IN CULLET MORE ACTIVE AS A FINING AGENT

M. Kawaguchi, T. Narita, J. Naka, H. Yamazaki and S. Yamamoto
Technical Division, Nippon Electric Glass Co.,Ltd.
7-1, Seiran 2-Chome
Otsu, Shiga 520-8639
Japan

ABSTRACT

A method to effectively activate arsenic ions in non-alkaline glass cullet with other multivalent ions was investigated. When a cullet containing arsenic ions was melted with Sb_2O_5, the mixed cullet showed a higher fining ability and has higher As^{5+} ratio than the cullet melted alone. This result was consistent with the prospects from a preliminary experiment, in which the redox states of arsenic, antimony and tin oxide in a non-alkaline glass melted at various temperatures were analyzed. It was found that Sb^{5+} could oxidize As^{3+} in the glass melt.

INTRODUCTION

Arsenic oxide is well known as the most effective fining agent. Arsenic oxide is still used for some kinds of glass with high viscosity, such as non-alkaline glass, although it has been replaced by antimony oxide in many kinds of glass. It is desirable to decrease the use of arsenic oxide, however, because of its toxicity. The activation of arsenic oxide in cullet would be an effective method to substantially decrease its use in the batch. Most arsenic ions in the cullet exist as As^{3+}, because As^{5+} is reduced during the melting process according to the following reaction, $As^{5+} + O^{2-} => As^{3+} + 1/2O_2$. Therefore, cullet would not have a fining ability even if the cullet were melted with nitrates, as reported by Krol et al., in the case of antimony ion[1]. However, if another redox ion is added to the cullet, and it oxidizes arsenic ion in the cullet during the primary melting process, the amount of arsenic oxide to be added to the batch could be decreased.

This report shows the fining abilities of cullet containing arsenic oxide when it is melted alone and with Sb_2O_5. It was presumed from the viewpoint of thermodynamics that Sb_2O_5 could oxidize As^{3+} in the cullet. A non-alkaline glass, OA-10, was used in this study.

EXPERIMENTS

Measurement of The Redox State of Fining Agents in OA-10

Industrial-grade sand, aluminum oxide, boric oxide, calcium carbonate, strontium nitrate, and barium nitrate were weighed to obtain 300 g of OA-10 glass with the composition, 60 SiO_2, 15 Al_2O_3, 10 B_2O_3, 5 CaO, 5 SrO, 2 BaO, and other minor components in mass per cent. As_2O_3, Sb_2O_5, or SnO_2 was added as a fining agent to each batch, so the glass contained 1 wt% of As_2O_3, Sb_2O_3, or SnO_2. The batch in a platinum crucible was melted in an electric furnace at various temperatures for 24 hours. After the glass melt was quenched, the glass was smashed into pieces with a size of about 10 mm and melted again at the same temperatures for 24 hours to accelerate reaching equilibrium. Then the ratios of the oxidized ion to the reduced ion of fining agent in the glass were analyzed by the combination of redox titration and ICP-AES. The equilibrium was confirmed by analyzing the redox ratios in the glass melted for various periods.

Fining Ability of The Cullet Containing Arsenic Oxide

The fining abilities of the cullet containing arsenic oxide were investigated when the cullet was melted alone and with Sb_2O_5. The glass OA-10 containing arsenic oxide was prepared by the same method as described above. The amount of additional As_2O_3 to the batch was 0.59 wt%, corresponding to 3×10^{-5} mol per 1g of glass. The batch was melted at 1923 K. Then the glass was smashed into pieces with a size of less than 1 mm. Sb_2O_5 was added to 50 g of cullet. The amount of additional Sb_2O_5 was 0.24, 0.49, or 0.73 g, which corresponds to 1.5×10^{-5}, 3×10^{-5}, and 4.5×10^{-5} mol per 1g of glass, respectively. The mixture of the cullet and Sb_2O_5 was melted in a cone-shaped platinum crucible in an electric furnace at 1673 K for 20 minutes and then heated up to 1873 K at the rate of 10 K/minute, and kept for 20 minutes at that temperature. After the melt was quenched, the number and diameter distribution of blisters in the glass were investigated, and the contents of As^{3+}, As^{5+}, Sb^{3+}, and Sb^{5+} were quantitatively analyzed by the combination of redox titration and ICP-AES.

RESULTS AND DISCUSSION

Comparison of ΔG^0 among Fining Agents

In general, redox reactions are discussed thermodynamically[2,3,4,5]. Redox reactions of arsenic and antimony ions in a glass melt are shown as follows:

$$As^{3+} + 1/2O_2 <=> As^{5+} + O^{2-} \qquad \Delta G^0_{As} \qquad (1)$$

and

$$Sb^{3+} + 1/2O_2 <=> Sb^{5+} + O^{2-} \qquad \Delta G^0_{Sb} \qquad (2)$$

where ΔG^0_{As} and ΔG^0_{Sb} are the standard Gibbs energy changes for the oxidation of arsenic and antimony ions, respectively. Combination of reaction (1) and reaction (2) gives

$$As^{3+} + Sb^{5+} \iff As^{5+} + Sb^{3+} \qquad \Delta G^0_{As-Sb} = \Delta G^0_{As} - \Delta G^0_{Sb} \qquad (3)$$

In the case of $\Delta G^0_{As-Sb} < 0$, the reaction (3) goes to the right, and $\Delta G^0_{As-Sb} > 0$, goes to the left. In other words, the direction of the reaction can be predicted from the comparison between ΔG^0_{As} and ΔG^0_{Sb}.

When the concentration of the species is low and the activity of O^{2-} is assumed as unity, the equilibrium constant of the reaction (1), K_{As}, is related to the standard Gibbs energy change as follows.

$$\ln K_{As} = \ln \frac{[As^{5+}]}{[As^{3+}]} - \frac{1}{2} \ln P_{O_2} = -\frac{\Delta G^0_{As}}{RT} = -\frac{\Delta H^0_{As}}{RT} + \frac{\Delta S^0_{As}}{R} \qquad (4)$$

where brackets represents concentration. R, T, ΔH^0_{As} and ΔS^0_{As} are gas constant, absolute temperature, the standard enthalpy change and the standard entropy change in the oxidation of As^{3+}, respectively. P_{O_2} is the oxygen partial pressure in the atmosphere above the glass melt ($P_{O_2} = 0.21$ in this study). By assuming that ΔH^0_{As} and ΔS^0_{As} are independent of temperature, the $1/T$ - $\ln K_{As}$ plot determines ΔH^0_{As} and ΔS^0_{As}, which in turn derives ΔG^0_{As}.

The contents of redox ions in the OA-10 glass melted at various temperatures are shown in Table I. The ratios of the oxidized ion to the reduced ion of the fining agents in logarithm were plotted against the reciprocal of the absolute temperature in Figure 1. The error bars correspond to the dispersion of the measured values. The data were fitted by least-squares in Figure 1. ΔH^0, ΔS^0 and ΔG^0 for each fining agent were calculated from the slope and the intercept using equation (4) and are given in Table II and Figure 2. In Table II, the values of ΔH^0 and ΔS^0 of arsenic and tin oxide are valid in comparison with the values reported previously[6], although they are not exactly the same due to the difference of the glass composition. On the other hand, ΔH^0 of antimony oxide is different from the previous data, apparently because the actual change in the ratio of Sb^{5+} to Sb^{3+} is smaller than the dispersion of the measurement as shown in Figure 1. However, the comparison of ΔG^0 among fining agents is considered to be possible.

The equilibrium constant of the reaction (3), K_{As-Sb}, is related to the standard Gibbs energy change as follows.

$$\ln K_{As-Sb} = \ln \frac{[As^{5+}][Sb^{3+}]}{[As^{3+}][Sb^{5+}]} = -\frac{\Delta G^0_{As-Sb}}{RT} \qquad (5)$$

In Figure 2, ΔG^0_{As-Sb} is always negative over the range of melting temperatures, and it can be stated that Sb^{5+} oxidizes As^{3+} in the glass melt. In addition, the difference increases with decreasing temperature, resulting in the fact that K_{As-Sb} increases, and the content of As^{5+} and Sb^{3+} increases with decreasing temperature. From this consideration it is expected that Sb^{5+} oxidizes more As^{3+} in the glass melt during the primary melting process. Then activated As^{5+} can release oxygen at a higher temperature.

Fining Ability of The Cullet Containing Arsenic Oxide

Figure 3 shows the cross sections of specimens melted at 1673 K for 20 minutes. The figure also gives the ratios of As^{3+} to total arsenic ion and of Sb^{3+} to total antimony ion in each specimen. The additional Sb_2O_5 decreased the As^{3+} ratio in the glass at 1673 K. This indicates that Sb^{5+} oxidized As^{3+} to As^{5+} in the glass melt. At that temperature, a foam layer is formed on the surface of the glass melt, and it becomes thicker as the additional Sb_2O_5 increases. This foam layer is due to the oxygen generation from the decomposition of excess Sb_2O_5. The results of chemical analysis showed that the ratio of Sb^{5+} used for the oxidation of As^{3+} to Sb^{5+} added to the cullet is estimated as 0.18, 0.17, and 0.11 for 0.24, 0.49, and 0.73 g of additional Sb_2O_5, respectively. In other words, the amount of Sb^{5+} not used for the oxidation increases as the additional Sb_2O_5 increases. The excess of additional Sb_2O_5 releases oxygen according to reaction (2) after oxidizing As^{3+} in the glass melt. These considerations are consistent with the foam-formation behavior. A certain amount of additional Sb_2O_5 to the cullet containing no arsenic oxide led to a much thicker foam layer than with the cullet containing arsenic oxide, which is also shown in Figure 3.

Figure 4 shows the cross sections of the specimens melted at 1673 K and then 1873 K. Figure 4 also gives the diameter distributions of blisters in the specimens. Even though the cullet contains arsenic oxide, blisters in the glass melt do not decrease so much by being heated to 1873 K without adding Sb_2O_5. On the other hand, the addition of Sb_2O_5 to the cullet containing arsenic oxide apparently decreased blisters at 1873 K. In this process it is assumed that As^{3+} is oxidized at 1673 K by Sb^{5+} and reduced at 1873 K. The oxygen released during reduction helps the fining. Finally, it is confirmed that the ratio of As^{3+} increases and the ratio of Sb^{3+} remains almost the same. When the cullet contains no arsenic oxide, additional Sb_2O_5 slightly decreases blisters at 1873 K, but the extent of the decrease is less than that in the specimens using the cullet containing arsenic oxide. From these results, it can be stated that arsenic oxide oxidized by additional Sb_2O_5 decreases blisters more efficiently than additional Sb_2O_5 itself.

CONCLUSION

It was confirmed that arsenic oxide in the cullet becomes more active as a fining agent by being melted with Sb_2O_5, and the number of blisters in the glass melt can be decreased. In this process it is assumed that arsenic oxide in the cullet is oxidized by Sb_2O_5 and then reduced by being heated. The oxygen released during the reduction helps the fining.

REFERENCES

[1] D. M. Krol and P. J. Rommers, "Oxidation-reduction behaviour of antimony in silicate glasses prepared from raw materials and cullet," Glass Technology, **25** [2] April 115-18 (1984).

[2-5] For example.

[2] W. D. Johnston, "Oxidation-reduction equilibria in molten $Na_2O/2SiO_2$ glass," J. Am. Ceram. Soc., **48**[4] April 184-90 (1965).

[3] A. Paul, "Effect of thermal stabilization on redox equilibria and colour of glass," J. Non-Cryst. Solids, **71** 269-78 (1985).

[4] H. D. Schreiber, L. J. Peters, J. W. Beckman and C. W. Schreiber, "Redox chemistry of iron-manganese and iron-chromium interactions in soda lime silicate glass melts," Glastech. Ber. Glass Sci. Technol., **69**[9] 269-77 (1996).

[5] H. M. Simon, "Electron exchange reactions between polyvalent elements in soda-lime-silica and sodium borate glasses," Glastech. Ber. Glass Sci. Technol., **69**[12] 387-95 (1996).

[6] P. Buhler, "Thermodynamics of the redox reactions between the oxygen and the oxides of polyvalent elements in glass melts," Glastech. Ber. Glass Sci. Technol., **72**[8] 245-53 (1999).

Table I.   Contents of redox ions in equilibrated OA-10

Temp. [K]	As^{3+} [wt%]	$As^{3+} + As^{5+}$ [wt%]	Sb^{3+} [wt%]	$Sb^{3+} + Sb^{5+}$ [wt%]	Sn^{2+} [wt%]	$Sn^{2+} + Sn^{4+}$ [wt%]
1623	-	-	0.778	0.824	0.038	0.803
1673	0.332	0.724	0.774	0.823	-	-
1723	0.375	0.701	0.763	0.832	0.077	0.831
1773	0.422	0.666	0.770	0.835	0.104	0.788
1823	0.408	0.598	0.778	0.816	0.151	0.772
1873	0.392	0.498	0.745	0.809	0.177	0.677
1923	0.343	0.382	-	-	0.178	0.575

Table II. ΔH^0 and ΔS^0 for each fining agent

Fining agent	ΔH^0 (kJ $\cdot$ mol^{-1})	ΔS^0 (J $\cdot$ K^{-1} $\cdot$ mol^{-1})
Arsenic oxide	-234	-131
Antimony oxide	14.9	-7.05
Tin oxide	-198	-90.0

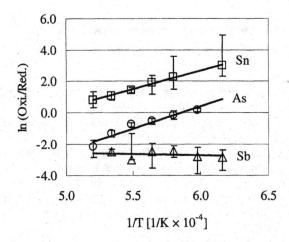

Figure 1. The ratios of oxidized ion to reduced ion of fining agents against the
reciprocal of absolute temperature.

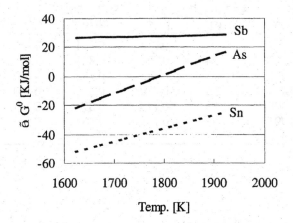

Figure 2. ΔG^0 for each fining agent against absolute temperature.

| Additional | As$_2$O$_3$ content in the cullet [wt%] | |
| Sb$_2$O$_5$ [g] | 0.42 | 0 |

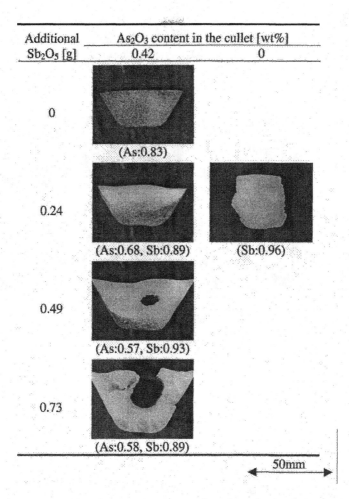

0	(As:0.83)	
0.24	(As:0.68, Sb:0.89)	(Sb:0.96)
0.49	(As:0.57, Sb:0.93)	
0.73	(As:0.58, Sb:0.89)	

50mm

Figure 3.   Cross sections of specimens made by mixing 50 g of OA-10 cullet with various amount of Sb$_2$O$_5$ followed by melting at 1673 K for 20 minutes. The figures in parentheses below each photo are the ratios of As^{3+} to total arsenic ion and of Sb^{3+} to total antimony ion in each specimen. The As^{3+} ratio in OA-10 cullet before being melted with Sb$_2$O$_5$ was 0.82. The scatter in the ratio is about ±0.03 and ±0.10 for As^{3+} and Sb^{3+}, respectively. As$_2$O$_3$ content in the cullet is smaller than the amount of additional As$_2$O$_3$ to the batch because of volatilization.

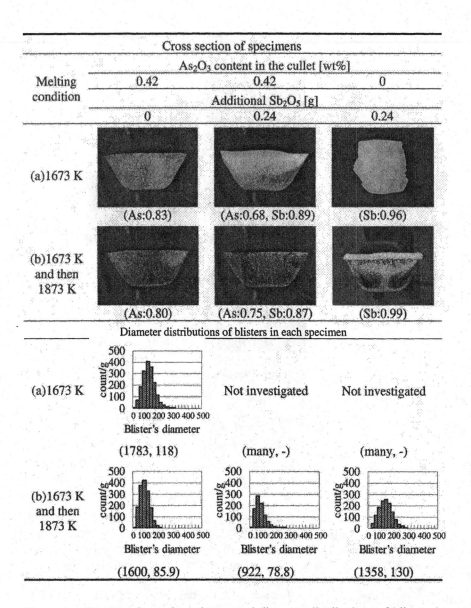

Figure 4. Cross sections of specimens and diameter distributions of blisters in the specimens. The left figures in parentheses under the bar charts are the numbers of blisters per 1g of glass, and the right figures are the average diameter in m. The range of each bar is 25 m. The melting conditions are (a) 1673 K for 20 minutes, (b) then heated to 1873 K at the rate of 10 K / minute, and kept at 1873 K for 20 minutes.

REDOX AND FOAMING BEHAVIOR OF E-GLASS MELTS

A.J. Faber and O.S. Verheijen
TNO Institute of Applied Physics
P.O. Box 595
5600 AN Eindhoven
The Netherlands

J.M. Simon
Borax Europe Ltd.
170 Priestley Road
Guildford GU2 5RQ
United Kingdom

ABSTRACT

The effect of the redox state of sulphate fined E-glass batches on the foaming behavior of the melt has been studied by experiments and by thermochemical modelling calculations. The objective of the study is to identify batch-related measures for foam control. It is found that the redox state and SO_3 content of the initial melt phase at 1250 °C are the key parameters for controlling secondary foam. Oxidized E-glass melts, with relatively high (> 0.1 wt%) SO_3-contents in the initial melt phase, are especially prone to excessive foaming. To reduce foam, while maintaining a sufficiently oxidised glass melt with adequate fining efficiency, the amount of sulphate added to the batch can be decreased by replacing a large part of the sulphate by an alternative oxidising agent. From a series of preliminary foaming experiments with alternative oxidants, sodium-perborate appears to have the strongest effect on the decrease of foam.

INTRODUCTION

E-glass is a calcium aluminoborosilicate glass, fined by sulphate, that is used in the manufacture of continuous filament fiber glass. In many E-glass tanks, fired by gas-oxy burners excessive foaming of the melt is observed.

A thick foam layer, covering the glass melt, will block the heat radiation from the combustion chamber into the melt. Thus, uncontrolled foaming has many detrimental consequences for the glass melting process, including: temperature instabilities in the tank, elevated crown temperatures, decreased melting capacity or increased energy consumption, extra evaporation of volatile glass components and accelerated attack of the refractory lining by the aggressive foam.

The intensified foaming behaviour in oxy-gas fired glass tanks is associated with the effects of the water-rich oxy-gas combustion atmosphere on the glass melt, resulting in an increased production of SO_2 fining gas and in a higher foam layer stability [1,2].

In this paper, the effects of the raw materials and the redox state of the glass melt on the foaming behavior are described in more detail. Here, the attention is focused on gas production in the melt, leading to the formation of bubbles and foam. The objective of this study is to identify batch-related measures for foam control. The backgrounds of foam stability and foam breakdown are outside the scope of this study (see e.g. [3]).

Two methods have been used to study fining and foaming of E-glass melts:
(a) an experimental approach, using a laboratory set-up to measure gas release, foam thickness and redox state of a glass melt during heating and melting of the batch and
(b) development of a thermo-chemical calculation model for predicting the gas production during fining, the SO_3 retention after melting and the evolution of the redox state during melting.

Results of both methods will be presented in the next sections.

EXPERIMENTAL

Set-up and measurement procedures

The laboratory set-up for studying the foaming behavior of glass melts is shown in Figure 1. A standard experimental procedure was developed, using well-defined conditions, which is described in the following:

To simulate batch heating kinetics in an industrial glass tank, the batch is heated rapidly from room temperature to the melting temperature. This is achieved by placing the quartz crucible containing the complete industrial batch sample, directly in a preheated laboratory furnace of around 800 °C. Immediately after introduction of the crucible, a controlled atmosphere (with about 2% oxygen and 55% water vapour), simulating an oxygen-gas fired combustion atmosphere, is applied above the batch sample. Next, the batch is ramp heated with 10 °C/minute to 1250 °C. The dwell time of the fresh, initial melt at 1250 °C is 20 minutes.

Each batch composition is subjected to two different maximum temperatures:
In a first experiment, the initial melt at 1250 °C is quenched to room temperature and the cooled glass is collected for further analysis.

In a second experiment, a similar batch sample is subjected to the same heating procedure; but this time, the initial melt is heated further to 1500 °C with 5 °C/min, after the dwell period at 1250 °C. After it has reached a temperature of 1500 °C, the second glass melt is quenched also, for further analysis.

During the heating process, the following measurements are carried out:
1. In-situ observation of batch melting and foam height on the glass melt in the quartz crucible by taking colour pictures of the batch/melt + foam layer, at every 25 °C temperature increase.
2. On-line analysis of the released gases (CO, CO_2, and SO_2), during the whole heating programme, using IR gas analysers.
3. Measurement of the volatile glass components (esp. S, B- and F-components) by leading a part of the off-gases through washing bottles (see Figure 1), during two temperature intervals: 800 – 1250 °C and 1250 – 1500 °C.

The residual SO_3-content of the two glass melts, heated to 1250 and to 1500 °C, respectively, is measured by X-Ray Fluorescence (XRF). Furthermore, the redox state of the glass samples is determined, either by determination of the Fe^{2+}/Fe_{tot}-ratio of the cooled sample or by measuring the partial oxygen pressure (pO_2) of the melt as a function of temperature with an oxygen sensor.

Using this experimental procedure, the effect of using different batch components on the foaming behavior has been compared. Here we present the effects of varying the redox-active additives, including: carbon, total sulphate addition and alternative oxidizing agents.

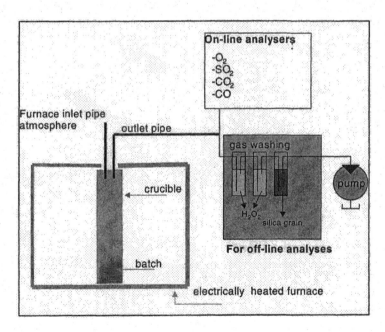

Figure 1 Experimental set-up for foaming experiments

Measuring results

Some typical results of the foaming experiments are presented in Figures 2 and 3. In Figure 2 the sequence of colour pictures illustrates the batch melting behaviour in the temperature interval 800 – 1250 °C; the lower part of Figure 2 shows the SO_2-release in this temperature interval. From this figure it is clear that around 1050 °C, some SO_2-gas is released due to batch reactions.

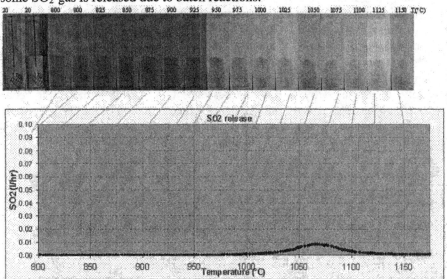

Figure 2 Batch heating process and graph of corresponding release of SO_2

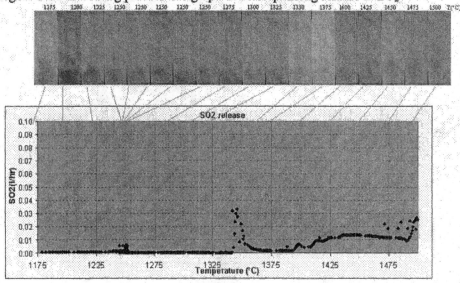

Figure 3 Foaming behavior of glass melts and corresponding release of SO_2 at temperatures > 1400 °C

In Figure 3, the foam formation, starting at $T_{foaming\ onset} = 1400\ °C$, clearly corresponds to a rise in the amount of SO_2 fining gas. Obviously, these experiments already provide a lot of information on the occurrence of batch reactions, the onset and progress of fining and foaming reactions, etc.

Even more valuable information can be obtained when the visual observations are compared to the analyses of the glass samples heated to 1250 °C (initial melt) and 1500 °C (melt after fining). This comparison is illustrated for two E-glass batch compositions, with different redox states, presented in Table I. The value for the total SO_3 content of the batches is obtained by adding the concentrations of S in the sulphate and in the other raw materials and by calculating this value as wt% SO_3. The total carbon content is a calculated value, estimated for the minerals in the batch.

Table I. E-glass batches with different redox states

	Oxidised batch	Reduced batch
Total SO_3 in batch	0.21 wt%	0.21 wt%
Total C in batch	0.09 wt%	0.16 wt%

The results of the foaming experiments and redox analyses carried out with the two batches of Table I are summarised in Table II.

Table II. Redox properties and foaming behavior of two similar E-glass melts, fused from raw materials with different redox state

	$LogpO_2$ at 1250 °C, with pO_2 in bar	SO_3 content of initial melt at 1250 °C (wt%)	SO_3 content of melt after fining at 1500 °C (wt%)	Foaming behavior
Oxidised batch	-1.8	0.090	0.033	Moderate foaming for T > 1350 °C
Reduced batch	-2.5	0.065	0.032	No foaming up to 1500 °C

As can be seen in Table II, the initial glass melt at 1250 °C, prepared from the more reduced batch has a lower pO_2 and a lower SO_3 content than the initial melt, prepared from the more oxidized batch. However, the residual SO_3 content of the final glass melts, after fining at 1500 °C, is practically similar for both batches: 0.033 wt% and 0.032 wt%, respectively. This implies that the difference in SO_3 content of the glass melt at 1250 and 1500 °C, ΔSO_3, is considerably larger for the oxidized glass batch than for the more reduced glass batch:

$$\Delta SO_3\ (oxidized\ melt) = 0.057\ wt\% > 0.033\ wt\% = \Delta SO_3\ (reduced\ melt)$$

As a consequence, the oxidised glass melt has released much more fining gases than the reduced glass melt in the temperature region between 1250 and 1500 °C, by the SO_3- dissociation reaction: SO_3 (m) → SO_2 (g) + ½ O_2 (g), where m = melt and g = gas. This observation was confirmed by the on-line gas analyses of both melts for the temperature region 1250 – 1500 °C (not shown here).

It is noted that the oxidized glass melt, with the higher ΔSO_3 and correspondingly higher SO_2 release for T > 1250 °C, forms a secondary foam layer for T > 1375 °C, while the more reduced glass melt with the lower ΔSO_3 does not show any foaming up to the maximum temperature of 1500 °C. Apparently, the fining gas production of the oxidized glass melt is so high that a foam layer is formed, in contrast to the more reduced glass melt.

This experiment explains the observation in industry that sulphate fined, oxidised E-glass batches, melted in oxy-fired glass tanks, are often prone to uncontrolled foaming behavior.

Effect of alternative oxidising agents on foaming

The most straightforward measure to reduce foam on E-glass melts is to decrease the amount of sulphate in the batch. However, the requirements of a stable redox state and sufficient fining efficiency of the melt will set a lower limit to the sulphate level in the batch. Still, an option for adequately lowering the sulphate addition would be to separate the low temperature oxidising role of sulphate from the role of fining agent in the melt.

In this scenario, the oxidizing role of sulphate in the batch is adopted by another oxidising agent, which releases oxygen during the batch reactions, but does not produce gas in the melt. Therefore, in a series of preliminary foaming experiments, the effect of partly replacing sulphate by an alternative oxidizing agent on redox and foaming behaviour of an E-glass melt was studied.

The basic E-glass batch consists of quartz sand, limestone, kaoline and boric acid. The alternative oxidizing agents that were studied, include: $NaNO_3$, MnO_2, Ca-peroxide (CaO_2) and Na-perborate monohydrate ($NaBO_3.H_2O$). A survey of the experiments is presented in table III.

It can be seen in this table that the original sulphate level in the batch of 0.20 wt% was reduced to a value of 0.05 wt%, corresponding to an SO_3 input of about 0.04 wt% on glass basis, which is lower than the maximum SO_3 solubility in an oxidized E-glass melt at 1250 °C (see next section).

Table III. Survey of studied E-glass batches with different types of oxidizing agents

No.	Oxidizing agents (wt% in batch)	Total SO$_3$ input (wt% in batch)	SO$_3$ input on glass basis (wt%)
1 ref.	-	0	0
2	0.20 Na$_2$SO$_4$	0.11	0.14
3	0.05 Na$_2$SO$_4$	0.03	0.04
4	0.05 Na$_2$SO$_4$ + 0.15 NaNO$_3$	0.03	0.04
5	0.05 Na$_2$SO$_4$ + 0.15 MnO$_2$	0.03	0.04
6	0.05 Na$_2$SO$_4$ + 0.15 CaO$_2$	0.03	0.04
7	0.05 Na$_2$SO$_4$ + 0.15 NaBO$_3$.H$_2$O	0.03	0.04

The most important results of this series of foaming experiments are summarised in Table IV. It is noted that the glass melt level in the quartz crucible after foaming is 2 cm only.

Table IV. Results of foaming experiments with batches containing alternative oxidizing agents, partly repacing sulphate

No	T$_{foaming}$ onset (°C)	Maximum foam height (cm) in quartz crucible	Maximum SO$_2$ release (l/hr) during fining/foaming	Residual SO$_3$ content in glass at 1500 °C (wt%)
1	No fining	no foam	< 0.005	0.025
2	1375	12	0.05 – 0.07	0.036
3	1425	8	0.01 – 0.02	0.037
4	1415	8	0.01 – 0.03	0.025
5	1395	11	0.01 – 0.03	0.027
6	1415	7	0.01 – 0.02	0.039
7	1440	5	0.01 – 0.02	0.040

From Table IV the following conclusions can be drawn:

- The foaming onset temperature for this E-glass batch with 0.2 wt% of Na$_2$SO$_4$ is around 1375 °C
- The foaming onset temperature can be shifted to higher temperatures > 1400 °C, by lowering the amount of sulphate in the batch and using an alternative oxidising agent
- A higher foaming onset temperature correlates with less foam formation, since at higher temperature the foam is less stable
- The sodium-perborate (NaBO$_3$.H$_2$O) appears to have the strongest effect in this E-glass batch: by replacing a major part (75%) of sulphate by this component, the foaming onset temperature is shifted from around 1375 °C to about 1440 °C and the foam formation is reduced considerably.

Except for the reference glass (no.1), melted without any sulphate or other oxidising agent, all glasses showed a good fining quality after the heating to 1500 °C. Industrial tests with alternative oxidants should be carried out to find out whether the above conclusions can be confirmed in a real glass melting tank.

REDOX MODELLING OF E-GLASS MELTS

A calculation model was developed, which describes the temperature dependent redox reactions and gas production in an oxidised E-glass melt, assuming thermodynamic equilibrium. For these calculations, the redox state of the initial glass melt at 1200 - 1250 °C, just after fusion of the raw materials, is treated as input parameter. It was not attempted to simulate redox reactions in the batch, since the kinetic effects occurring in the batch are too complex to accommodate for in a predictive model.

In the redox model of the melt, the following reactions are taken into account:

Once the initial glass melt has been formed, the sulphate in the melt starts to oxidize Fe^{2+}, according to:

$$SO_4^{2-}(m) + 2\ Fe^{2+}(m) \qquad \Leftrightarrow SO_2(g) + 2\ O^{2-}(m) + 2\ Fe^{3+}(m)$$

With increasing temperature, the dominating fining reactions, causing secondary foaming, are the thermal decomposition of sulphate:

$$SO_4^{2-}(m) \qquad \Leftrightarrow SO_2(g) + \tfrac{1}{2}\ O_2(g) + O^{2-}(m)$$

and the reduction reaction of Fe^{3+}:

$$Fe^{3+}(m) + O^{2-}(m) \qquad \Leftrightarrow Fe^{2+}(m) + 1/4\ O_2(g)$$

The temperature dependent equilibrium constants of the latter two reactions can be written as:

$$K_S = \frac{p_{SO_2}\ p_{O_2}^{0.5}}{c_{SO_4^{2-}}} = e^{\frac{-(\Delta H_S - T\Delta S_S)}{RT}}$$

and:

$$K_{Fe} = \frac{c_{Fe^{2+}}\ p_{O_2}^{0.25}}{c_{Fe^{3+}}} = e^{\frac{-(\Delta H_{Fe} - T\Delta S_{Fe})}{RT}} \quad , \text{ respectively,}$$

where the c_i represent the concentrations of the dissolved components and the p_i represent the partial pressure of the gaseous species. ΔH_S is the reaction enthalpy and

ΔS_S the reaction entropy of the thermal decomposition reaction of sulphate and ΔH_{Fe} is the reaction enthalpy and ΔS_{Fe} the reaction entropy of the iron redox reaction. The values of these thermodynamic quantities are specific for each glass type and can be determined experimentally.

Using the thermodynamic properties of a specific glass type, the redox behavior and gas production during heating can be predicted by solving a set of non-linear equations, involving the above redox reaction equilibria and the mass balances of the redox active species.

In the present study the thermodynamic properties of a typical E-glass composition from [2] were used for calculating the following redox related properties:

- Residual sulphate content in the glass melt as a function of temperature (see Fig. 4)
- The total volume of gas production in the melt as a function of temperature (Fig. 5)

Considering the Figures 4 and 5, the following comments can be made: In Figure 4 it can be seen that at temperatures above 1500°C, the residual SO_4^{2-} content in the glass melt approaches a constant value. Measurements of the redox state in the forehearth of an E-glass furnace by use of an oxygen sensor confirm this observation by showing a practically constant redox state. From these calculations and observations in practice, it is concluded that the redox of E-glass fiber products is largely determined by the maximum temperature in the glass melting tank.

Fig. 5 illustrates the effect of the oxidation state of the initial, fresh melt at 1200 °C on the total gas release. It appears that, especially for more oxidized E-glass melts with an initial iron redox ratio < 0.20, the gas release rises steeply in a very small temperature interval, due to the sharply decreasing solubility of sulphate with temperature (cf. Fig. 4). This strong temperature dependency was also observed in the foaming experiments, and is confirmed by observations in industry: oxidized E-glass melts often show an excessive foaming behavior in response to a slight temperature increase.

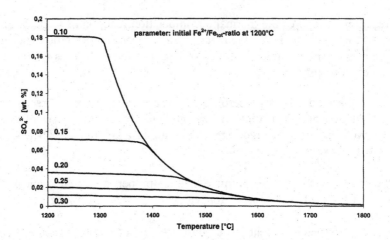

Figure 4 Residual sulphur content expressed as wt.% SO_4^{2-} during heating of an E-glass melt as a function of temperature for different oxidation states of the initial melt phase.

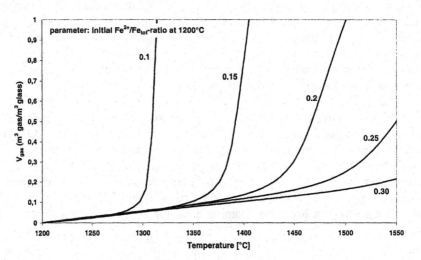

Figure 5 Gas production during heating of an E-glass melt as a function of temperature for different oxidation states of the initial melt phase.

CONCLUSIONS

From the combined experimental and modelling studies it appears that the foam formation on sulphate fined E-glass melts is governed by redox reactions during batch heating and melting. These redox effects can be understood by splitting up the whole melting process in two stages:

- The batch fusion reactions up to 1250 °C
- The fining and foaming processes between 1250 and 1500 °C

The redox state and residual SO_3-content of the initial melt phase at 1250 °C, just after completion of the batch fusion reactions, are the key parameters determining the fining and foaming behavior of the melt at higher temperatures. For oxidised glass batches, containing sulphate as oxidizing and as fining agent, the SO_3 content of the initial glass melt at 1250 °C is, typically, around 0.1 wt% or higher. The SO_3 solubility of E-glass melts strongly decreases with temperature in the region 1300 – 1500 °C, and reaches values of only 0.01 – 0.03 wt% at temperatures above 1500 °C, practically independent of the starting redox conditions. ΔSO_3, the difference in SO_3 content of the melt between 1250 °C and the maximum melt temperature, is released from the melt in the form of fining gas. A foam layer is formed when the production rate of gas bubbles exceeds the breakdown rate of the foam bubbles. According to the laboratory experiments, foaming of E-glass melts will occur for onset temperatures of fining/foaming up to around 1440 °C, under a simulated oxy-gas fired atmosphere. The tendency for foaming grows at lower fining/foaming onset temperatures, resulting in more stable foam, and also increases at higher values of $\Delta SO_3(1250 – 1500 °C)$, leading to higher gas production.

On the basis of the above findings, foam control of E-glass melts should focus on measures to decrease $\Delta SO_3(1250 – 1500 °C)$ to a minimum value, but at the same time maintaining an acceptable fining efficiency. It is noted that a ΔSO_3 of merely 0.01 wt% corresponds to a fining gas volume of more than 400 liter per m^3 of glass melt at 1500 °C, which already may be sufficient for fining.

To minimise the SO_3 content of the initial melt phase in the industrial practice, the following measures are to be taken:

- Control of the S- and C-contamination of the minerals in the batch (clay, boron-compounds, ..)
- Control of the firing conditions (reducing/oxidising) above the batch blanket, since less oxidising firing conditions will result in early loss of sulphate
- Reduce the amount of sulphate addition to the batch, while maintaining a sufficiently oxidised glass, by replacing part of the sulphate by an alternative oxidizing agent. From a series of preliminary foaming experiments with alternative oxidants, sodium-perborate appears to have the strongest effect on the decrease of foam.

It is emphasized that the experimental set-up and modelling calculations described in this paper are useful tools in finding the correct trends and directions for foam control measures. However, industrial tests are still necessary for fine-tuning the optimum redox conditions for a specific glass tank.

REFERENCES

[1] R.G.C. Beerkens, P. Laimbock, *Ceram. Eng. Sci. Proc.* **21** (1) (2000) 41
[2] P.R. Laimbock, *PhD Thesis, Technical University of Eindhoven*, 1998
[3] A.G. Fedorov, L. Pilon, *J. Non-Cryst. Sol.* **311** (2002) 154 - 173

COMPARISON OF MEASURED AND CALCULATED GAS RELEASE BY FINING AGENTS

Detlef Koepsel, Olaf Claussen, Werner Rausch
Schott Glas
Hattenbergstrasse 10
D-55122 Mainz
Germany

ABSTRACT

The efficiency of fining agents is determined by the temperature range in which fining gas is released. This temperature range should usually correspond to a viscosity of less than 100 dPas.

On one hand, the evolution of fining gas can be calculated as a function of temperature, if thermodynamic data for the investigated glass composition is available. As an example, these data can be determined by means of square-wave-voltammetry. For this calculation it is assumed that equilibrium is immediately reached at each temperature.

On the other hand, the evolution of fining gas can be quantitatively measured by means of evolved gas analysis (EGA) using a mass spectrometer. In contrast to the thermodynamic calculation, EGA is a dynamic method taking into account kinetic effects of reactions in the glass melt. The gas evolution is measured during heating of a glass batch up to 1630°C under an atmosphere of helium, which is used as a carrier gas.

For some fining agents, e.g., Sb_2O_3, As_2O_3, and SnO_2, both methods are compared. If the gas evolution occurs in a temperature range in which batch reactions and sand grain dissolution are almost completed, it can be shown that the results are in good agreement. Otherwise the glass has not reached the final composition, and the calculation using thermodynamic data of the final glass composition is not valid.

INTRODUCTION

The characterization of fining agents is necessary, especially when the most efficient agent in new glass compositions has to be found. Crucible tests followed by counting of bubbles are not very precise and reliable. A better way consists in calculating or measuring the gas evolution by fining agents. If the fining gas is released in a temperature range corresponding to viscosities higher than 100 dPas, the fining efficiency of the investigated agents is usually limited.

For the theoretical calculation, thermodynamic data should be available. This data can be determined, for example, by square-wave voltammetry. Though the results usually are very precise, this method is time consuming. The direct measurement of this gas release by EGA is fast in comparison to the determination of thermodynamic data. Another advantage of EGA is that it takes into account batch reactions and sand-grain dissolution. These reactions may strongly influence the gas evolution, if both processes occur at the same time.

SQUARE-WAVE-VOLTAMMETRY

This method is well known [1,2]. Therefore, only one example and some results are shown (Figure1, Table I) for two different glass types.

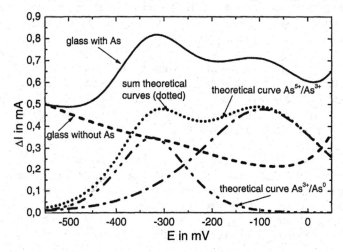

Figure 1. Square-wave voltammogram for a neutral borosilicate glass with As

Table I. Thermodynamic data of redox reaction

	ΔH^0, KJ/mol	ΔS^0, J/(mol K)	Glass type
As^{5+}/As^{3+}	194	120	neutral borosilicate
As^{5+}/As^{3+}	172	96	alkali free
Sb^{5+}/Sb^{3+}	160	114	alkali free
Sn^{4+}/Sn^{2+}	156	74	alkali free

CALCULATION OF THEORETICAL GAS EVOLUTION

For a redox reaction written in the general form:

$$M^{(m+z)+} + {}^z/_2 O^{2-} \rightleftharpoons M^{m+} + {}^z/_4 O_2 \qquad (1)$$

with the corresponding equilibrium constant K_M:

$$K_M = \frac{[M^{m+}]}{[M^{(m+z)+}]} p_{O_2}^{z/4} = \exp\left(-\frac{\Delta H^0}{R \cdot T} + \frac{\Delta S^0}{R}\right) \qquad (2)$$

the theoretical gas evolution can be calculated by equations (3, 4):

$$\frac{dn_{O_2}}{dT} = \frac{z \cdot [M]_{tot} K_M \cdot p_{O_2}^{z/4} \cdot \Delta H^0}{4 \cdot R \cdot T^2 \left(K_M + p_{O_2}^{z/4}\right)^2}, \qquad \left[\frac{moles}{K}\right] \qquad (3)$$

$$\frac{dV_{O_2}}{dT} = \frac{R \cdot T}{P_{tot}} \cdot \frac{dn_{O_2}}{dT} \qquad (4)$$

This calculation is based on the assumption that equilibrium of reaction (1) is immediately reached at any temperature. An example of the calculated oxygen evolution[*] is shown in Figure 2.

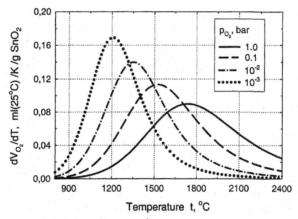

Figure 2. Calculated oxygen evolution[*] in an alkali-free glass with SnO_2 glass

---

[*] The volume of oxygen (Equation 4) in all following figures and tables is calculated for a temperature of 25°C.

EVOLVED GAS ANALYSIS

The gas analysis can be done either with gas chromatography [3] or with mass spectrometry [4]. In the current case a mass spectrometer is used (Figure 3).

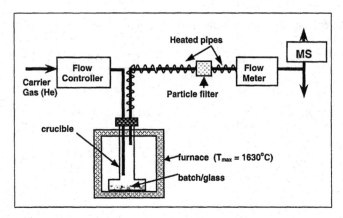

Figure 3. Evolved gas analysis

To determine accurate results some procedures have to be followed. The gas released from the glass in the crucible needs some time to arrive at the mass spectrometer. If this retention time is significant, the gas-release peaks are apparently shifted to higher temperatures. By minimizing the dead volume of the crucible and gas pipes, by maximizing the carrier gas flow (without worsening the detection limits), and by moderate heating rates, this retention time can be reduced to a negligible value. The heating rate in the current case is 8 K/min. The thickness of the glass layer in the crucible should not exceed 2-3mm; otherwise, the released gas first diffuses into bubbles, which need time to rise to the surface. This rising time can also contribute to an apparent peak shift to higher temperatures. If the glass layer is too small, the glass equilibrates with the carrier gas atmosphere above the melt; and, therefore, the oxygen partial pressure (Equation 3) is not dominated by the fining agent in the glass. In this case, the peaks are shifted to lower temperatures, as illustrated in Figure 1.

FINING WITH SnO_2

For an alkali-free batch with nitrate and 1 wt% SnO_2, the measured and calculated gas evolution are shown in Figure 4. In addition, the same glass composition, but previously equilibrated with pure oxygen atmosphere at 1350°C, is shown (Figure 4). The theoretical gas evolution in Figure 4 is calculated for two cases:

- For a constant oxygen partial pressure of 1 bar and
- For oxygen partial pressure, measured in the glass with a ZrO_2 sensor (Figure 5).

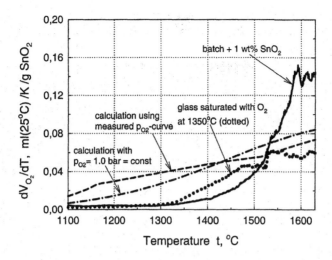

Figure 4. Oxygen evolution by SnO_2 (1 wt%) in an alkali-free glass

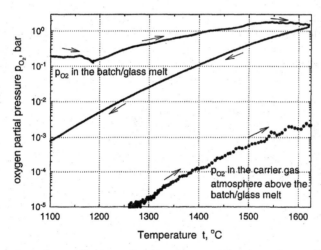

Figure 5. Experimentally determined oxygen partial pressure in an alkali-free batch with 1 wt% SnO_2 and in the carrier gas atmosphere above the batch/glass

The difference between the theoretically calculated and experimentally determined curves at lower temperatures is caused by incomplete batch reactions and sand-grain dissolution (Figure 6); i.e., the molten glass has not yet reached its final composition. In this case, the thermodynamic properties, which are valid only for the final composition, should not be used. Therefore, the prediction in Figure 5 is not accurate.

Considering that only 87% of the total Sn is oxidized to Sn^{4+} (analyzed with Mössbauer spectroscopy) during the equilibration under oxygen at 1350°C, the agreement between the measurement for this glass and the theoretical prediction is better than that for the batch curve.

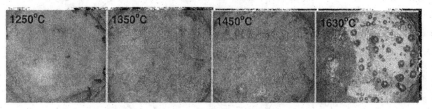

Figure 6. Melting state of the batch at different temperatures

If for the calculation the oxygen partial pressure in the carrier gas atmosphere above the melt is used, the peak should be observed in the temperature region below 1300°C. In reality, the peak is found at t >1550°C. It can therefore be concluded that the pO_2 is controlled by the fining agent in the glass melt. The assumption that pO_2 =1 bar for the theoretical calculation is not far from reality.

FINING WITH Sb_2O_5

For an alkali-free batch with nitrate and 0.96 wt% Sb_2O_3, the measured and calculated gas evolution are shown in Figure 7. The difference between the theoretical and the measured gas evolution is greater than that for SnO_2. As mentioned before, it is caused by the incomplete batch reactions.

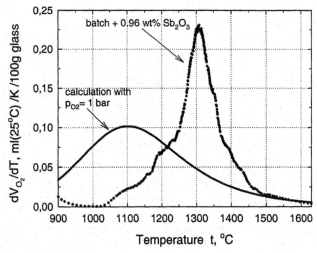

Figure 7. Oxygen evolution in an alkali-free batch with nitrate and 0.96wt% Sb_2O_3

FINING WITH As₂O₃

FINING WITH As$_2$O$_3$

The measured and calculated gas evolutions for an alkali-free and a neutral borosilicate batch with nitrate and As$_2$O$_3$ are shown in Figure 8. For the alkali-free glass the maxima of the calculated and measured gas release peaks are nearly at the same temperature, but the absolute values of gas evolution rates are different.

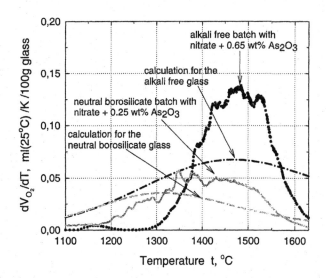

Figure 8. Oxygen evolution in an alkali-free batch with nitrate+0.65wt% As$_2$O$_3$ and in a neutral borosilicate batch with nitrate+0.25wt% As$_2$O$_3$

GAS BALANCE

While the agreement between theoretical and measured oxygen release rates is not satisfactory in many cases, the agreement between measured and calculated total oxygen volume is rather good (Table II). The experimental determination includes the gas evolution during heating up to the maximum temperature of 1630°C and a following dwell time at this maximum temperature until the gas evolution is finished (Figure 9).

Table II. Calculated and measured total oxygen volume

fining agent	$V_{O_2}^{tot}$, ml (25°C) /100g glass		glass type
	EGA	Calculation	
1.0 wt% SnO$_2$	25.8	22.8	alkali free
0.65 wt% As$_2$O$_3$	30.7	27.1	alkali free
0.96 wt% Sb$_2$O$_3$	44.1	39.9	alkali free
0.25 wt% As$_2$O$_3$	15.7	14.0	neutral borosilicate

The agreement is not surprising, because at the maximum temperature of 1630°C, all batch reactions and the sand-grain dissolution are complete. At this point, the glass with its polyvalent species is in equilibrium with temperature.

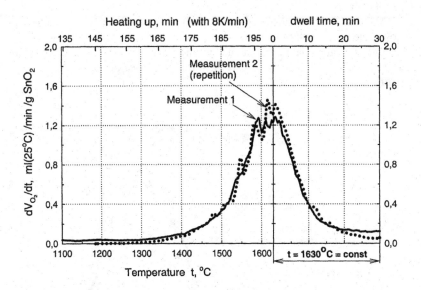

Figure 9. Oxygen evolution in an alkali-free batch +1 wt% SnO_2 as a function of time (two measurements under the same conditions)

CONCLUSION

The evolved gas analysis is a fast quantitative method to characterize fining agents. A theoretical calculation with available thermodynamic data provides reliable results if the evolution of the fining gas takes place in a temperature range within which all batch reactions and the sand-grain dissolution are almost completed.

REFERENCES

[1] C. Rüssel, E. Freude, "Voltammetric studies of the redox behaviour of various multivalent ions in soda-lime-silica glass melts," Phys. Chem. Glasses **30** [2] 62-68 (1989)
[2] O. Claußen, C. Rüssel: "Voltammetry in silicate and borosilicate melts," Ber. Bunsenges. Phys. Chem. **100**, 1475-1478 (1996)
[3] S. Kawachi, M. Kato, "Evaluation of reaction rate of refining agents," Glastechn.Ber.Glass Sci.Technol. **72** [6] 182-187 (1999)
[4] F. Krämer, "Gasprofilmessungen zur Bestimmung der Gasabgabe beim Glasschmelzprozeß," Glastechn.Ber. **53** [7] 177-188 (1980)

BUBBLE CONTINUUM MODEL

Andreas Bensberg and Christian Breitbach
SCHOTT Glas
Hattenbergstr. 10
55122 Mainz
Germany

ABSTRACT

A bubble continuum model is presented which considerably enhances the simulation capabilities for the fining process in glass tanks. The basic assumptions are that bubbles can be treated as a continuum phase and their size distribution can be approximated by a lognormal function. Experimental observations in particular confirm the latter assumption. The dynamic behaviour of the bubble phase consisting of a single gas is calculated by solving the transport equations for three moments of the size distribution. For each additional gas species to be considered in the bubbles, one additional transport equation has to be solved. The model is capable of covering several orders of magnitude for the number density of bubbles in a glass tank. It also takes into account that, in general, there are bubbles of different sizes with different rising velocities within the same location. Mass exchange between the liquid and the bubble phase, including degassing of the liquid phase, is considered, as well as the repercussion of the bubble phase on the mean velocity field due to a reduction of the mean density of the suspension. The novel model allows incorporation of all relevant phenomena of the fining process without the need for any further material and process parameters beyond the usual, such as solubility and diffusion coefficients of the gases.

The model is implemented in Fluent, via user-defined functions. Glass tank simulations are presented to illustrate its capability and to identify sensitive parameters.

INTRODUCTION

A glass melt consists not only of a liquid, but also of a suspension of particles and bubbles. For the fining process, the behaviour of the bubbles is critical. Due to buoyancy and gas exchange with the liquid phase, the size and the number density of the bubbles can vary by several orders of magnitude. In aerosol physics, a method for describing these phenomena has been derived and successfully applied which is based on transport equations for the leading moments of an assumed size distribution [1, 2, 3, 4]. The main benefit of this method is that usually three transport equations suffice to cover the whole size spectrum, including nucleation, condensation, and coagulation. This paper briefly outlines the application of this method to bubbles in a liquid. After the basic ideas of the method have been outlined the method is applied to a 2D model of a glass tank. Parameter variations are carried out to demonstrate the capability of the model.

BASIC IDEAS

The fundamental assumption of the model is firstly, that the bubbles can be described as a continuum phase by forming of mean values in a similar way as in the kinetic theory of gases. Further, experimental observations [5] indicate that bubble size distributions fit a lognormal function. The size distribution can then be written as

$$N(r_\alpha) = \frac{N_B}{\sqrt{2\pi} \ln \omega} \exp\left\{-\frac{1}{2}\left(\frac{\ln(r_\alpha / r_g)}{\ln \omega}\right)^2\right\} \frac{1}{r_\alpha} \tag{1}$$

$N(r_\alpha)dr_\alpha$ is the number density of bubbles with a radius between r_α and $r_\alpha+dr_\alpha$. The right-hand side contains the three parameters

- Total number density N_B
- Geometric mean value r_g of the radius
- Standard deviation ω

Although the assumption just employed restricts the size distribution to the form of a lognormal function, it should be noted that the three parameters are allowed to vary in space and time. Possible shifts of the function are sketched in Figure 1. As shown in the literature on aerosol physics cited above, moments M_l, l=0,1,2,... of the size distribution function are defined as mean values of a size parameter to the power of l. Transport equations for these moments can be derived to give

$$\frac{\partial M_l}{\partial t} + \nabla \cdot \left(M_l \vec{v}_s + M_l \vec{V}_{Auf,l} - D_l \nabla M_l\right) = P_l \tag{2}$$

$\vec{v}_s$ is the mean velocity vector of the melt, $\vec{V}_{Auf,l}$ the buoyancy velocity vector of

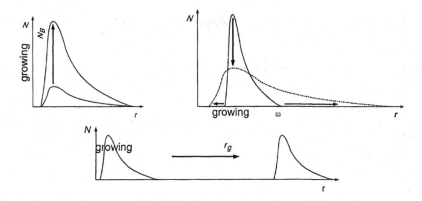

Figure 1. Possible shifts of the lognormal size distribution.

the lth moment and D_l the diffusion coefficient of the bubbles due to their Brownian motion. P_l is a source term to take into account mass exchange between the phases. A corresponding source term has been formulated for the transport equations of the gas species in the liquid phase to account for the chemical interaction with the liquid. If the values of the first three moments M_l are known, not only the parameters N_B, r_g und ω can be determined, but also the volume fraction of the bubble phase ε. If the bubbles contain more than one gas species, further moments have to be defined. The resulting number of equations is equal to the number of parameters. On the other hand, no additional material parameters beyond diffusion coefficients, solubilities, and free-reaction enthalpies of the gas species, and initial concentrations of dissolved gases are needed.

The density ρ_s and the viscosity μ_s [6] of the melt, i. e., of the suspension, depend on the volume fraction ε by

$$\rho_s = (1-\varepsilon)\rho_F$$
$$\mu_s = (1+\varepsilon)\mu_F \tag{3}$$

ρ_F and μ_F are the corresponding values of the bubble-free liquid glass. This relation produces the mechanical interaction in the melt.
The equations are formulated such that they can be coupled with conventional simulation tools. In particular, the CFD software Fluent offers, via user-defined functions, the possibility to implement the model.

EXAMPLE: 2D MODEL OF A GLASS TANK
The bubble continuum model has been applied to a 2D model of a glass tank. The material parameters are similar to those of a neutral borosilicate glass. The bubbles contain the gas species H_2O and O_2 at an initial mole fraction of 0.8. To accelerate the numerical calculations, the diffusion coefficient of O_2 has been decreased approximately by one order of magnitude compared to the usually high values in such glasses. This causes a reduction of the mass exchange between the phases, and damps the fining efficiency. Further, H_2O is used as a dummy gas,

which cannot leave or enter the bubbles. In the liquid phase, O_2 is subject to a reaction according to

$$MeO_2 \leftrightarrow MeO + \tfrac{1}{2}O_2 \qquad (4)$$

where MeO_2 is an arbitrary metal oxide.

The initial total number density of the bubbles is 200 cm^{-3}, the initial mean bubble radius is 100 μm, and the initial standard deviation is 1.7. Despite these simplifications, the example is suitable to demonstrate the capabilities of the model.

Figure 2 shows the streamlines of the flow. Since no batch model is used, the flow just drops down at the inlet. The bubblers produce a large recirculation between the inlet and the bubblers. Three more recirculation zones occur between bubblers and dam wall. Behind the dam wall, part of the glass flows down along the wall due to its cooling and enters a recirculation zone. Another part of the glass flows along the surface towards the back wall of the tank, where another recirculation zone occurs. At the bottom in front of the throat, a stagnation point can be observed. The temperature distribution in the tank is shown in Figure 3. It reaches its maximum at the surface near the dam wall.

Figure 4 shows the distributions of the total number density and the mean radius of the bubbles. While the total number density decreases only slightly across the bubblers, the major reduction of the total number density occurs beyond the dam wall. In particular, near the surface towards the back wall, this reduction is at its maximum. Due to the high temperature, the solubility of O_2 is

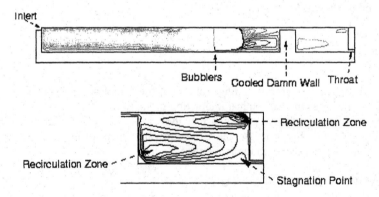

Figure 2. Streamlines in the glass tank.

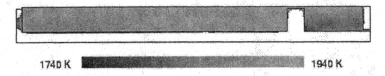

Figure 3. Temperature in the glass tank.

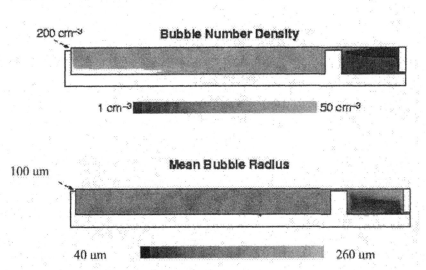

Figure 4. Total number density and mean radius of the bubbles in the glass tank.

low and the diffusion coefficient is high. The bubbles grow and ascend faster, and finally leave the melt. Here, the bubble radius increases to 2.6 times the initial value, as shown in Figure 4 also. At the bottom behind the dam wall, the temperature is low, and the hydrostatic pressure high. So the bubbles tend to release O_2. The bubble radius is smaller, and the reduction of the total number density is less efficient. Particularly at the stagnation point at the bottom near the throat, the residence time of the bubbles is high, and the bubbles have a lot of time to release O_2. Accordingly, the mean bubble radius reaches its minimum here.

The mole fractions of O_2 in the bubbles and in the liquid behave correspondingly, as shown in Figure 5. In the top right part behind the dam wall, where the bubbles reach their maximum size, their O_2 mole fraction is also maximal and reaches a minimum value of about 0.4 at the stagnation point at the bottom near the throat. The O_2 mole fraction in the liquid is minimal in the top right part behind the dam wall and increases again towards the bottom; however, it does not attain a local maximum at the stagnation point where the bubbles release a lot of O_2. The reason is that only small and few bubbles arrive there. They have not passed a region where they could absorb O_2. Hence they do not transport much gas.

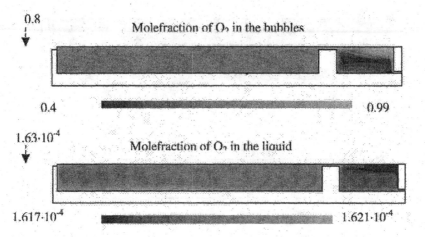

Figure 5. Mole fraction of O_2 in the bubbles and in the liquid.

Now we examine the influence of variations of the initial parameters of the size distribution on their values at the throat. Figures 6 through 8 show the distributions of the total bubble number density and the mean bubble radius over the height of the inlet to the throat. A height of 0.49 m corresponds to the bottom of the throat, and a height of 0.55 m to the top. First we compare simulations for the two initial standard deviations of 1.7 and 1.4. Both are realistic values. In Figure 6 we see that the curves for the total number density and the mean radius are in both cases qualitatively equal. The total number density decreases from the bottom to the top of the throat, while the mean radius increases along this line. Those bubbles which crossed the dam wall reached the top of the throat, moved along the surface towards the back wall, and absorbed a lot of O_2. Most of them left the melt, via the surface. Only few, but large bubbles, reach the top of the throat. The bubbles at the bottom of the throat moved down the dam wall and did not absorb much O_2. They are smaller and do not ascend very fast. At the bottom of the throat, the total number density is higher, but the mean radius is smaller. For the lower value of the initial standard deviation, the total number density is more than 50 % higher than for the larger initial standard deviation, while the mean radius does not differ much. Although the interpretation of this result is difficult, it is clear that even moderate variations of the initial standard deviation significantly impacts the total number density of the bubbles which leave the tank. It is obviously important to consider a bubble size distribution rather than a monodisperse bubble phase.

The influence of the initial mean bubble radius is shown in Figure 7. If this value is halved, the total number density at the throat is almost three times bigger than before. While the mean radius at the bottom of the throat is nearly halved also, the bubbles at the top of the throat are not affected much by this variation.

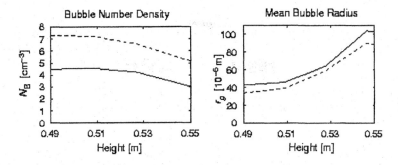

Figure 6. Total number density and mean radius of the bubbles at the throat for initial values of $\omega = 1.7$ (—) and $\omega = 1.4$ (---).

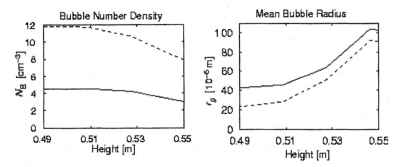

Figure 7. Total number density and mean radius of the bubbles at the throat for initial values of $r_g = 100$ μm (—) and $r_g = 50$ μm (---).

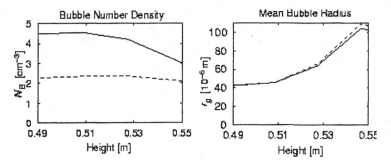

Figure 8. Total number density and mean radius of the bubbles at the throat for an uncoupled (—) and a coupled (---) simulation.

This value is determined mainly by the processes at the surface near the back wall. The importance of the mechanical interaction with the melt via the Equations (3) is demonstrated in Figure 8. While the mean radius is almost unaffected, the total number density at the throat reduces to about half the value as before, if the coupling is taken into account.

CONCLUSION

The bubble continuum model based on a moment method is designed such that it provides all important information at a minimum effort, and the presented results demonstrate the complexity of this information. The authors are convinced that the model implemented in a commercial CFD code represents an optimal tool for the design and optimisation of melters.

To become efficiently applicable some points have to be improved:

- More gas species (N_2, CO_2, ...) will be incorporated.
- Due to the sensitive coupling of the model equations, the numerical behaviour is unstable and has to be improved.
- The concept of a dummy gas has to be abandoned.
- A batch model has to be implemented for a more realistic flow simulation and to provide source terms for the formation of bubbles.

REFERENCES

[1] K.W. Lee, H. Chen and J.A. Gieseke "Log-normally preserving size distribution for Brownian coagulation in the free-molecule regime," *Aerosol Sci. and Technol.*, **3** 53-62 (1984).

[2] E.R. Whitby, P.H. McMurry, U. Shankar and F.S. Binkowski, "Modal Aerosol Dynamics Modeling," *EPA reports*, 600/3-91/020 (1991).

[3] K.W. Lee, Y.J. Lee and D.S. Han, "The log-normal size distribution theory for Brownian coagulation in the low Knudsen number regime," *J. Colloid Interface Sci.*, **188** 486-492 (1997).

[4] A. Bensberg, P. Roth, R. Brink and L. Lange "Modeling of Particle Evolution in Aerosol Reactors with Coflowing Gaseous Reactants," *AIChE Journal*, **45** [10] (1999).

[5] T. Roi, O. Seidel, G. Nölle and D. Höhne "Modeling of the bubble population in glass melts," *Glastec. Ber.* **67** [10] 263-271 (1994)

[6] M. Ishii and T.C. Chawla "Local Drag Laws in Dispersed Two-Phase Flow," *Argonne National Laboratory, Report No. NUREG/CR-1230, ANL-79-105* (1980)

SELECTIVE BATCHING FOR IMPROVED COMMERCIAL GLASS MELTING

Ungsoo Kim and William M. Carty
Whiteware Research Center
New York State College of Ceramic at Alfred University
2 Pine Street
Alfred, NY 14802

Christopher W. Sinton
Center for Environmental and Energy Research
Alfred University
2 Pine Street
Alfred, NY 14802

ABSTRACT

Current methods of continuous glass production rely on the introduction of mixed batch materials into a furnace. It is proposed that initial reactions between alkali and alkaline earth carbonates create a low viscosity melt that segregates flux from quartz, resulting in a mixture that must then be re-homogenized in the tank. In this study two alternative batching methods, selective pre-reaction (1000°C and 1400°C) and spray drying, are examined for changes in the batch free time relative to the current process. Two intermediate materials of $\{Na_2CO_3+SiO_2\}$ and $\{CaCO_3+MgCO_3 \cdot CaCO_3 +SiO_2\}$ are prepared using both methods, subsequently mixed mechanically, and introduced into the furnace. When the pre-reacted materials were blended together and heat-treated, batch-free glass was obtained in extremely short times, on the order of 20 minutes at 1350°C compared to 240 minutes for a typical batch. Similar results were obtained from the selective spray drying of slurries with batch-free times at 1350°C of 45 minutes.

INTRODUCTION

The batching and melting process used in the manufacture of modern commercial glass (float glass, container glass, and fiber glass) has remained basically unchanged since the 1960s. The batching process consists of weighing raw materials directly from storage silos into a weigh hopper, adding to the weighed materials a specified amount of cullet (ground, recycled glass of similar composition to the batch), subsequently mixing the raw batch and the cullet in a large scale mixer, then transferring the mixed batch to storage hoppers located directly adjacent to the "dog house," or the end of the glass furnace where the batch is introduced and melting commences. Commercial batching and mixing processes are reasonably well understood and have been adequately addressed in several review articles.[1-5] Nevertheless, batching and mixing studies are still of great interest, considering that over the last ten years, an average of five papers have been given each fall at the Conference on Glass Problems (CGP), each addressing batching or melting problems (amounting to 27% of the total number of papers presented).[6] Clearly, a large amount of research has been conducted correlating the quality of the mixed batch with the quality of the resulting glass, and a variety of studies have addressed the cullet particle size, the use of water or oil to minimize segregation of the batch prior to its introduction into the furnace, and the various properties of pelletizing (or briquetting) and preheating (of either the cullet, the batch, or the entire mixture). Regardless of the changes introduced to improve the melting process, the energy efficiency is still far behind the calculated energy necessary to melt the batch. The calculated energy required to melt a glass batch is 2.2 million Btu/ton – a level that is approximately two to three times less than the energy expended to melt glass (4.7 to 6.9 million Btu/ton).[7]

Pelletizing the batch has been evaluated as a batch preparation route. This approach has been addressed and is periodically revisited by glass researchers. There are documented melting efficiencies observed, but remaining problems, such as the potential for contamination from the pelletizing equipment, minimize its use in the current industry.[8]

Once the batch is introduced to the furnace, several reactions take place that almost immediately segregate the batch. As Na_2CO_3, $CaCO_3$, $CaCO_3 \cdot MgCO_3$, Na_2SO_4, and SiO_2 (quartz) are the most commonly used major raw materials in float glass production, and assuming that water has not been added to reduce segregation of the batch in the storage hopper, the first reaction is usually the formation of a eutectic liquid (785°C) by the reaction of Na_2CO_3 and $CaCO_3$.[9] Similar reactions are observed between Na_2CO_3, $CaCO_3$, and Na_2SO_4. The eutectic liquid, in this case composed of molten salts, will have a very low viscosity, similar to that of water (viscosity of 1 to 4 mPa·s, or 1 to 4 centipoise),[1] that then reacts with the quartz to eventually provide a homogeneous glass of the desired composition. It is therefore proposed that the formation of this eutectic liquid immediately promotes batch segregation, reversing the efforts of batch mixing. Similar arguments can be made for container glass compositions, and in the case of fiber glass production, borates exhibit similar problems in the initial stages of melting. This process of segregation leads to large-scale domains, or agglomerates, of nearly pure silica that then require excessively long residence times for dissolution into the surrounding liquid melt.

To prevent segregation in the furnace, it is proposed that substantial improvements in glass batch melting can be accomplished by decreasing the viscosity differences between the reaction products and the glass melt. Namely, segregation of the batch constituents during the initial stages of melting can be limited by promoting the formation of higher viscosity intermediate compounds during heating through the control of the reaction pathways. This can conceptually be accomplished in two ways, listed in order of decreasing anticipated energy demand: Approach I, pre-react selective batch components to form an intermediate feedstock; and Approach II, selectively batch and pelletize to form small "reaction volumes" that react initially to form an intermediate reaction product. This differs from palletizing in the traditional sense in that these pellets would not contain the entire batch, but selected batch constituents.

EXPERIMENTAL PROCEDURES

To control the melting sequence and consequently the viscosity of the molten phase(s), raw materials were selectively batched into mixtures and the melting behaviors were compared to the standard batches, in which all the batch components were simply mixed together prior to charging. Two proposed approaches were investigated with the typical float glass composition as denoted in Table I. The composition for standard batch without dolomite is formulated by substituting on a molar basis the divalent cation contribution from dolomite with limestone. Raw materials were provided by Guardian Industries (Geneva, NY) and the measured properties are summarized in Table II. Quartz (325 mesh) was provided by U.S. Silica.

Table I. Chemical composition of typical float glass.

Name	Chemical Formula	(w/o)	(w/o)
Soda ash	Na_2CO_3	19.61	19.37
Limestone	$CaCO_3$	5.90	21.05
Dolomite	$CaCO_3 \cdot MgCO_3$	14.19	
Sand	SiO_2	60.30	59.58

Table II. Physical properties of raw materials used in this investigation.

Name	Chemical Formula	Density (g/cm^3)	Specific Surface Area (m^2/g)	Mean Diameter (μm)
Soda ash	Na_2CO_3	2.53	0.54	185
Limestone	$CaCO_3$	2.72	0.49	516
Dolomite	$CaCO_3 \cdot MgCO_3$	2.86	0.23	199
Sand	SiO_2	2.65	0.13	323
Quartz	SiO_2	2.65	1.05	45

As a preliminary study mixtures of soda ash+quartz and limestone+quartz were prepared (100 g) and fired at 1200°C with dwell times of 2, 4, 6, and 8 hours. The weight ratio for soda ash+quartz was 38.75:61.25 to form the Na_2O-SiO_2 eutectic. The weight ratio for limestone+quartz was 42.09:57.91, corresponding to the composition of a line drawn from Na_2O-SiO_2 eutectic to the CaO-SiO_2 boundary passing through the typical float glass composition on Na_2O-

CaO-SiO$_2$ phase diagram. X-ray diffraction analysis (XRD, XRG 3100, Philips, Natick, MA) on fired samples indicated that the mixture of Na$_2$CO$_3$-SiO$_2$ formed glass at all the dwell times. The mixture of limestone-quartz showed that limestone formed CaO by the dissociation of CO$_2$ and two intermediate compounds (CaSiO$_3$ and Ca$_2$SiO$_4$) were formed, but the relative amounts did not change significantly with longer dwell times. XRD analysis was performed from $2\theta = 10°$ to $70°$ with a $0.04°$ step size and a two second dwell time.

Based on the previous experiments intermediate materials were prepared in larger scale to investigate the melting behavior of the pre-reacted batch materials. Float glass batch samples were prepared by mixing the Na$_2$O-SiO$_2$ eutectic with two intermediate materials; 1) limestone and quartz and 2) limestone, dolomite, and quartz. The melted Na$_2$O-SiO$_2$ eutectic was ground into powders with particle size <180μm before mixing. Two sets of intermediate materials were prepared by heating the intermediate RO+SiO$_2$ materials at 1000°C ("Unreacted") and 1400°C ("Reacted") for 4 hours, respectively. Each batch was made with two different sizes of quartz; 325 mesh fine quartz and coarse sand. The aggregates formed after firing were grounded in a dry ball mill. In total, eight batches were prepared for this study using the batch formulas summarized in Table I and II. The final mixtures were dry ball milled for 30 minutes with nylon balls to homogenize the batches. Samples were weighed into mullite crucibles to make 25g of glass and heat treated to determine batch free time, which is defined as the time required to dissolve the entire batch. Samples were placed in a gas furnace at 1350°C initially for 60 minutes. If the batch was completely dissolved, the time was decreased; if not completely dissolved, the time was increased.

To investigate the melting behavior of the selectively batched and pelletized materials, three 25 volume percent slurries were prepared: {soda ash + quartz}, {limestone + quartz}, and {limestone + dolomite + quartz}. These were prepared without any polymeric additives (dispersant or binders) due to concerns regarding redox conditions within the batch. Only 325 mesh quartz was used for the spray drying studies and the other constituents were ball milled to reduce particle size before addition into the slurries. After dry ball milling for 48 hours the specific surface area for limestone and dolomite were measured as 3.39 and 2.25 m^2/g, respectively, indicating a particle size similar to the 325 mesh quartz. Spray drying was performed using a laboratory scale mixed flow spray dryer (BE 985, Bowen Engineering, North Branch, NJ). The slurries were introduced into a spray dryer under the conditions of 350°C inlet temperature, 145°C outlet temperature, and 20psi atomizing air pressure. The mean granule size was determined to be 80 μm. The batch mixtures were prepared by mixing the {soda ash + quartz} granules separately with the two other granules and dry ball milling for 30 minutes with nylon balls to homogenize the batches. Samples were weighed into mullite crucibles to make 25g of glass and heat treated (as described above) to determine the batch free time.

Table III. Batch formula including Na_2CO_3, $CaCO_3$, $CaCO_3 \cdot MgCO_3$, and quartz. $Na_2O\text{-}SiO_2$ or Eutectic indicates the eutectic intermediate material between Na_2CO_3 and quartz. MCQ stands for the mixture of $CaCO_3$, $CaCO_3 \cdot MgCO_3$, and quartz. Unreacted and reacted indicates intermediate materials fired at 1000°C and 1400°C, respectively. Fine and coarse in parenthesis indicates 325 mesh quartz and sand from Guardian Glass, respectively.

Batch ID	Description	Na_2CO_3	$Na_2O\text{-}SiO_2$	$CaCO_3$	$CaCO_3 \cdot MgCO_3$	CaO	$CaO \cdot MgO$	Fine Quartz	Sand
A	Standard batch (Coarse)	x		x	x				x
B	Standard batch (Fine)	x		x	x			x	
C	Eutectic / Unreacted MCQ (Coarse)		x			x	x		x
D	Eutectic / Reacted MCQ (Coarse)		x			x	x		x
E	Eutectic / Unreacted MCQ (Fine)		x			x	x	x	
F	Eutectic / Reacted MCQ (Fine)		x			x	x	x	
G	Spray dried standard	x		x	x			x	

Table IV. Batch formula including Na_2CO_3, $CaCO_3$, and quartz. $Na_2O\text{-}SiO_2$ or Eutectic indicates the eutectic intermediate material between Na_2CO_3 and quartz. CQ stands for the mixture of $CaCO_3$ and quartz. Unreacted and reacted indicates intermediate materials fired at 1000°C and 1400°C, respectively. Fine and coarse in parenthesis indicates 325 mesh quartz and sand from Guardian Glass, respectively.

Batch ID	Description	Na_2CO_3	$Na_2O\text{-}SiO_2$	$CaCO_3$	CaO	Fine Quartz	Sand
H	Standard batch (Coarse)	x		x			x
I	Standard batch (Fine)	x		x		x	
J	Eutectic / Unreacted CQ (Coarse)		x		x		x
K	Eutectic / Reacted CQ (Coarse)		x		x		x
L	Eutectic / Unreacted CQ (Fine)		x		x	x	
M	Eutectic / Reacted CQ (Fine)		x		x	x	
N	Spray dried standard	x		x		x	

RESULTS and DISCUSSION

The batch free time for all the tested samples are summarized in Figures 1 and 2. Several experimental conditions were evaluated, including selective pre-reacting and spray drying, and the role of quartz particle size. When the pre-reacted materials were blended together and heat-treated, batch-free glass was obtained in extremely short times, on the order of 20 minutes at 1350°C (Figure 1). Similar results were obtained from the selective spray drying sample, with batch-free time of 45 minutes. Samples made with reacted intermediate materials, which were fired at 1400°C, showed shorter batch free time than those with unreacted intermediate materials, fired at 1000°C. A standard commercial batch exhibited a batch-free time on the order of 240 minutes for coarse (323 μm) and 120 minutes for fine (45 μm) quartz, demonstrating that the melting efficiency improvements are not simply an artifact of particle size. Compared to a batch-free time of 240 minutes, the pre-reacted materials exhibit a batch-free time reduction of 92%; the selectively spray dried batch a reduction of 81%.

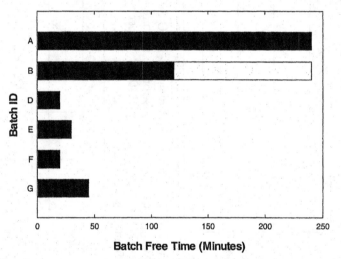

Figure 1. Batch free time for the batches including Na_2CO_3, $CaCO_3$, $CaCO_3 \cdot MgCO_3$, and quartz. Black bar indicates batch free time and gray portion on the bar indicates state with un-melted materials on the rim. (Batches are identified in Table III.)

Similar results were repeated with pre reacted and selective spray drying samples for the samples without dolomite (Figure 2). The batch free time for the standard batch matched that of the dolomite-containing batch at 240 minutes, while the batch free times of 20 and 30 minutes were obtained for the pre-reacting and selective spray drying samples, respectively. Again, the benefit of reduced batch free time is not shown to be simply an artifact of particle size (as illustrated by batches H and I).

As noted above, the substantial reduction in batch free times is not solely a function of selective batching, but is also a result of the reduction in particle size-a necessary condition spray drying. The selective spray drying approach has therefore the added benefit of allowing finer particle size raw materials to be used. Overall, conservative energy savings using the selective batching approach compared to current methods are on the order of 20-33% for the entire process, assuming that melting times can be reduced by 50%. Additional observations during these experiments suggest that additional energy savings may be possible by reducing the needed fining times (the time to remove bubbles and other defects); the efficiency of melting obtained from this approach may allow a reduction in the peak temperature in the melting portion of the furnace potentially increasing refractory lifetimes; and improve the overall quality of the glass produced may increase by reducing the tendency to convey un-reacted batch materials out of the melting portion of the furnace into the forming sections by improving homogeneity earlier in the furnace. Further environmental benefits may be gained by a reduction in the use of Na_2SO_4 (salt cake) currently added in the float glass industry to assist in the fining operation.

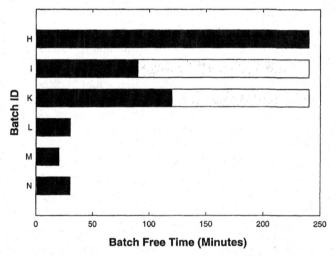

Figure 2. Batch free time for the batches including Na_2CO_3, $CaCO_3$, and quartz. Black bar indicates batch free time and gray portion on the bar indicates state with un-melted materials on the rim. (Batches are identified in Table IV.)

CONCLUSIONS

These experimental results demonstrate that significant improvements to commercial glass melting may be possible through the use of a selective batching method, leading to substantial energy savings and potentially reduced emissions of criteria air pollutants. In the selective batching process, {soda ash + silica} are blended to make one granulated feedstock, and in a separate step, {limestone + dolomite + silica} are blended to make another feedstock. The feedstocks will be produced by spray drying, subsequently mixed mechanically, and introduced into the furnace. The process eliminates low viscosity liquid formation by preventing the 785°C eutectic reaction of {soda ash + limestone} that is proposed to lead to gross segregation of flux from quartz in the early stages of melting.

ACKNOWLEDGEMENTS

This study was funded by the U.S. Department of Energy (DOE) under Award Number: DE-PS36-02GO90014. This support does not constitute an endorsement by DOE of the views expressed in this paper. Also, the donation of raw materials from Guardian Industries and the assistance of Hyo Jin Lee and Brian Tanico are gratefully acknowledged.

REFERENCES

1. F. E. Woolley, "Melting/Fining," *Ceramics and Glasses*, Engineered Materials Handbook, Volume 4, Eds.: S. Schneider, Jr., J. R. Davis, G. M. Davidson, S. R. Lampman, M. S. Woods, T. B. Zorc, and R. C. Uhl, ASM International, 386-393, 1987.

2. M. Cable, "Principles of Glass Melting," *Glass Science and Technology, Vol. 2*, Eds. D. R. Uhlmann and N. J. Kreidl, Academic Press, 1-44, 1984.

3. A. G. Pincus and David H. Davies, *Batching in the Glass Industry*, (Books for Industry and *The Glass Industry Magazine*), 1981, Section II, "The Batching Process," pages 28-105, and Section IV, "Batch Plants," pages 132-205.

4. W. C. Bauer, F. V. Tooley, and W. M. Manring, "Raw Materials (Section 2)," pp 57-90; F. V. Tooley, M. A. Knight, A. K. Lyle, and V. C. Swicker, "The Glass Preparation Process (Section 9)," pp 517-580; *The Handbook of Glass Manufacture*, Volume 1, Editor F. V. Tooley, Books for Industry and The Glass Industry Magazine, 1974.

5. J. J. Hammel, "Some Aspects of Tank Melting," *Commercial Glasses, Advances in Ceramics Volume 18*, American Ceramic Society, pp. 177-186, 1986.

6. *Conference on Glass Problems,* published yearly as Issue #1 of *Ceramic Engineering and Science Proceedings*, American Ceramic Society, edited by C. H. Drummond, III in odd numbered years and J. Kieffer in even numbered years.

7. J. Eisenhauer, P. Donnelly, S. McQueen, A. Monis, J. Pellegrino, J. Julien, *Report of the Glass Technology Roadmap Workshop*, Prepared by Energetics, Inc., Columbia, Maryland, Sponsored by the Department of Energy, 1997.

8. R. Miller and H. Moore, "Compacted Batch—Will it Make a Difference?" (Chapter 23, pp. 87-89); B. Bansal, K. Jones, P. M. Stephan, and J. R. Schorr, "Batch Pelletizing and Preheating" (Chapter 24, pp. 90-95); H. W. Morelissen, A. H. M. Rikken, and A. J. M. van Tienen, "Pelletized Batch: Its Manufacture and Melting Behavior" (Chapter 24, pp. 90-95); in Reference 2.

9. P. Niggli, "K_2CO_3-Na_2CO_3-$CaCO_3$" Figure 1016 in *Phase Diagrams for Ceramists*, Eds. E. M. Levin, C. R. Robbins, and H. F. MacMurdie, American Ceramic Society, (p 324) 1964.

OBSERAVTION AND ANALYSIS OF DISSOLUTION KINETICS, SUPPORTED BY MICROSCOPY

Ann-Katrin Glüsing
RWTH University Aachen,
Department of Glass and
Ceramic Composites,
and Chair of Mineral Engineering,
Mauerstrasse5, D-52064 Aachen

Reinhard Conradt
RWTH University Aachen,
Department of Glass and
Ceramic Composites,
and Chair of Mineral Engineering,
Mauerstrasse5, D-52064 Aachen

ABSTRACT

The impurities in recycled cullet are distinguished by their potential risk to impare glass quality. It is the objective of this study to identify the time laws by which individual types of impurities are dissolved and to translate the findings to the industrial scale. By this procedure, the potency of individual types of impurities to reach the product, i.e., to cause defects, is quantified. This yields most valuable criteria for the acceptable upper treshold of particle sizes, which may be used to optimize recycling and glass processing techniques.

MOTIVATION

In container glass melting in Europe, recycled cullet has become the main batch component. The cullet ratio depends on glass color, on availability, and on the level of impurities imported into the material via the public collection system. Typical impurities are crystalline ceramics, quartz stones, porcelain, and metals. Quality defects caused by such impurities are usually discovered at the end of the working line only. Therefore, any assistance for an effective input quality control of recycled cullet is highly appreciated. Emphasis must be laid on a safe separation of the most critical species prior to melting. Until today, recyclers and glass melters discuss quite controversially the acceptable thresholds of impurities in terms of size and total amount.

Importance of cullet as raw material for the glass industry

Waste reduction has been a leading theme since the energy crisis in the 70ties. As a bonus, waste reduction measures often result in a payback in terms of energy and raw materials savings. More recently, legislation has been focusing on the reduction of CO_2 emission. The use of large amounts of recycled cullet has a very positive influence on both CO_2 emission and waste reduction. Beyond this, the amount of CO_2 generating carbonates is significantly reduced, as described for soda ash by de Waal [1]. Each kg of primary raw materials replaced by cullet helps to save an amount of energy of 2 MJ per kg glass. Table I illustrates the development in glass industry during the past 80 years.

Table I. Development of the energy demand, average cullet content, CO_2 output, and other production parameters during the past 80 years.

Year		1928	1968	1990	1998
spec. energy demand H_{in}[*]	[kWh/t]	5600	2600	1550	1100
pull rate r[*]	[t/m²·d]	0.2	1.1	3.0	3.5
tank operation time[*]	[d]	300	2100	3000	4500
melting temperature[*]	[°C]	1370	1450	1500	1500
heat exploitation	[%]	9	18	26	36
exploited heat H_{ex}	[kWh/t]	504	471	410	396
cullet content	[%]	11	35	78	88
CO_2 output	[kg/t]	1320	700	400	270

[*] data for 1928 – 1980 after Gebhardt [2]

Presently, a cullet ratio of 30 to 90 % is reached depending on product colour and type. The relative amount of energy saving by cullet use versus the exploited heat is calculated as 0.36 kWh per wt. % of recycled cullet and t of produced glass. Beyond this, the use of cullet accelerates the melting process, thereby improving the efficiency of heat exploitation. As to the CO_2 emission: The output is reduced in direct proportion to the energy input ($\approx$ 0.2 kg CO_2 per kWh), and beyond this, to the reduction of primary raw materials ($\approx$ 190 kg CO_2 per t of primary batch).

OBJECTIVE

The impurities in recycled cullet are distinguished by their potential risk to impare glass quality. In recycling cullet, emphasis must be laid on a safe separation of the most critical species. Nowadays, recycling techniques are highly discriminative for particle sizes above 10 mm. In finer cullet fractions, however, comparatively high levels of ceramic and metallic impurities are accumulated.

It is the objective of this study to identify the time laws by which individual types of impurities (such as: alumina rich ceramics, hard porcelain, quartz stones, glass ceramics, metallic aluminium, etc.) are dissolved, and to translate the findings to the industrial scale.

EXPERIMENTAL PROCEDURE

The behavior of impurities is examined in detail by two types of experiments. In the first type of experiments, individual particles (porcelain, dense and porous stone ware, quartz stones, glass ceramics, and metallic aluminum) of a few mm size are dipped into a flint glass melt of a given temperature (1200 to 1500 °C).

Quantitative data on the dissolution kinetics are measured by in-situ observation, video recording and image analysis. The decrease of particle size is evaluated by a regression analysis helping to identify the type of time law of dissolution and to quantify it in terms of empirical parameters (effective dissolution velocity v, effective diffusion coefficient D).The identification of the predominant type of dissolution is supported by microscopy of partially reacted particles revealing the occurrence of reaction layers, intermediate phases, enriched or depleted zones. The findings are checked against thermodynamic data and phase diagrams.

In complementary experiments, individual particles are mixed into a batch and charged on top of a hot melt. This results in an assessment of the particle sizes actually introduced into the bulk of the glass melt.

The procedure leads to the formulation of a ranking of different oxide and metallic impurities (with respect to their time demand for total dissolution), and in turn, to the formulation of an acceptance threshold for their sizes. This information is combined with the mean and critical residence times of the glass melt in industrial glass furnaces.

INDENTIFICATION OF THE INDIVIDUAL TIME LAWS OF PARTICLE DISSOLUTION

The dissolution of individual particles is recorded by a video camera as a function of time and temperature. The results present a quantitative view of local kinetics. Along with the local kinetics of the materials, the time demand t* for total dissolution is determined, the so-called lifetime. Industrial relevance of these results is obtained by discussing the lifetime with respect to the residence time of the glass melts in individual furnaces.

We distinguish between linear and diffusion controlled types of kinetics, and different combinations of both types. With the identification of the local mechanism, kinetic parameters v and D denoting an effective dissolution velocity and diffusion coefficient, respectively, are determined. The kinetic equations used to describe the dissolution progress are compiled in Table II. They are given in terms of: the local reaction progress $\Delta r(t)/r_0$, the particle lifetime t*, and the diameter d_0 associated with the given lifetime.

Table II. Time laws of different types of dissolution kinetics, see also in [3]

Type of local kinetics	Reaction progress $\Delta r(t)/r_0=$	Particle lifetime $t^*=$	Critcal particle size $D_0^*(t^*)=$
A: Linear (congruent) dissolution	$\dfrac{v \cdot t}{r_0}$	$\dfrac{r_0}{v}$	$2 \cdot v \cdot t^*$
B: Diffusion controlled dissolution	$\dfrac{\sqrt{4 \cdot D \cdot t}}{r_0}$	$\dfrac{r_0^2}{4 \cdot D}$	$\sqrt{16 \cdot D \cdot t^*}$
C: Diffusion controlled dissolution through a reaction product layer	$\dfrac{2 \cdot D}{v \cdot r_0} \cdot \left[\sqrt{1 + \dfrac{v^2}{D} \cdot t} - 1 \right]$	$\dfrac{r_0^2}{4 \cdot D} + \dfrac{r_0}{v}$	$\dfrac{4 \cdot D}{v} \cdot \left[\sqrt{1 + \dfrac{v^2}{D} \cdot t^*} - 1 \right]$
D: Incongruent dissolution (slower species diffusion controlled)	$\dfrac{\sqrt{4 \cdot D \cdot t}}{r_0} + \dfrac{v}{r_0} \cdot t$	$\dfrac{2 \cdot D}{v^2} \cdot \left[1 - \sqrt{1 + \dfrac{v \cdot r_0}{D}} \right] + \dfrac{r_0}{v}$	$\sqrt{16 \cdot D \cdot t^*} + 2 \cdot v \cdot t^*$

The decrease of the size of individual particles is recorded and evaluated by an image analyzer. Among the different evaluation methods, the analysis of the apparent area A(t) of the particle silhouette turned out to be the most reliable way. When put into relation to the initial value A_0, the average decrease of the particle radius $\Delta r(t)$, is received by:

$$\frac{\Delta r(t)}{r_0} = \sqrt{\frac{A(t)}{A_0}} \qquad (1)$$

As an example, Figure I shows the immediate results for porcelain at 1500 °C.

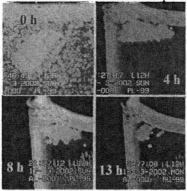

Figure I Video recording of dissolution kinetics
(particle decrease) of a porcelain at 1500 °C

DISSOLUTION BEHAVIOUR OF INDIVIDUAL PARTICLES DURING MELTING

In a next step, the data for porous earthenware and dense porcelain are analysed with respect to the most probable type of dissolution kinetics. The results for T = 1400 °C are shown by Figure II.

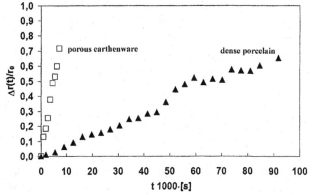

Figure II. Decrease of size of a porous earthenware
and a porcelain particle at 1400 °C

It is obvious that the long term dissolution of the porcelain has a higher defect potential than the porous earthenware. The infiltration of melt and the early weakening of the ceramic matrix in the latter case increases the dissolution velocity.

In order to identify the appropriate time law of dissolution, the relative size decrease $\Delta r(t)/r_0$ is checked for its compatibility with all equations compiles in Table II. This is done by a multiple regression of $\Delta r(t)/r_0$ vs. t an $\sqrt{t}$, or of t vs. Δr and Δr^2. Then the experimental data are plotted versus the normalized time t*/t and contrasted to the regression curves.

This results in highly discriminative plots which allow to select the type of kinetic equation best representing the data set. As an example, Figure III shows the immediate results for natural quartz stone at 1300 °C. Types A and D can be excluded right away based on the obvious mismatch between data points and curves.

Finally, type C is preferred to type B because of its better representation of the initial behavior. At higher temperatures the contribution of the diffusive part of the combined dissolution predominates.

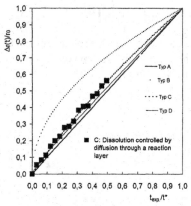

Figure III. Decrease of particle size versus normalized time t*/t
of a natural quartz stone at 1300 °C

Further particles typically dissolving via a B type kinetics are: porcelain (at
T < 1400 °C only), and metallic aluminium. Particles porcelain, dense stone ware,
and porous stone ware (at T > 1400 °C), by contrast, dissolve via an A type kinetics, while particles glass ceramics, and alumina dissolve via a D type kinetics.

OPTIMAL BATCH-TO-MELT CONVERSION OF IMPURITIES

The behavior of individual particles in the batch is examined, too. For this
purpose, impurities are mixed into batch samples and charged on top of a cullet
melt. The observation is done in different observation furnaces from above as well
as through the melt. The following qualitative results are obtained:
Coarse ceramic particles segregate; fragments with high aspect ratios are vertically orientated. Aluminium foils roll up and adhere to the surface of the melt.
Quartz stones behave like coarse sand grains forming agglomerates and bubbles in
the batch. Porous ceramics are quickly transformed by dissolution of the bonding
phase. Coloured glass ceramics particles sink directly into the melt and do not
participate in the batch reactions.

In batches containing fine cullet fractions, as described in [4-5], the particles
are stagnant; thus even dense impurity particles are dissolved prior to reaching the
melt. In batches containing coarse cullet fractions, by contrast, there is an enhanced relative motion of the batch constituents. This supports gas release and
mixing. However, dense particles are more easily conveyed to the melt underneath the batch.

MICROSCOPICAL IDENTIFICATION OF SPECIFIC DISSOLUTION
KINETICS

The examples described in this paragraph show the relevance of observing the
microstructure of the impurities. Figure IV shows the transformation of metallic
aluminium as observed by light microscopy and SEM.

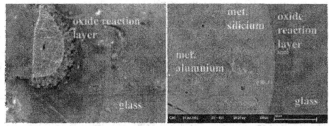

Figure IV. Dissolution of a aluminium bottle cap into a Si sphere with an iron penetration phase surrounded by alumina oxide barrier

The microscopical observations support the result of the kinetic analysis as presented in the previous section. The time law of dissolution is essentially constituted by the microstructure of the impurities. For example, dense alumina rich ceramics dissolves incongruently while metallic aluminum builds up a reaction layer imposing a diffusion controlled kinetics. An aluminium cap dissolves under formation of a Si sphere with an iron penetration phase. The massive oxide alumina barrier surrounding the metallic Si sphere dissolves upon long time exposure. The formation of carnegeit or nephelin takes place by a reaction of alumina with sodium disilicate.

LIFETIME DETERMINATION OF IMPURITIES WITH RESPECT TO THE MASS FLOW BEHAVIOR OF INDIVIDUAL TANKS

The data on particle dissolution determined of individual particles are transformed into plots of particle size versus lifetime d (t*). Here, d(t*) denotes the particle size of impurities requiring a time t* for total dissolution. So, for a given lifetime t*, the corresponding particle size can be determined. This is shown in Figure V for natural quartz stones at several temperatures.

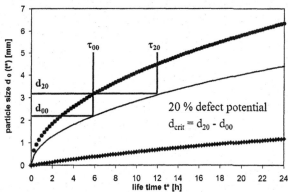

Figure V. The lifetime t* and the particle size d_{00} for a natural quartz stone at 1300 – 1500 °C including the residence time and the critical paticle size.

As seen from Figure V, a residence time of 10 h requires a particle size until total dissolution of 0.6 mm at 1300 °C, 2.8 mm at 1400 °C and 4.0 mm at 1500 °C. In comparison to this, a 10 h exposure at 1400 °C leads to a complete dissolution of a hard porcelain particle of 1.2 mm, a metallic aluminium particle of 1.6 mm, a glass ceramics particle of 3.1 mm, and a porous stone ware particle of 22.2 mm.

In order to determine the particle sizes acceptable for industrial glass melting, the individual life times t* have to be evaluated with respect to their residence time in the glass tank furnace. Only for the (rather unrealistic) case of a plug flow behavior of the tank is the particle size d_0 identical with the acceptable size. The evaluation proceeds as follows: For each kind of impurity, temperature dependent families of curves $d_0(t^*)$ are plotted and compared to the residence time spectrum of a given furnace determined, e.g. by a tracer experiment.

By a previous investigation [6], the following characteristic times were determined: τ_{00} = 6 h (first detection of tracer), τ_{05} = 8 h, τ_{20} = 12 h. Here, τ_{00} (see Figure V) denotes the time of which a nn % throughput of the tracer has taken place. The times correspond to critical particle sizes d_{crit}. The highest security level is related to the time τ_{00}. There is no tracer response, hence the entire volume is still contained in the furnace, with no defects being able to reach the end of the feeder. At the time τ_{20}, 80 % of particles with the size d_{20} are still contained in the furnace. These will dissolve for sure. But 20 % of the said size are put through the furnace and may cause defects of sizes ranging up to $\Delta d = d_{20} - d_{00}$.

CONCLUSIONS

The paper comprehensively treats the problem of impurities in cullet. This includes the local reaction mechanisms (dissolution kinetics, life time, recristallisation) as well as participation in the global mass flow in the furnace.

The time demand for the dissolution of individual impurity particles is determied as a function of the kind of material, the initial particle size, and the exposure temperature. The most critical particles are alumina, hard porcellain, metallic aluminium, glass ecramics; upon a 10 h exposure to a 1400 °C flint glass melt, only particles with diameters smaller than 3.0 mm dissolve. By contrast, porous ceramics dissolves comparatively fast; the corresponding "safe" particle size is 22.0 mm, respectively.

The results are put into relation to their physical and chemical properties, and evaluated for industrial relevance with respect to the residence time distribution of individual tanks. This can be used to formulate acceptance thresholds for particle sizes of ceramic and metallic impurities. By this, interesting problem solutions are derived, such as a combination of particle sizes, fine and coarse fractions of cullet, to minimize the glass defects according to the cullets.

REFERENCES

[1] H. de Waal, "Technological aspects of recycling glass melting"; pp. 87 – 97 in *El vidro reciclado en la fabricación de envases, Red ibereoamericana sobre cienca y technología de materiales vítreos*, Valencia, 2000

[2] F. Gebhardt, „Feuerfeste Werkstoffe für den Betrieb von Glasschmel-zöfen", VGT-Dyko Industriekeramik GmbH, Düsseldorf, 1996

[3] A.-K. Glüsing and R. Conradt, "Dissolution kinetics of ceramic and metal-lic impurities of cullet"; pp. 29 – 42 in *Recycling and reuse of glass cullet*, 1st. ed. Edited by R.K. Dhir, M.C. Limbachiya and T.D. Dyer. Thomas Telford Publishing, London, 2001

[4] A.-K. Glüsing and R. Conradt, "Melting behavior of recyceld cullet; effects of impurity dissolution, pulverisation and solution pre-treatment", Verre, **9** [1] 16-20 (2003)

[5] Research project on reuse of the fine rest fraction, GHI, 2001 and V. Roumeas, „Prozesstechnische Bewältigung der Verunreinigungen im Altglas durch Feinvermahlung", Diploma thesis RWTH, Aachen, 2001

[6] D. Schippan, Tracerversuche diskreter Verweilzeiten an einer Hohlglas-wanne, Diploma thesis RWTH, Aachen, 1999

Characterization of Glass Melts/Glass Melt Properties

INERT GAS SOLUBILITY IN GLASSES AND MELTS OF COMMERCIAL COMPOSITIONS

C. C. Tournour, M. G. Mesko, & J. E. Shelby
New York State College of Ceramics
Alfred University
Alfred, NY 14802

ABSTRACT

Inert gas solubilities (He, Ne, and Ar) have been measured in melts of several commercial glass compositions, including those commonly used for windows, containers, TV panels, fiberglass, and borosilicate tubing. Results for each gas are combined with data from the literature to produce curves showing the general effect of changes in melt composition on inert gas solubility in silicate and borosilicate melts. These curves can be used to estimate the solubility of each of these gases in such melts where no experimental data exist. Data from the literature were used to construct a similar curve for nitrogen solubility in silicate melts.

He solubilities were also determined over the temperature range from below T_g to 1500°C for the same compositions. Results are compared with the temperature dependence of helium solubility previously reported for vitreous silica and for a binary sodium silicate composition. A minimum in solubility as a function of temperature occurs for the float and several borosilicate compositions in the region slightly above T_g. The low diffusivity of helium in the other glasses at low temperatures prevents extension of the data to sufficiently low temperatures to determine if similar minima exist for all silicate compositions, but it is clear that the solubility of helium in the melts is always greater than in the glasses near T_g.

INTRODUCTION

Inert gas solubility in glasses and melts is important in many applications [1,2]. Trapping of atmospheric gases during the initial stages of batch melting can result in bubbles containing argon and/or nitrogen [2,3]. Bubbling helium through melts can be used to speed fining by producing large, rapidly rising bubbles which sweep up the smaller seeds in the melt [3]. Removal of any residual helium bubbles is dependent on the solubility of helium in the melt. The presence of all of these gases in natural glasses and melts is routinely used to determine the origin, age, and history of these materials [2]. Diffusion of all of these gases through glasses can be detrimental to vacuum systems, vacuum tubes, and other devices which depend on the exclusion of gases [1].

Discussion of inert gases is restricted in this paper to the most common gases which do not chemically react with glasses or glassforming melts [1-3]. While gases such as O_2, CO_2, and SO_2 can exist in melts in the molecular form, their primary interactions with melts involve chemical reactions [2,3]. Nitrogen, on the other hand, which can react with melts to form "nitrided" glasses, usually exists only as the physically dissolved nitrogen molecule in most glasses and melts [2].

This paper presents new data for the solubility of helium, neon, and argon in glasses and melts, along with a review of the literature dealing with the solubility of these gases and N_2 in melts. Since the emphasis of this paper lies in the compositions commonly used to produce commercial glasses, data for binary alkali silicate, borate, and similar melts are not included in most of the figures. Data for some ternary soda-lime-silicate melts are included since these compositions are near those of many commercial glasses. In addition, data for solid glasses are restricted to helium, since no solubility data for other gases exist for the types of glasses considered here, while data for melts are presented for all four of these gases.

EXPERIMENTAL PROCEDURES
Samples of most of the materials used in this study were produced by remelting commercial glasses in Pt crucibles. Glass studies were carried out on plates cut from either the remelted glasses or on slices from plates or tubing of the as-received glass. Since sulfate-fined glasses foam during remelting under vacuum, which is necessary for the solubility measurements on melts, commercial glasses could not be used for melt studies in some cases. Under those circumstances, "mock" glasses were made based on the nominal compositions reported elsewhere [4].

Details of the measurement technique have been discussed in several other papers [5-7]. Helium solubility measurements on glasses are made by exposing the plates of glass to helium at a known pressure and temperature until the glass is saturated with the gas. The sample is then transferred to a residual gas analyzer system and reheated under vacuum to 500°C to drive off the dissolved gas. The amount of helium released from the sample is determined by comparison with the integrated signal obtained from a known amount of helium contained in a standard vial. A similar procedure is used for measurements on melts, where the melt is cooled in the crucible, which is then placed in the outgassing system and outgassed at 900°C. In either case, the solubility is determined by dividing the amount of gas released, expressed as a number of atoms, by the mass of the sample in grams and the soaking pressure in atmospheres.

The temperature dependence of gas solubility was determined by soaking the sample at ≈700 torr of the desired gas at different temperatures. In those cases where the data are compared at either 600°C for glasses or at 1300°C for melts, several measurements were made under identical conditions. The value reported is the average of these measurements.

RESULTS
The effect of temperature on helium solubility in commercial silicate glasses and melts over a broad temperature range has been determined for several borosilicate glasses [7] and for float, container, TV panel, wool, and E-glass. Examples of the results for a borosilicate and a float glass are shown in Figure 1. These results are similar to those reported earlier for a sodium silicate composition in that the solubility of helium decreases with increasing temperature in the glass, passes through a minimum somewhat above the glass transformation temperature, and then

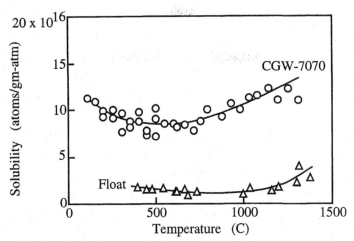

Figure 1: Effect of temperature on helium solubility in CGW-7070 borosilicate and soda-lime-silicate float glasses and melts.

increases with increasing temperature in the melt [6]. The same general behavior is observed for all of the glasses studied.

It is generally assumed that the concentration of dissolved helium and other inert gases in glasses and melts will increase linearly with pressure. While this has been shown to be true at high pressures for argon solubility in boric oxide melts [8] and at lower pressures in some feldspar melts [9], there has been no proof that this behavior is observed for silicate glasses and melts with compositions near those of commercial glasses. The data shown in Figure 2 for dissolved helium concentration as a function of helium partial pressure in glasses and melts of a $20Na_2O-80SiO_2$ glass and in a float glass at 500°C demonstrate that the solubility of He in these glasses and melts increases linearly with increasing partial pressure of the gas.

Isothermal helium solubility data at 600°C for vitreous silica, several commercial borosilicate glasses, and for our measurements on float, container, TV panel, wool, and E-glasses, along with data for some other commercial glasses taken from Ref. 10, are shown in Figure 3. Helium solubility in typical commercial soda-lime-silicate glasses is about 10% of that in vitreous silica at this temperature. The solubility of helium decreases rapidly with initial reduction of the concentration of glassforming oxides in the glass, followed by a large compositional range where helium solubility decreases only gradually with continued reduction in the total glassforming oxide concentration.

The effect of composition on the solubility of neon and argon in melts has also been measured for several of the compositions discussed above, along with a few other commercial compositions. The gas solubility always decreases in the order He>Ne>Ar. Typical results for the temperature dependence of the solubility of all three of these gases in a borosilicate melt are shown in Figure 4. The large range of solubilities requires use of a log axis for the solubility in order to adequately show these data on a single plot.

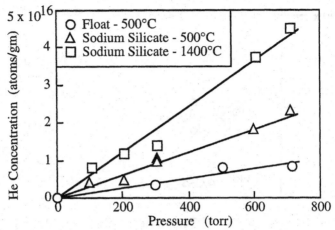

Figure 2: Effect of helium partial pressure on the concentration of helium in solution in sodium silicate and float glasses at 500°C and in sodium silicate melts at 1400°C.

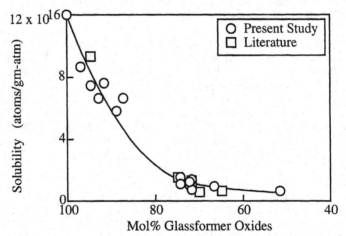

Figure 3: Solubility of helium in complex silicate glasses at 600°C. Literature data are from Ref. 10.

The compositional dependence of helium, neon, and argon solubility at 1300°C as a function of the total glassforming oxide content of the melt is shown in Figures 5, 6, and 7, respectively. In some cases, the values indicated are taken from a best fit curve to our data taken over a range of temperatures. In other cases, we only determined solubilities at 1300°C. In the latter case, the values shown here represent the average of a number of repetitions of the measurement under identical conditions. The scales used on the ordinates of these plots are progressively changed in order to provide a better visual presentation of the data. It should be noted that vitreous silica is within its glass transformation range at 1300°C, while all of the other compositions are melts, albeit very viscous melts, at this temperature.

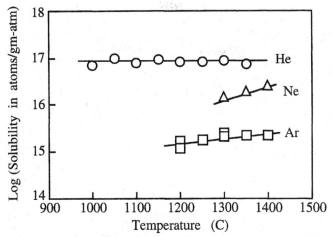

Figure 4: Effect of temperature on He, Ne, and Ar solubility in melts of a borosilicate glass.

Although no measurements of nitrogen solubility were made during this study, the high incidence of nitrogen bubbles in commercial melts justifies an attempt to include information regarding nitrogen solubility in this paper. Examination of the literature reveals very little data for nitrogen solubility in silicate melts [11, 12, 16-19]. These results are shown as a function of glassforming oxide content, which is entirely silica in this case, in Figure 8. Most of the existing data are found in the work of Mulfinger, et al. [11,12] from the early 1960s. These studies can be subdivided into measurements on lithium and potassium silicate melts, indicated as "alkali silicates" on the figure. Mulfinger, et al. [11] did report the results for one ternary soda-lime-silicate melt containing 74 mol% silica. The other data shown in this figure are for commercial melts [17-20]. In most cases, the glass studied is

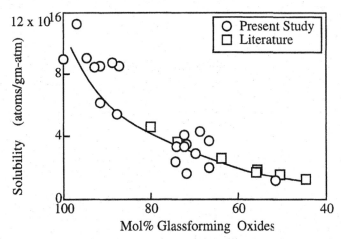

Figure 5: Solubility of helium in complex silicate melts at 1300°C. Literature data are from Refs. 9, 11 and 12.

designated by a generic term such as TV glass, with no indication of the exact composition of the glass. Note that the data for the binary alkali silicate melts lie on a completely different line from that for the more complex compositions. It is not known if this represents a true difference or is simply indicative of the difficulty in measuring nitrogen solubility at low pressures in silicate melts.

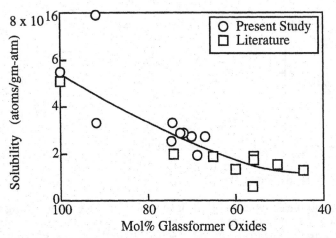

Figure 6: Solubility of neon in complex silicate melts at 1300°C. Literature data are from Refs. 9, 11, and 13.

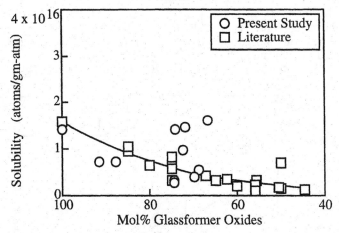

Figure 7: Solubility of argon in complex silicate melts at 1300°C. Literature data are from Refs. 9, 13, 14, and 15.

The trend in inert gas solubility as a function of the identity of the dissolved gas in a number of the melts studied here is shown in Figure 9. Shelby [1], for example, has analyzed the literature data for a wide range of compositions, including not only the complex silicates considered here, but also alkali silicates and borates, alkali fluoride, sodium fluorozirconate, and natural silicate melts. He found that the log of the solubility decreases linearly with increasing atomic diameter of the inert gas atom or equivalent spherical diameter of diatomic inert gas molecules, such as N_2. Results of the present study (Figure 9) indicate that the solubility, rather than the log solubility, is a linear function of the diameter of the dissolved entity. The results presented in Figure 9 should prove useful in estimating the solubility of He, Ne, Ar, and N_2 in complex silicate melts.

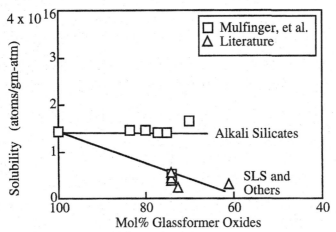

Figure 8: Solubility of nitrogen in silicate melts at 1300°C.
Literature data are from Refs. 11, 12, and 16-19.

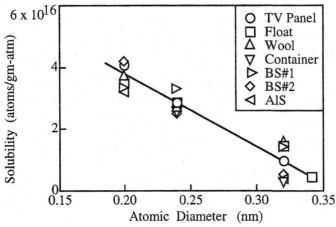

Figure 9: Effect of molecular diameter on gas solubility in silicate melts at 1300°C. All data are from the present study.

ACKNOWLEDGMENTS

Portions of this work were funded by the NSF Industry/University Center for Glass Research at Alfred University (CGR) and portions by the US Department of Energy, Office of Industrial Technologies, grant #DE-FG07-96EE41262.

REFERENCES

[1]J. E. Shelby, "Gas Solubility in Inorganic Glasses"; pp. 55-70 in *Handbook of Gas Diffusion in Solids and Melts*, by J. E. Shelby, ASM International, Materials Park, OH, 1996.

[2]J. E. Shelby, "Gases in Melts"; pp. 161-200 in *Handbook of Gas Diffusion in Solids and Melts*, by J. E. Shelby, ASM International, Materials Park, OH, 1996.

[3]J. E. Shelby, "Glass Melting," pp. 25-47 in *Introduction to Glass Science and Technology*, by J. E. Shelby, Royal Society of Chemistry, Cambridge, England, 1997.

[4]M. G. Mesko and J. E. Shelby, "Water Solubility and Diffusion in Melts of Commercial Silicate Glasses," *Glastech. Ber.*, **73** [C2] 13-22 (2000).

[5]M. G. Mesko and J. E. Shelby, "Helium Solubility in Ternary Soda-Lime-Silica Glasses and Melts," *Phys. Chem. Glasses*, **43** [2] 91-96 (2002).

[6]M. G. Mesko, K. Newton and J. E. Shelby, "Helium Solubility in Sodium Silicate Glasses and Melts," *Phys. Chem. Glasses*, **41** [3] 111-16 (2000).

[7]M. G. Mesko, B. E. Kenyon and J. E. Shelby, "Helium Solubility in Commercial Borosilicate Glasses and Melts," *Glastech. Ber.*, **73** [C2] 33-42 (2000).

[8]S. P. Faile and D. M. Roy, "Solubilities of Ar, N_2, CO_2, and He in Glasses at Pressures to 10 kbars," *J. Am. Ceram. Soc.*, **49** [12] 638-43 (1966).

[9]G. Lux, "Behavior of Noble Gases in Silicate Liquids: Solution, Diffusion, Bubbles, and Surface Effects, with Application to Natural Samples," *Geochim. Cosmochim. Acta*, **51** 1549-60 (1987).

[10]A. Suckow, P. Schlosser, H. Rupp and R. Bayer, "Diffusion and Permeation Constants of Helium in Various Glasses," *Glass Technol.*, **31** [4] 160-64 (1990).

[11]H. O. Mulfinger, A. Dietzel and J. M. F. Navarro, "Physical Solubility of He, Ne, and N_2 in Glass Melts," *Glastechn. Ber.*, **45** [9] 389-96 (1972).

[12]H. O. Mulfinger and H. Scholze, "Solubility and Diffusion of Helium in Glassforming Melts, I. Solubility," *Glastechn. Ber.*, **35** [11] 466-78 (1962).

[13]T. Shabata, E. Takahashi and J. Matsuda, "Solubility of Ne, Ar, Kr, and Xe in Binary and Ternary Silicate Systems: A New View on Noble Gas Solubility," *Geochim. Cosmochim.* Acta, **62** [7] 1241-53 (1998).

[14]B. S. White, M. Brearley and A. Montana, "Solubility of Argon in Silicate Liquids at High Pressures," *Am. Mineral.*, **74** 513-29 (1989).

[15]M. R. Carol and E. M. Stolper, "Noble Gas Solubilities in Silicate Melts and Glasses: New Experimental Results for Argon and the Relationship between Solubility and Ionic Porosity," *Geochim. Cosmochim.* Acta, **57** 5039-51 (1993).

[16]S. Kawachi, unpublished data.

[17]G. H. Frischat, O. Buschmann and H. Meyer, "Diffusion of Nitrogen in Glass Melts," *Glastechn. Ber.*, **51** 321-27 (1978).

[18]E. L. Swarts, "Gases in Glass," *Ceram. Eng. Sci. Proc.*, **7** [3-4] 390-403 (1986).

[19]H. Meyer, G. H. Frischat, and F. W. Kramer, "Diffusion of Nitrogen in Glass Melts," pp. 317-26, *Proceedings XI ICG, Vol. 2*, Prague, 1977.

WATER DIFFUSION AND SOLUBILITY IN GLASSES AND MELTS OF FLOAT, CONTAINER, AND OTHER COMMERCIAL COMPOSITIONS

Peter B. McGinnis, Melissa G. Mesko, Douglas B. Rapp, and James E. Shelby
New York State College of Ceramics
Alfred University
Alfred, NY 14802

ABSTRACT

Water diffusion coefficients have been determined over the temperature range from below the glass transformation temperature to 1500°C for glasses and melts of float, container, TV panel and funnel, and wool and E-glass compositions. Diffusivities in melts were determined during both the formation and removal of hydroxyl. Most diffusivities at temperatures below 800°C were determined by dehydroxylation of thin plates of glass, but a few measurements were also made during hydroxyl formation in similar plates. Hydroxylation and dehydroxylation methods yield identical results for these materials. Diffusion coefficients lie on a smooth Arrhenian curve over the entire temperature range studied, with no indication of a change in the glass transformation region. Water solubilities were also determined for float and container melts. Results indicate that the temperature dependence of water solubility in these melts is very small.

INTRODUCTION

Water enters the structure of silicate glasses and melts by reacting to form hydroxyl [1,2]. Since replacement of a bridging oxygen by a hydroxyl reduces the connectivity of the network, hydroxyl significantly reduces the glass transformation temperature (T_g) of silicate glasses, with variations of 20 to 30 K for soda-lime-silicate glasses [2] for changes in water concentration of 1000 wtppm. The transformation region viscosity is altered by more than an order of magnitude by varying the hydroxyl concentration over this range [2]. Crystallization rates, which are very sensitive to viscosity, can be altered by orders of magnitude by changes in hydroxyl content [3,4]. Changes of this magnitude are particularly important in annealing and reshaping glasses and in the prevention of crystallization during processing.

Although the product of the reaction of water with oxide melts is hydroxyl, the reaction is the result of exposure to water vapor. Since the reaction rate is diffusion-controlled, this process is usually termed "water diffusion". While the diffusing species is usually assumed to be molecular water, details of the mechanisms involved in the production of hydroxyl have not been established [5]. Use of the term "water" here only indicates that the species in the atmosphere surrounding the melt is molecular water and does not imply any particular diffusion mechanism.

Water solubility in melts is determined by exposing the melt to a known water vapor pressure until it reaches equilibrium and then using one of several methods to determine the actual amount of water in the glass [2]. The most common method involves measurement of the infrared spectrum of the glass and calculation of the water concentration using the Beer-Lambert Law, i.e. the expression

$$A = \varepsilon cL, \tag{1}$$

where A is the absorbance measured at a given wavelength, c is the concentration of hydroxyl in the sample, L is the path length of the light through the sample, and ε is a proportionality constant known as the extinction coefficient. Since A and L are easily measured, the concentration of hydroxyl can be obtained if ε for the specific material studied is known. If ε is not known, it must be measured before quantitative determination of the solubility using the Beer-Lambert Law is possible.

Unfortunately, the extinction coefficient for water in glasses has been measured for only a few compositions. Values reported for commercial soda-lime-silica glasses [2] lie in the range of ≈ 40 L/mol-cm. Large differences among values reported for different glasses suggest that the extinction coefficient varies over a wide range with glass composition, so that no single value can be applied to glasses of significantly different composition.

The diffusivity of water has usually been determined by measuring the rate of removal of hydroxyl from a glass or melt, i.e. during dehydroxylation, utilizing either direct measurements of water content or infrared spectral measurements of the hydroxyl concentration [6,7]. Very few studies have measured the diffusivity of water in glasses or melts during the hydroxyl formation process. Since studies of hydroxyl formation and removal in vitreous silica [8] yield different diffusivities, there is some question regarding the validity of comparison of results obtained during hydroxylation with those obtained during dehydroxylation.

Only one study has considered the solubility and diffusivity of water in float glass melts [9]. No reports of water solubility and diffusion in container melts were found. The solubility of water in float glass melts in the previous study was determined using infrared spectroscopy and the 3-band method of Scholze [10].

This paper presents the results of a study of the hydroxylation and dehydroxylation of glasses and melts of several commercial compositions at temperatures ranging from below Tg to 1500°C. The extinction coefficient for hydroxyl in float and container glass was also determined.

EXPERIMENTAL PROCEDURES

Samples used in this study were produced by remelting commercial glasses in Pt/5Au crucibles. Nominal compositions of these glasses have been reported elsewhere [11]. Since details of the measurement technique have been discussed in several other papers [8,9,11-15], they will only be reviewed briefly here.

Extinction coefficients for water in float and container glasses were determined using the vacuum heat treatment technique reported earlier for TV panel, wool, and E-glass [12,13]. Weight losses during heat treatment were measured using a microbalance, while changes in the infrared spectra were measured using a FTIR spectrometer. The extinction coefficient was determined from the slope of a plot of change in absorption versus change in weight, with the assumption that all of the weight loss was due to removal of water from the glass.

Diffusivities are determined by measuring the rate of change of absorption at the maximum of the infrared band due to hydroxyl, which occurs at approximately 3500 cm^{-1} for these glasses. Since most of the procedures used for both melts and glasses are discussed in detail elsewhere [11-15], only those which are specific to this study will be given here. In general, hydroxyl is added to melts by exposure to an atmosphere enriched in water vapor. Since laboratory air has a lower water vapor partial pressure than that found in glass tanks, however, remelting of commercial glasses in air will cause the melt to release water vapor. The final hydroxyl content of the glass will be determined by the humidity at the time of the study. If the glass or melt is exposed to vacuum, it is usually assumed that the water vapor partial pressure is zero, so that all of the hydroxyl would be removed from the glass if sufficient time were available. This assumption is necessary for the determination of the diffusivity in glasses at the lower temperatures since the time to actually reach the final water concentration would be prohibitive.

The change in relative hydroxyl concentration from that of the initial condition is obtained by subtracting the value for the absorption/mm for the initial state (C_0) from each of the values obtained during the heat treatments (C_t). Dividing the change in absorption/mm by the value obtained by subtracting the initial melt value from that of the saturated melt yields the fractional change, F, in concentration for each treatment time. The fraction of saturation is given by the expression

$$F = (C_t - C_0)/(C_\infty - C_0). \tag{2}$$

The effective diffusivity for hydroxyl formation, D, into or out of a semi-infinite plate, where diffusion occurs from both surfaces, is given by the expression

$$\frac{M_t}{M_\infty} = \frac{4}{\sqrt{\pi}} \sqrt{\frac{Dt}{L^2}} \tag{3}$$

where M_t is the concentration after treatment time t, M_∞ is the total change in concentration ($M_t/M_\infty = F$), and L is the thickness of the sample. If the data are plotted as F versus $\sqrt{t}/L$, it can be shown that the diffusivity is given by

$$D = (\pi/16)R^2. \tag{4}$$

where R is the slope of the plot of F versus $\sqrt{t}/L$ at short times (the plot is essentially linear up to a value of F = 0.5). This expression is used to obtain D for water vapor diffusion from the plate samples used in the transformation range.

Since Eq. 3 assumes that the sample is exposed to the diffusing species on both surfaces, Eq. 4 must be corrected by a factor of 4 to yield the expression for diffusion into or out of a melt, where the sample is only exposed to the atmosphere on one surface, or

$$D = (\pi/4)R^2. \tag{5}$$

RESULTS

Results of determinations of the extinction coefficient for water in float and container melts are shown in Figures 1 and 2. The extinction coefficients for the band at ≈3500 cm⁻¹ are 41.3 and 42.4 L/mol-cm, respectively, for float and container glasses. These values compare well with other reported values in the range of 39 to 43 L/mol-cm for soda-lime-silica glasses [2,10].

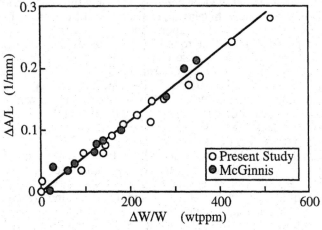

Figure 1: Determination of the extinction coefficient for water in a float glass.

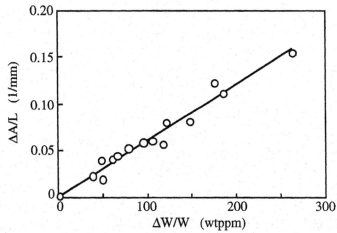

Figure 2: Determination of the extinction coefficient for water in a container glass.

Water solubilities have been determined [9,11] in our laboratory for float and container glasses for melting temperatures ranging from 1000 to 1400°C. The results, as shown in Figure 3, are about 50 wtppm higher than values reported by others [16,17] for 16/10/74 soda-lime-silica melts. The slight dependence of the solubility on temperature reported here is similar to that reported by Franz and Scholze for a simple ternary soda-lime-silica melt [16].

The effect of temperature on the diffusivity of water in float and container glasses and melts is shown in Figures 4 and 5, respectively. Measurements were made during addition of water by exposing the samples to flowing air saturated with 1 atmosphere of water vapor [9,11] and during dehydroxylation by exposing plates to vacuum at lower temperatures or by remelting the glass in ambient air at the higher temperatures. Since the original glasses were produced in air/fuel fired furnaces, they contain more water than glasses which are remelted in an electrical furnace. As a result, the glasses lose about 2/3 of their original water content during remelting.

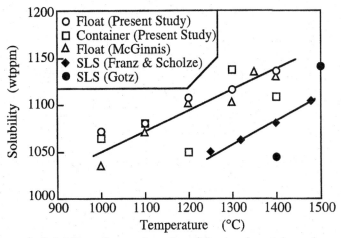

Figure 3: Solubility of water in melts of float and container glasses.

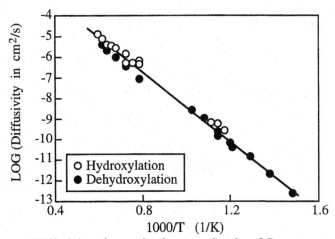

Figure 4: Diffusivity of water in glasses and melts of float composition.

Results of similar studies for compositions used for television tube (panel and funnel) glasses and melts are shown in Figure 6 and for glasses and melts of compositions used to produce glass fibers (wool and E-glass) are shown in Figure 7. Data for hydroxylation of the melts of the TV panel glass are from Ref. 12, while those for the wool and E-glasses are from Ref. 13. All other data, including those for the TV funnel melts and for all of the glasses at lower temperatures, are from the present study.

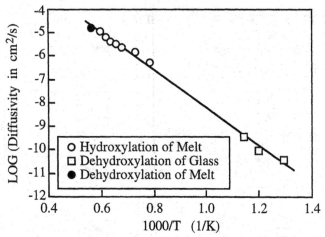

Figure 5: Diffusivity of water in glasses and melts of container composition.

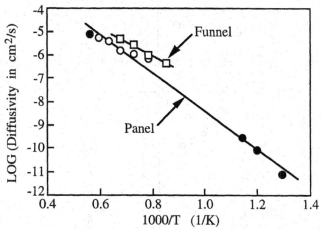

Figure 6: Diffusivity of water in glasses and melts of glasses used in the production of television tubes.

DISCUSSION

Results of the present study indicate that the commonly used approximation of 41 L/mol-cm for the extinction coefficient for water in commercial glasses is very close to the experimentally determined values of 41.3 and 42.4 L/mol-cm for float and container compositions, respectively. These values differ substantially from the extinction coefficient for other commercial glasses of 46.8, 49.1, and 18.2 L/mol-cm for wool, E-glass, and TV panel glass, as reported elsewhere [11-13].

Water solubilities in float and container melts are only a small function of melt temperature over the range from 1000 to 1500°C. Values reported here are about 50 wtppm greater than those reported by Franz and Scholze [16]. The melts considered by Franz and Scholze were actually treated at 0.5 atmosphere of water vapor, while those studied here were exposed to a full atmosphere of water vapor. The values reported in the earlier study were multiplied by a factor of 1.414 to correct for the square root dependence of water solubility in oxide melts on water vapor partial pressure [11-15]. Franz and Scholze state that their error in the determination of the water concentration is ±10%, which is considerably larger than the ≈5% difference between their results and those found here. The temperature dependence found in the present study is very similar to that reported by Franz and Scholze, as indicated by the approximately parallel lines through the data shown in Figure 3.

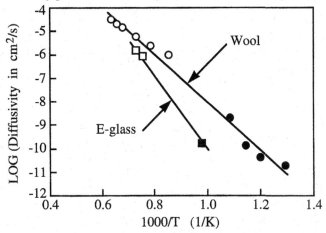

Figure 7: Diffusivity of water in glasses and melts of glasses used in the production of commercial glass fibers.

While a number of studies have separately considered the diffusion of water in the melt and glass transformation ranges [2,6-9], this is the first report of a study which covers both temperature regions using the identical glasses as samples. The results indicate that there is no significant deviation in the log(D) versus reciprocal temperature plot over a range of 1000 K. In particular, results for float glass convincingly show that no unusual behavior in water diffusivity occurs in the glass transformation range. This result is rather surprising in that other diffusion-related phenomena, e.g. sodium [18] and helium [2] diffusion [18] and ionic conductivity [18], have been reported to show significant deviations in the glass transformation range from the simple behavior shown here.

CONCLUSIONS

Water solubility in commercial glasses and melts lies in the range of 1000 to 1200 wtppm for most compositions, with a very small dependence on temperature. Water diffusivities in the melts of these glasses lie in the range of 10^{-7} to 10^{-5} cm^2/s, with a large temperature dependence. The diffusivities of water in glasses are much lower, but lie on the same Arrhenian line as the melts. There is no discontinuity in the log(D) vs. reciprocal temperature plot in the glass transformation region.

ACKNOWLEDGMENTS

This work was supported by the NSF Industry/University Center for Glass Research at Alfred University.

REFERENCES

[1]J. E. Shelby, "Glass Melting"; pp. 25-47 in *Introduction to Glass Science and Technology*, by J. E. Shelby, Royal Society of Chemistry, Cambridge, 1997.

[2]J. E. Shelby, "Water in Glasses and Melts"; pp. 217-234 in *Handbook of Gas Diffusion in Solids and Melts*, by J. E. Shelby, ASM International, Materials Park, OH, 1996.

[3]M. R. Heslin and J. E. Shelby, "Effect of Hydroxyl Content on the Nucleation and Crystallization of Lithium Disilicate Glass, *Ceram. Trans.*, **30** 189-96 (1993).

[4]C. J. R. Gonzalez-Oliver, P. S. Johnson and P. F. James, "Influence of Water Content on the Rates of Crystal Nucleation and Growth in Lithia-Silica and Soda-Lime-Silica Glasses," *J. Mater. Sci.*, **14** 1159-69 (1979).

[5]R. H. Doremus, "Speciation of Water in Silicate Glasses and Melts"; pp. 129-47 in *Diffusion of Reactive Molecules in Solids and Melts*, by R. H. Doremus, Wiley Interscience, New York, 2002.

[6]H Scholze and H. O. Mulfinger, "The Influence of Water in Glasses V. The Diffusion of Water in Glasses at High Temperatures," *Glastech. Ber.*, 32 [9] 381-86 (1959).

[7]B. J. Todd, "Outgassing of Glass," *J. Appl. Phys.*, **26** [10] 1238-43 (1955).

[8]P. B. McGinnis and J. E. Shelby, "Diffusion of Water in Vitreous Silica," *J. Non-Cryst. Solids*, **179** 185--93 (1994).

[9]P. B. McGinnis and J. E. Shelby, "Diffusion of Water in Float Glass Melts," *J. Non-Cryst. Solids*, **177** 381-88 (1994).

[10]H. Scholze, "The Influence of Water in Glasses I. Influence of Dissolved Water in Glass on the IR Spectra and the Quantitative IR Spectroscopic Determination of Water in Glasses," *Glastech. Ber.*, **32** [3] 81-8 (1959).

[11]M. G. Mesko and J. E. Shelby, "Water Solubility and Diffusion in Melts of Commercial Silicate Glasses," *Glastech. Ber. Glass Sci. Technol.*, **73** [C2] 13-22 (2000).

[12]M. G. Mesko and J. E. Shelby, "Solubility and Diffusion of Water in Melts of a TV Panel Glass," *Phys. Chem. Glasses*, **42** [1] 17-22 (2001).

[13]M. G. Mesko and J. E. Shelby, "Solubility and Diffusion of Water in Melts of E and Wool Glasses," *Phys. Chem. Glasses*, **42** [6] 389-96 (2001).

[14]M. G. Mesko and J. E. Shelby, "Water Solubility and Diffusion in Alkali Silicate Melts," *Phys. Chem. Glasses*, **42** [3] 173-8 (2001).

[15]M. G. Mesko, P. A. Schader and J. E. Shelby, "Water Solubility and Diffusion in Sodium Silicate Melts," *Phys. Chem. Glasses*, **43** [6] 283-90 (2002).

[16]H. Franz and H. Scholze, "Solubility of Water Vapor in Glass Melts of Varying Basicity," *Glastech. Ber.*, **36** [9] 347-56 (1963).

[17]J. Gotz, "Influence of Melting Parameters on Water Solubility in Glass, *Glastech. Ber.*, **45** [1] 14-18 (1972).

[18]R. H. Doremus, "Electrical Conductivity and Ionic Diffusion," pp. 271-6 *Glass Science*, *2nd Ed.*, by R. H. Doremus, Wiley Interscience, New York, 1994.

THE EFFECTS OF VANADIUM ADDITIONS ON THE SURFACE TENSION OF SODA LIME SILICATE MELTS

Douglas R. Wing, Alexis G. Clare & Linda E. Jones
NYS College of Ceramics
Alfred University
Alfred, NY 14802, U.S.A.

ABSTRACT

The density and surface tension of (mol %) 15 Na_2O 11 CaO 74 SiO_2 glass melts containing additions of 0.0, 0.1, 0.2 0.5, 1.0, and 1.5 mol% V_2O_5 were measured using the sessile and pendant drop techniques at temperatures between 1200 and 1400°C. Surface tension measurements were carried out in reducing (96%Ar/4%H_2), oxidizing (Dry Air) and wet (Wet Air) atmospheres. The density of the melts increased with increasing vanadium content and decreasing melt temperature. The surface tension of the melts was found to increase between 10 to 30 mN/m at low additions of vanadia (0.1 to 0.2 mol%), above which the surface tension decreases. The surface tension was found to decrease approximately 50mN/m with an addition of 1.5 mol% V_2O_5. This behavior can be explained using the transitional structure theory. At low concentrations, vanadium ions accumulate in alkali-rich regions. The vanadium ions have a higher field strength than the other modifiers in the melt and therefore, the vanadium strengthen the structure. At higher concentrations, four fold coordinated V^{5+} replaces silica in the structure and weakens the network. The surface tension of vanadium containing soda lime silica melts was found to be highest in a reducing environment. Little difference in surface tension was detected for melts in dry versus wet atmospheres. The surface tension of the vanadia-containing melts was not greatly influenced by temperature.

INTRODUCTION

Studies have shown that the addition of vanadium to glass melts leads to a decrease in the melt's surface tension.[1,2] Badger, Parmelee and Williams[1] showed that the surface tension of a soda lime silica melt dropped from 304 mN/m to 235 mN/m with the addition of V_2O_5. The authors also observed that the surface tension of vanadium-containing soda lime silica melts increases with temperature, though this increase was within the experimental error. Badger and Pinnow[2] also showed that the addition of V_2O_5 to soda lime silica glass lowered the surface tension of the melt. Mackenzie[3] noted a decrease in the high temperature viscosity of silica glass when V_2O_5 was added.

Volf[4] attributed the decrease in surface tension to the large deformability of V^{5+}. Vanadium has a high atomic refraction, i.e. a strong polarizability. It also has a small formal lattice energy. These factors lead to a decrease in surface tension when vanadium is added to a glass. This decrease is a result of vanadium accumulating at the surface to compensate for charge balance.[4]

Vanadium can be present in either the pentavalent (V_2O_5), tetravalent (VO^{2+} (vanadyl)) or trivalent (V_2O_3) state in silicate glasses.[5,6] V^{5+} can have coordination numbers of 4 or 6, while V^{4+} and V^{3+} have a coordination number of 6.[5] The bond strength decreases with decreasing valence. This decrease is the result of increasing ionic radius (r_i) and decreasing charge (z).[4] V^{3+}, V^{4+} and V^{5+} have been shown to co-exist in sodium disilicate glasses.[6] However, Leister, Ehrt, von der Gönna, Rüssel, and Breitbarth [5] showed that the valence state of vanadium in silicate glasses is heavily dependant on the forming conditions of the glass. Under oxidizing conditions, the majority of vanadium is in the V^{5+} state with some V^{4+} and a negligible amount of V^{3+}, whereas, when melted under reducing conditions, the glass contains a large amount of V^{3+} a small amount of V^{4+} and a trace amount of V^{5+}.[5]

This study investigates the effects of vanadium content, temperature and atmosphere on the density and surface tension of soda lime silicate melts. Additionally, measurements of the glass transition temperature and the UV-vis spectra are used to help elucidate the role of vanadium in the melt structure.

EXPERIMENTAL PROCEDURE

The glasses in this study were batched from analytical grade oxides and carbonates and were of the nominal composition (mol%) 15 Na_2O 11 CaO 74 SiO_2 with additions of 0.0, 0.1, 0.2, 0.5, 1.0 and 1.5 mol % V_2O_5. The as-batched compositions are listed in Table I. The glasses were prepared by melting the 100g batches in slip cast silica crucibles at 1450°C for 30 min. The melts were quenched by pouring them onto a steel plate. The resulting glasses were subsequently annealed in a furnace at 600°C for 60 min, then slow cooled to room temperature.

A Perkin-Elmer Lambda 900 UV-Vis-NIR spectrometer was used to measure the optical spectra of the glasses. Samples of the annealed glasses were cut, ground parallel and polished. The resulting thickness of the specimens was approximately 1mm. The spectra were measured from 190 to 1900nm. The percentage of V^{4+} was calculated using the Beer-Lambert Law:

$$A = \alpha c d \tag{1}$$

where A is the absorbance, α is the extinction coefficient, c is the concentration of the absorbing species and d is the path length through the sample. The extinction coefficent used for the V^{4+} band at 622nm was 3.0×10^{-4}/ppm•cm, as previously determined by Leister et al..[5]

Table I. As-batched Compositions of Experimental Glasses.

Mol % V$_2$O$_5$ Added	Na$_2$O	CaO	SiO$_2$	V$_2$O$_5$
0.0	15.00	11.00	74.00	0.00
0.1	14.99	10.99	73.92	0.10
0.2	14.97	10.98	73.85	0.20
0.5	14.93	10.94	73.63	0.50
1.0	14.85	10.89	73.27	0.99
1.5	14.78	10.84	72.91	1.48

The densities of the melts were measured at 50K intervals from 1200 to 1400°C using the sessile drop technique. The measurements were made with the drops on a graphite substrate in a flowing argon atmosphere. The surface tension of the melts containing additions of 0.0, 0.1, 02, and 0.5 mol% V$_2$O$_5$ were measured over the range of temperatures (1200-1400°C) using the pendant drop technique. These measurements were conducted in reducing (96% Argon/ 4% Hydrogen), oxidizing (Dry Air) and wet (Wet Air) environments. The Wet Air environment was created by bubbling dry air through a room temperature water bath. The melts with 1.0 and 1.5 mol% additions of V$_2$O$_5$ had surface tensions that were too low to be measured using the existing pendant drop arrangement. The drops would not stay suspended from the platinum loop. Therefore, a second set of measurements was made using the sessile drop technique at 1400°C (on graphite, under flowing argon) for the all of the melts in the study. Details of both the sessile and pendant drop techniques have been described in detail in a prior publication.[7]

The glass transition temperature, T_g, of the glasses was measured using a TA Instruments DSC 2910 Differential Scanning Calorimeter. Approximately 20mg of glass was heated at 20K/min over a range of 450 to 650°C in Pt pans.

RESULTS

Figure 1 shows the UV-Vis-NIR spectra for the glasses in this study. The V^{4+} bands occur near 400, 625 and 1020nm and V^{5+} has 2 bands near 300nm.[5] The V^{4+} bands at 625 and 1020 nm are labeled in the inset. The V^{5+} bands near 300nm are evident though they are not individually resolved. As the vanadium content increases, the V^{4+} peak at 400nm becomes more prominent and the V^{5+} peaks at 300 nm become skewed towards 400nm. V^{3+} peaks, which would occur near 450 and 700nm,[5] are not readily apparent in these glasses

The percentage of vanadium in the V^{4+} state as a function of total vanadia is shown in Figure 2. The plot shows that the V^{4+} concentration rapidly decreases to 0.2 mol% V_2O_5, then begins to level off. It can be reasonably assumed that the rest of the vanadium is in the pentavalent state.[5]

Figure 3 shows the density of the melts as a function of temperature. The density is seen to increase with increasing vanadia content and decreasing temperature. The relationship between density and temperature appears to be linear in this temperature range. The temperature coefficients, i.e. the slopes, are similar for each of the vanadium containing melts. The isothermal density and molar volume, at 1400°C, as a function of V_2O_5 addition, are shown in Figure 4. The density of the melt at 1400°C increases rapidly to 0.2mol% V_2O_5, then tapers off with further addition of vanadia. An inflection in the molar volume is also present at 0.2mol% V_2O_5.

Figure 5 shows the surface tension as a function of temperature for the 4 melts in this study. The plots show that there is little difference in surface tension between the melts in dry air versus wet air atmospheres. The vanadia-containing melts in the reducing atmosphere have this highest surface tension. This becomes more pronounced as the amount of vanadia
added to the system increases. The surface tension of these melts remains fairly constant over the range of temperatures in each atmosphere.

Isothermal plots of the surface tension as a function of vanadia addition are shown for each of the three atmospheres in Figure 6. In each of these series, the surface tension is seen to increase at low additions of V_2O_5 up to 0.1 to 0.5 mol%. In this region, a maximum occurs and the surface tension decreases with further increases in vanadia.

Figure 7 shows the glass transition temperature (T_g) and the surface tension at 1400°C as function of vanadia content. The surface tension measurements were made via the sessile drop technique, in an argon atmosphere, for this series of data. Both the T_g and the surface tension exhibit a maximum near 0.1-0.2mol% V_2O_5. The surface tension drops approximately 50 nm/m after an addition of 1.5 mol % vanadia.

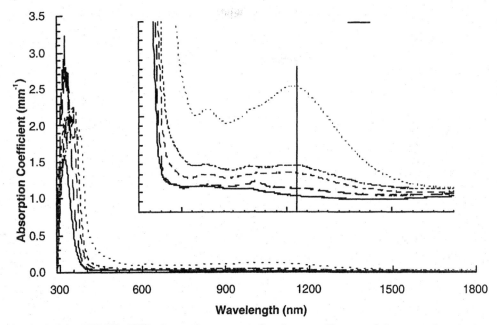

Figure 1. The UV-Vis-NIR absorption spectra for the vanadia-containing samples. The inset shows the details of the V^{4+} peaks.

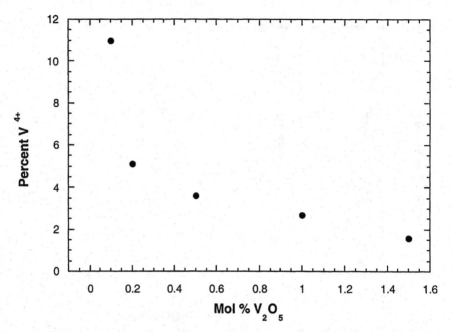

Figure 2. The percentage of vanadium in the 4+ state as a function of the total amount of V_2O_5 added to the melt, as determined from the UV-Vis-NIR spectra.

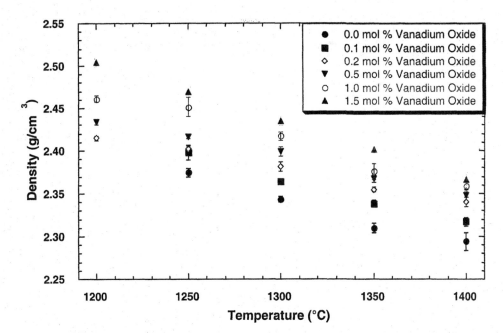

Figure 3. The densities of the melts as a function of temperature. The densities were measured using the sessile drop technique in an argon environment.

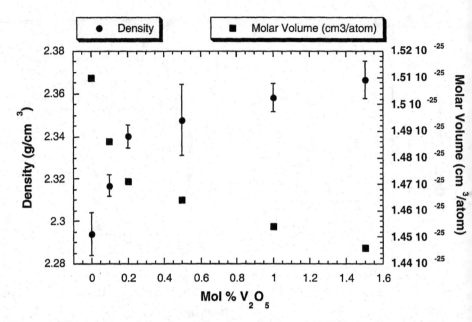

Figure 4. The isothermal density and molar volume at 1400°C as a function of the amount of vanadia added to the melt.

a)

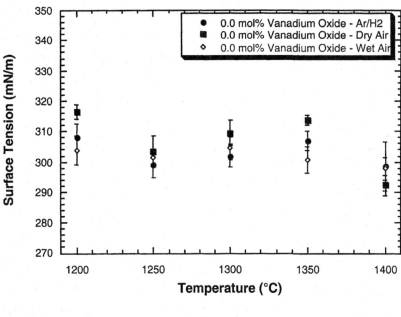

b)

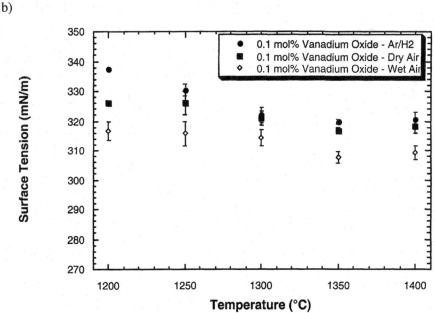

Figure 5. Surface tension as a function of temperature for the a) 0.0, b) 0.1, c) 0.2, and d) 0.5 mol% V$_2$O$_5$ containing melts.

c)

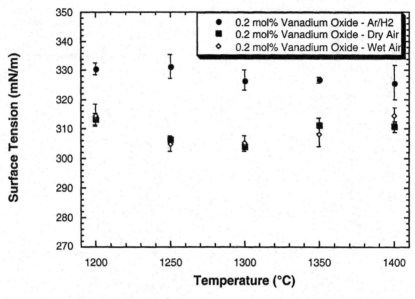

d)

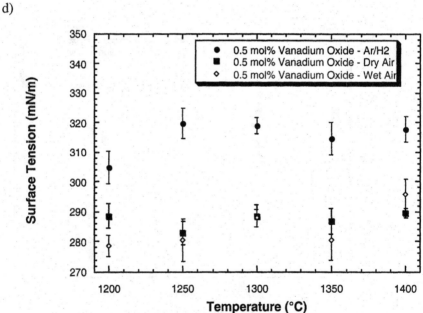

Figure 5. Surface tension as a function of temperature for the a) 0.0, b) 0.1, c) 0.2, and d) 0.5 mol% V$_2$O$_5$ containing melts.

Advances in Fusion and Processing of Glass III

a)

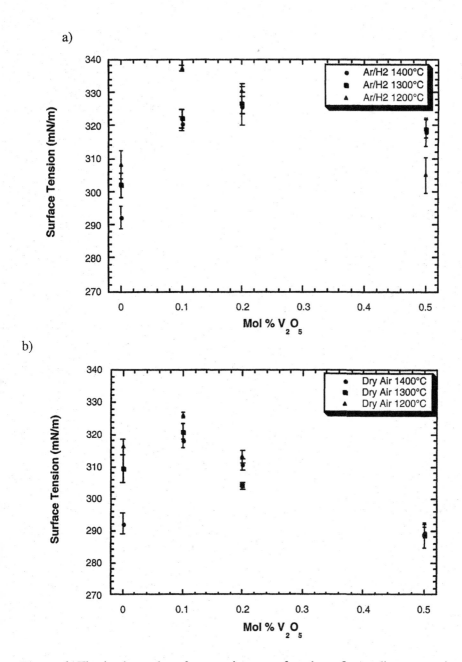

b)

Figure 6. The isothermal surface tension as a function of vanadia content in a) Ar/H₂, b) Dry Air, and c) Wet Air atmospheres.

c)

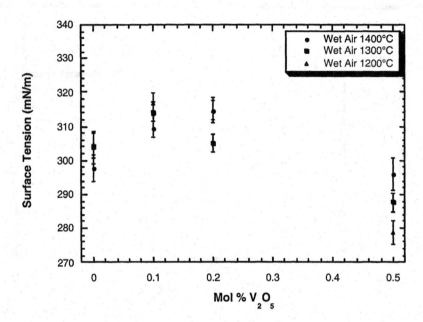

Figure 6. The isothermal surface tension as a function of vanadia content in a) Ar/H₂, b) Dry Air, and c) Wet Air atmospheres.

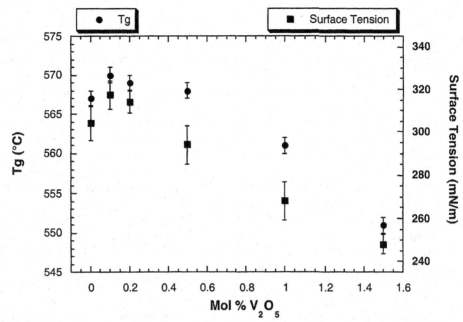

Figure 7. The glass transition temperature (T_g) and the surface tension at 1400°C (as measured via the sessile drop technique in Ar) as function of vanadia content.

DISCUSSION

The surface tension of soda lime silicates was observed to decrease with addition of V_2O_5, as reported previously by others,[1,2] but only after an initial increase at low V_2O_5 content. In this study, the surface tension increased between 10 and 30 mN/m after the addition of 0.1 mol % V_2O_5. The amount of increase varied with atmosphere and temperature. A maximum in surface tension occurred around 0.1 to 0.2 mol % V_2O_5, after which the surface tension decreased. The surface tension had significantly decreased after an addition of 1.5 mol% vanadia.

The data for the redox state, density, molar volume, surface tension and T_g all exhibit an inflection around 0.1 to 0.2 mol% V_2O_5 addition. This information suggests that vanadium is serving one role in the structure at low concentrations and a different role at larger concentrations. This is consistent with the transitional structure theory.[8] The transitional structure theory postulates that modifier ions may exist in a melt or glass in random clustered regions.[8] These modifier-rich regions are on a scale of 1-2nm (smaller than the traditional phase separation definition). As the modifier content increases, the clusters begin to interconnect. In this "transitional range," the properties of the glass change from being controlled by silica-rich regions to the modifier-rich region.[8] Figure 8 represents structures of alkali silicates using the cluster model. The left-hand figure represents the structure of a glass containing less than 10% alkali, while the right-hand figure represents a high alkali glass with continuous alkali-rich regions.

It is proposed that the changes in properties observed in the vanadium-containing melts are the result of a transition in the melt structure. Commercial soda lime silicates with compositions close to that of the base glass used in this study have transitional structures.[8] According to the transitional structure theory, the base composition has interconnecting regions of both modifier-rich and silica-rich regions. As vanadia is added to the melt, the vanadium will tend to congregate or accumulate in the modifier-rich regions. The six fold coordinated V^{4+} and V^{5+} will act as modifiers in the modifier-rich region, however since these ions have a relatively high field strength, they will help collapse the local structure. Meanwhile, four fold coordinated V^{5+} will act as a network former in this region. The four fold coordinated V^{5+} increases the size of the network-rich regions. The modifier-rich regions are thus divided up to the point at which they are no longer interconnected (near 0.2 mol% V_2O_5.) At this point the properties become dominated by the interconnected silica-rich network and the surface tension is highest. Above 0.2 mol% V_2O_5, the melt is less segregated and the vanadium impacts the silica-rich network by replacing silica as four fold coordinated V^{5+}. Since V^{5+} has a much lower field strength than Si^{4+},[3,9] the average strength of the bonds in the former-rich region decreases. This theory explains the increase in T_g and surface tension up to 0.2 mol% V_2O_5, and the subsequent decrease in these properties.

■ Si-Rich Region ▨ NBO⁻Na⁺-Rich Region

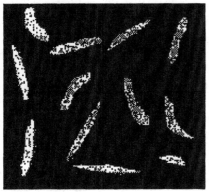 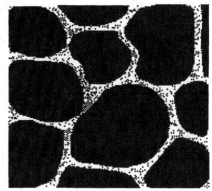

Figure 8. Structures of alkali silicates using the cluster model. The left-hand figure represents the structure of a glass containing less than 10% alkali, while the right-hand figure represents a high alkali glass with continuous alkali-rich regions.[8] [Pending permission of The American Ceramic Society, PO Box 6136, Westerville, Ohio 43086-6136. Copyright 1998. All rights reserved.]

In order to further validate this theory, a second series of vanadium-containing melts was prepared. These melts had a base composition (mol%) of 20 Na_2O 6 CaO 74 SiO_2. The as-batched compositions as listed in Table II. According to the transitional structure argument, one would expect that these melts would have a larger proportion of alkali-rich regions compared to the 15 Na_2O 11 CaO 74 SiO_2 melts. The melts with larger regions of alkali would accommodate more vanadium as a modifier before reaching the transition point where the properties are network dominated. At the transition point, the vanadium starts to impact the structure as a network former in the silica rich region. Hence, in this composition, the same modest increase, maximum, and subsequent decrease of the glass transition temperature and surface tension would be expected to occur as higher values of vanadia addition.

Table II. As-batched Composition of High-alkali Glasses.

Mol % V_2O_5 Added	Na_2O	CaO	SiO_2	V_2O_5
0.0	20.00	6.00	74.00	0.00
0.2	19.96	5.99	73.85	0.20
0.5	19.90	5.97	73.63	0.50
1.0	19.80	5.94	73.27	0.99
1.5	19.70	5.91	72.91	1.48

Figure 9 shows the T_g as a function of vanadia addition to glasses of composition (mol%) 20 Na_2O 6 CaO 74 SiO_2. The T_g remains fairly constant up to 0.5 mol% V_2O_5, after which it rapidly decreases. This data illustrates the shift in the transition point as predicted by the transitional structure theory.

The surface tension of the 15 mol% Na_2O glass in reduced environments tended to be higher than those in inert or oxidizing environments. The reducing environment of Ar/H_2 would have 2 possible effects. The first effect would be to reduce the vanadium in the melt, the second would be the introduction of OH ions at the melt surface. At no point was V^{3+} evident in the UV-vis specta of the glasses. Therefore, the effect of a reducing atmosphere would have been to keep the vanadium redox ratio (V^{4+}/V^{5+}) higher, which would raise the surface tension.

CONCLUSIONS

The addition of vanadium to soda lime silica melts was found to result in a decrease in melt surface tension, but only after an initial increase. The point of onset is dependant on the overall composition of the glass. Inflection points were also found in the data for density, molar volume and T_g. The transitional structure theory can be used to explain these inflections. At low concentrations, vanadium congregates in modifier-rich regions within the melt. The six-fold V^{4+} and V^{5+} act as a high bond strength modifiers. This results in the initial increases in surface tension, T_g, and density. Meanwhile, four fold coordinated V^{5+} enters the

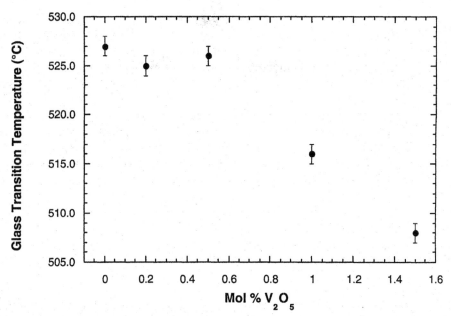

Figure 9. Glass transition temperature, Tg, as a function of vanadia addition to glasses of mol% composition 20 Na_2O 6 CaO 74 SiO_2.

structure as a network former and increases the size of the network-rich region. The modifier-rich regions are broken up to the point at which they are no longer interconnected, this is the transition point. Above this point, the four fold coordinated V^{5+} begins to replace silica in the network structure. The low field strength of vanadium, compared to silica, results in decreases in surface tension and T_g and a leveling off of density.

The surface tension of vanadium-containing soda lime silica melts was highest in a reducing environment. There was little difference in surface tension for melts in dry versus wet atmospheres. Temperature did not greatly influence the surface tension of the melts over the range of measurements. The concentration of vanadium oxide in a melt was determined to have a large influence on the surface tension, regardless of the environment.

ACKNOWLEDGEMENTS

This research was sponsored by the NSF University-Industry Center for Glass Research at Alfred University.

REFERENCES

[1]A.E. Badger, C.W. Parmalee, and A.E. Williams, "Surface Tension of Various Molten Glasses," *J. Am. Ceram. Soc.*, **20** [10] 325-29 (1937).

[2]A.E. Badger and H.R. Pinnow, "Vanadium and Chromium Oxides in Refractories as Aids in the Elimination of Cords in Glass," *Glass Ind.*, **22** [4] 161-2 (1941).

[3]J.D. Mackenzie, "The Refractory Behavior of Vanadia-Silica Glass" *Phys. Chem. Glasses*, **3** [2] 50-53 (1962).

[4]M.B. Volf, *Chemical Approach to Glass*; pp. 326-333. Elsevier Amsterdam, 1984.

[5]M. Leister, D. Ehrt, G. von der Gönna, C. Rüssel, and F.W. Breitbarth, "Redox States and Coordination of Vanadium in Sodium Silicates Melted at High Temperatures," *Phys. Chem. Glasses*, **40** [6] 319-25 (1999).

[6]W.D. Johnston, "Optical Spectra of the Various Valence States of Vanadium in $Na_2O \cdot 2SiO_2$ Glass," *J. Am. Ceram. Soc.*, **48** [2] 608-611 (1965).

[7]D.R. Wing, A.G. Clare, and L.E. Jones, "The Effects of Redox State and Atmosphere on the Surface Tension of Iron-Containing Soda Lime Silicate Melts" (to be published).

[8]W.C. LaCourse and A.N. Cormack, "Glasses with Transitional Structures," *Ceram. Trans.*, **82**, 273-9 (1998).

[9]Dietzel, "Die Kationenfeldstärken Und Ihre Beziehungen Zu Entglasungsvorgängen, Zur Verbindungsbildung Und Zu Den Schmelzpunkten Von Silicaten," *Z. Elektrochem.*, **48** [1] 9-23 (1942).

MODELING OF GLASS MAKING PROCESSES FOR IMPROVED EFFICIENCY: HIGH TEMPERATURE GLASS MELT PROPERTY DATABASE FOR MODELING. US DOE PROJECT DE-FG07-96EE41262

T. Vascott, A.K. Varshneya, T.P. Seward,
R. Karuppannan, J.M. Jones and D. Oksoy.
New York State College of Ceramics
Alfred University
Alfred, NY 14802

ABSTRACT

Electrical boosting of glass melters is commonplace and many specialty and optical glasses are melted completely by electricity. Increased fossil fuel prices and/or improved designs for electrically heated finers could extend this trend. Knowledge of melt electrical resistivity is essential for effective furnace modeling and design. We have measured electrical resistance at a frequency of 2 KHz in the range of 950 to 1450°C for about 150 glass compositions of container, float, low-expansion borosilicate, TV panel, wool and textile fiberglass-types using a simple two-probe method, and report here the ranges of resistivity and general composition-dependent behavior observed. Details of the specific glass compositions will be withheld, pending evaluation of the data by Center for Glass Research member companies.

INTRODUCTION

The overall objective of this study is the development of a comprehensive and reliable data base for glass forming melts that will allow full use of numerical simulation models by a broad cross-section of the glass industry for the purpose of achieving energy savings, improving product quality, increasing productivity, and meeting present and future environmental regulations. This poster focuses on electrical resistivity in the glass melt (900-1450°C) for six families of glass compositions. Twenty-four compositional variations were developed around each of six commercial base glasses: container glass, E glass, float glass, low-expansion borosilicate glass, TV panel glass, and wool glass. Glass composition ranges are shown below in Table I.

Table I: DOE Glass-Melt Properties Project; Glass Composition Ranges (wt. %)

DOE Glass-Melt Properties Project			Glass Composition Ranges (wt. %)		
	Container	E	Float	Low-exp. B-Si	TV Panel	Wool
SiO_2	65-75	52-60	70-74	70-81	58-64	56-66
B_2O_3		0-9		10-15		3-9
Al_2O_3	1-3	12-16		2-7	1.3-3.5	0-6
MgO	0-3	0.5-4.5	3-4		0-1.5	1-5
CaO	7-12	16-24	7-9	0-2	0-3.5	5-11
SrO					1-10	
BaO				0-2	2-13	
Li_2O	0-1				0-0.5	
Na_2O	11-15	0-2	12-15	4-8	6-9	13-17
K_2O	0-2	0-0.5	0.05-0.8	0-3	6-9	0-2
Fe_2O_3	0-0.4	0-0.8	0.1-1.5			0-0.6
Cr_2O_3	0-0.3					
TiO_2	0-0.5	0-1			0.1-0.5	
CeO_2					0-0.7	
ZrO_2					0-3	
PbO					0-3	
ZnO					0-1.5	
As_2O_3					0-0.3	
Sb_2O_3					0.2-0.6	
SO_3	0-0.3					0-0.2
F		0-0.6			0-0.7	0-0.6

EXPERIMENTAL

Measurements were made using a two-probe method. Platinum electrodes were lowered to a depth of 1.0 cm into 70 ml of melt. The glass was melted in air in a Pt-3.5%Rh crucible in an electric furnace. Resistance was measured at 2 kHZ from 1450°C to 950-900°C at 50-degree descending intervals. The instrument is calibrated by measuring the cell constant for a desired configuration using a KCl solution (0.1 Demal: 7.4745g KCl + 1000g DI H_2O and 1.0 Demal: 76.583g KCl + 1000g DI H_2O). Verification is made by measuring the resistance of a standard glass melt, in this case SRM1414. Resistivity (ρ) was calculated at each temperature using the cell constant (L/A) determined from the Demal solution measurement and the resistance (R) measured at temperature: ρ (ohm-cm) $=R/(L/A)$. The measurement system is illustrated in Figure 1.

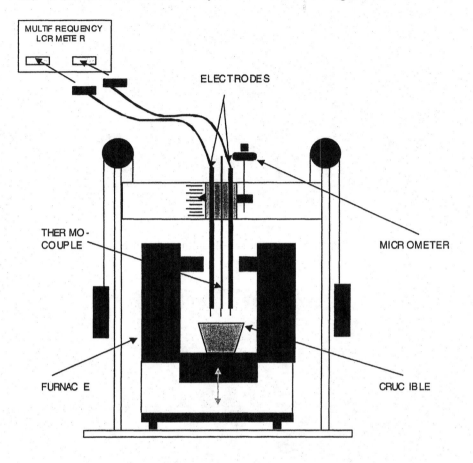

Figure 1: Electrical Resistivity Measurement System

RESULTS

Resistivity measurement results are shown below in Figures 2-8. Resistivity of the six commercial base glasses as a function of mol % total alkali at 1200°C and 950°C is shown in Figure 9.

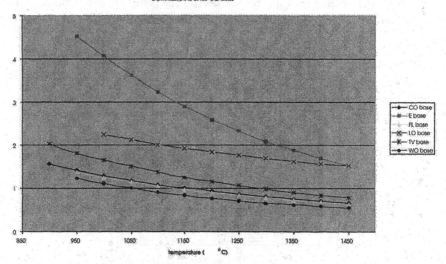

Figure 2: Log Resistivity vs. Temperature—Commercial Base Glasses

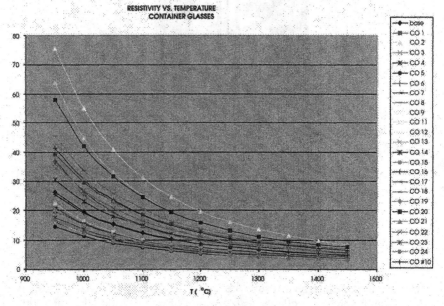

Figure 3: Resistivity vs. Temperature—Container Glasses

Advances in Fusion and Processing of Glass III

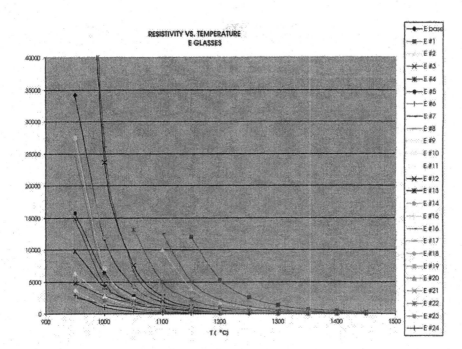

Figure 4: Resistivity vs. Temperature—E Glasses

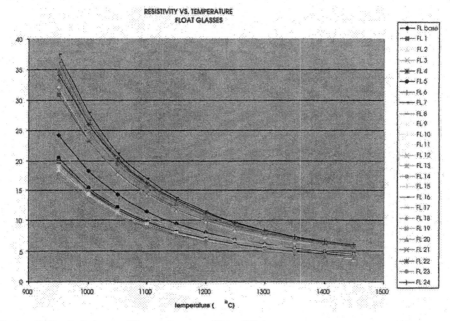

Figure 5: Resistivity vs. Temperature—Float Glasses

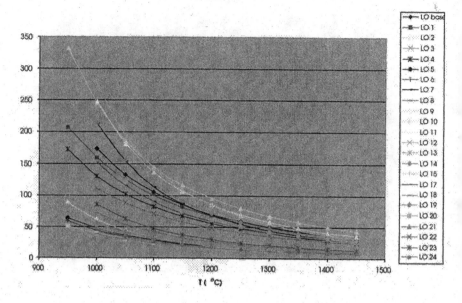

Figure 6: Resistivity vs. Temperature—Low-exp. Borosilicate Glasses

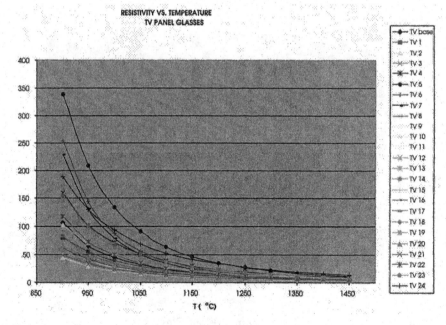

Figure 7: Resistivity vs. Temperature—TV Panel Glasses

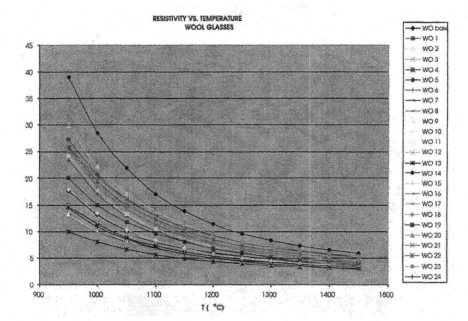

Figure 8: Resistivity vs. Temperature—Wool Glasses

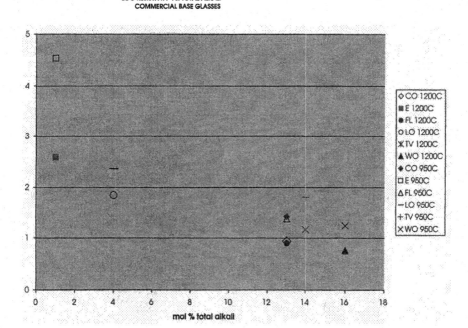

Figure 9: Resistivity vs. Mol % Total Alkali—Base Glasses

Materials for Glassmaking

ANALYTICAL MODELS FOR HIGH-TEMPERATURE CORROSION OF SILICA REFRACTORIES IN GLASS-MELTING FURNACES

M. D. Allendorf, R. H. Nilson
Sandia National Laboratories
Livermore, CA 94551-0969

K. E. Spear
Penn State University
University Park, PA 16802

B. Bugeat, M. U. Ghani, O. Marin
American Air Liquide
Countryside, IL 60525

P. M. Walsh
University of Alabama at Birmingham
Birmingham, AL 35294-4461

A. Gupta
Monofrax Inc.
Falconer, NY 14733-1797

H. E. Wolfe
ANH Refractories
West Mifflin, PA 15122

G. A. Pecoraro[*]
PPG Industries, Inc.
Pittsburgh, PA 15238-0472

ABSTRACT

Corrosion of refractory silica crown brick in glass-melting furnaces is a serious problem in oxy-fuel furnaces. To better understand and to quantify this process we utilize analytical models to evaluate the importance of three potential rate-limiting processes: 1) gas-phase transport of NaOH to the crown surface; 2) diffusion of sodium-containing reactants through a liquid product layer that forms on the brick face; and 3) gas-phase diffusion of NaOH into refractory pores. Predictions are compared with reported corrosion rates and product compositions previously determined by post-mortem analysis of refractory samples. We conclude that corrosion occurs largely by reaction and removal of material from the exposed brick face, rather than by transport of reactants into the porous bricks. The rate is limited by process (1). The observed presence of corrosion products deep within the brick pores is consistent with capillary suction of liquid products from the hot face into the interior. Although computed corrosion rates based on

---

[*] Retired.

mass transport through a gas boundary layer are somewhat greater than those observed, the results are very sensitive to the gas-phase concentration of NaOH and to the refractory temperature, both of which contain significant uncertainties. We also present a new thermodynamic analysis of the SiO_2-CaO-Na_2O that provides an explanation for the increased corrosion rates observed in lime-containing silica relative to low- or no-lime silica. Finally, we combine our analytical model with results of the *Athena* CFD furnace code to predict corrosion rates in an oxygen-fuel furnace. The results are encouraging, but suggest that accurate knowledge of both NaOH(gas) concentrations and refractory temperatures is needed to validate these models.

INTRODUCTION

Corrosion of silica-containing refractories used to line the crown of glass-melting furnaces is a problem that has attracted considerable attention in recent years due to the increased use of oxy-fuel melting techniques. Higher water vapor concentrations produced in such furnaces relative to air-fuel furnaces leads to increased concentrations of vapor-phase alkali hydroxides. The hydroxide can react with silica to form a low-melting alkali-silicate product:

$$2MOH(gas) \leftrightarrow M_2O(\text{dissolved in liquid } SiO_2) + H_2O(gas); M = Na \text{ or } K \qquad (1)$$

The quasi-steady surface corrosion process may be viewed as a series of distinct steps:

(1) transport of NaOH(gas) from the furnace gas to the external surface of the liquid product layer that coats the crown surface.
(2) formation of Na_2O at the external surface of the liquid product layer.
(3) diffusion of Na_2O through the liquid product layer to the SiO_2 surface.
(4) dissolution of silica to form a sodium-rich silicate at the interface between the liquid product and the unreacted silica grains.
(5) transport of liquid reaction product from the reaction surface either by suction into the porous bricks or by dripping or flowing away.

Although the corrosion phenomenon has been qualitatively characterized, the identity of the step that controls the rate of corrosion has not been firmly established.

As a result of an extensive analysis of the thermochemistry[1,2] and kinetics[3] of the corrosion of refractories in glass-melting environments, we conclude that steady-state corrosion of typical low-density silica refractories is controlled by the transport of gas-phase NaOH from the furnace atmosphere to a liquid product layer formed on the outer (hot face) surface of the furnace refractory crown. The chemical reactions occurring at the outer refractory surface appear to reach equilibrium on the time scales typical of refractory lifetimes (years) and other potential transport mechanisms are either very fast (diffusion of sodium through the liquid product layer) or ineffective (gas-phase diffusion of NaOH through pores in the refractory). This paper provides a concise summary of the results of these investigations; complete details can be found in the relevant references.[2-4] We also discuss new results concerning the effects of calcium oxide on the SiO_2/Na_2O equilibrium. Finally, we combine our model of transport-limited silica corrosion with predictions of crown temperature, gas-phase NaOH concentration, and gas velocity from the *Athena* furnace code to predict crown corrosion rates for an entire furnace. The results are qualitatively consistent with observation, but demonstrate that model predictions are highly sensitive to the refractory temperature and the NaOH(gas) concentration, pointing to the need for more accurate data that can be used to validate both the corrosion model and heat transfer models in furnace codes.

THERMOCHEMISTRY OF CORROSION PRODUCTS

Development of a model that can predict silica corrosion rates requires knowledge of the chemical reactions and their associated rates leading to the consumption of refractory. Although there is some reason to believe that, during the early stages of corrosion the rate may be limited by chemical reaction rates, over the long time scales typical of refractory lifetimes it is reasonable to assume that chemical equilibrium is achieved. In fact, Na_2O concentrations found in corroded silica samples taken from a glass furnace are consistent with the predictions of equilibrium calculations (see below). Although a thermodynamic analysis can predict the amount of NaOH(gas) in equilibrium with silica as a function of temperature, and thus identify NaOH(gas) concentration regimes in which corrosion is energetically favored to occur, thermodynamic data for all species involved must be available first. In the case of the SiO_2/Na_2O system, data for the variable-composition liquids that form when alkali reacts with the refractory (Reaction 1) were not readily available. These liquids are non-ideal solutions (i.e., they do not follow Raoult's law of vapor pressures). This behavior can be ascribed to strong negative free-energy of interaction terms between "species" in solution. The exact identity of these species is unknown, but the concept can be used to develop a model that can account for the interactions and thus predict the heat of formation, entropy, and heat capacity for these liquids.

One approach to generating the needed data is the Associate Species Model (ASM), developed originally by Bonnell and Hastie to model metal slags[5] and then extended by Spear and Besmann for application to nuclear waste glass.[6] In this modified ASM model, the negative free-energy terms that produce non-ideal solution behavior are accounted for by treating the mixture as an ideal solution of "associate species." Thermodynamic properties for these species are obtained by fitting their values to achieve agreement with published phase diagrams and activity data. For positive mixing energies such as encountered with liquid immiscibility, positive regular solution constants between associate species are used. To date we have applied the ASM to the following binary systems: SiO_2/Na_2O,[2] SiO_2/K_2O,[2] Al_2O_3/Na_2O,[1] and Al_2O_3/K_2O.[1] Data for these systems and for additional oxide systems can be found at a web site and are available for download.[†] In the SiO_2/Na_2O case, the liquid associate species used are $(2/5)Na_4SiO_4$, $(2/3)Na_2SiO_3$, $(1/2)Na_2Si_2O_5$, and the end members Na_2O and Si_2O_4. An excellent fit to the published phase diagram was achieved.[7]

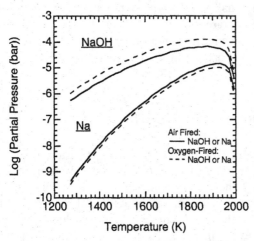

Combining the thermodynamic data for the associate species with data from standard sources allows the concentration of NaOH(gas) in equilibrium with crystalline SiO_2 and sodium-containing liquid SiO_2 to be predicted, as well as the amount of Na_2O dissolved in silica when NaOH reacts with the silica grains. The predicted NaOH(gas) partial pressures are shown in Figure 1 for both air- and oxy-fired cases. It is seen that the NaOH(gas) concentration increases with temperature, reaching a maximum at ~ 1873 K, then decreases as temperatures approach the melting point of crystalline silica. In contrast, the

Figure 1.   Partial   pressures   of NaOH(gas) and Na(gas) for air- and oxy-fired furnace conditions, as predicted by equilibrium. Data are from Ref. 2.

amount of Na_2O that can dissolve in liquid SiO_2 decreases continuously with temperature and is zero at the melting point of silica (1996 K), which has consequences for the corrosion rates predicted by a gas-transport-limited model, as will be seen below. These predicted NaOH(gas) concentrations define the minimum amount of NaOH(gas) required to corrode the refractory, assuming that the chemical reaction is at equilibrium.[2] We use these results in the analysis of potential rate-controlling corrosion mechanisms, as described in the sections below.

[†] www.ca.sandia.gov/HiTempThermo/index.html

Advances in Fusion and Processing of Glass III

Thermodynamics of the SiO$_2$/Na$_2$O/CaO system

CaO as a binder to provide structural integrity. Recently, some short-duration experiments in an oxygen-fuel glass furnace indicate that removing some or all of this calcium can reduce the rate of corrosion.[8] Since thermodynamic data for the SiO$_2$/Na$_2$O/CaO system are now available (see the web site cited above) it is possible to evaluate the causes of this effect from a thermodynamic point of view. Equilibrium calculations were thus performed for a typical oxygen-fuel mixture (O$_2$/CH$_4$ = 2.05; 0.1 mole Na$_2$O; activity of SiO$_2$(cristobalite) fixed at 1.0). The CaO/Na$_2$O ratio was varied from 0 to 1.0, using temperatures expected for the crown hot face. The results are shown in Figure 2.

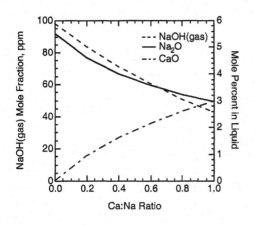

Figure 2. **Equilibrium predictions of the effect of CaO addition on the amount of NaOH(gas) and the concentration of Na$_2$O that can dissolve in liquid SiO$_2$ that is in equilibrium with crystalline SiO$_2$.**

The results suggest two reasons why the presence of CaO in the refractory could increase corrosion rates. First, CaO addition decreases the partial pressure of NaOH(gas) that can be in equilibrium with the liquid corrosion product and crystalline silica. As will be seen below, this increases the corrosion rate by increasing the magnitude of the concentration gradient between the ambient NaOH(gas) in the furnace and that at the surface of the refractory. Second, the concentration of Na$_2$O that can dissolve in the liquid corrosion product decreases, which accelerates corrosion since less Na$_2$O is required to form a liquid corrosion product (see equation 2 below). Both factors support arguments presented below that corrosion occurs primarily at the front (hot) face of the refractory.

ANALYTICAL MODELS OF TRANSPORT PROCESSES

As discussed in the introduction, there are at least five processes that may be occurring during the corrosion of silica by vapor-phase alkali, four of which could be rate-limiting. The fifth, capillary suction, can transport liquid corrosion products to the interior of a porous refractory brick. Each of these is discussed briefly below; complete details can be found elsewhere.[3]

Gas-phase transport

The recession rate of the brick face, dl/dt, is equal to the areal reaction rate of SiO_2, $\dot{n}_{SiO_2}$ (moles/m^2/s), divided by the molar density of SiO_2 in the unreacted brick, ρ_1. In addition, the number of moles of SiO_2 eroded by one mole of NaOH(g) is $(1-x)/x/2$ where x is the mole fraction of Na_2O in the melt and the factor of two accounts for the fact that two moles of NaOH(g) are reacted to form each mole of Na_2O. Assuming steady-state conditions, the reaction rate of the sodium must be in balance with the transport of NaOH(g) from the furnace gas to the refractory surface. It can thus be shown that

$$\frac{dl}{dt} = \frac{\rho_{gas}}{\rho_{SiO_2}} \Delta X \; Sh \frac{D}{L} \left(\frac{1-x}{2x} \right) \tag{2}$$

where ΔX is the difference between the ambient mole fraction outside the boundary layer and that immediately adjacent to the gas-liquid interface, L is the length of the flow path over the surface and Sh is the Sherwood number (essentially the ratio of the actual transport through the boundary layer to that which would occur by diffusion through the length scale, L), ρ_{gas} and ρ_{SiO2} are the densities of the gas and the silica refractory, respectively, and D is the diffusivity of NaOH(gas). Sh is typically much greater than unity. Three different models are used to estimate the magnitude of the Sherwood number: (i) laminar flow over a flat surface, (ii) turbulent flow over a flat surface, and (iii) analysis of numerical CFD results.

Figures 3 and 4 show computed recession rates versus crown temperature for various NaOH(g) concentrations in oxy- and air-fuel furnaces, respectively. In the oxy-fuel case, the H_2O concentration is 65%, while in the air-fuel case it is ~18%. Recession rates indicated by solid lines utilize the turbulent flow estimate of the Sherwood number. Computed recession rates based on laminar flow (dotted) and the CFD friction velocities (chain-dotted) shown in the two figures differ by less than 30%, mainly because the flow is transitional, on the borderline between laminar and turbulent flow (Reynolds number Re = 30,000). The estimates based on the CFD friction velocities are about twice as great, perhaps because they better reflect the furnace-scale turbulence induced by the burner jets.

The most important conclusion to be drawn from Figure 3 is that computed recession rates for large but credible NaOH(g) concentrations of 150-200 ppm are generally greater than those observed in oxygen/fuel furnaces. For example, at a crown temperature of 1823 K, concentrations of 150 – 200 ppm produce computed recession rates of 40 – 100 mm/year (60 – 150 mm/yr based on CFD friction) as compared with anecdotal observations of about 25 mm/year (although

Advances in Fusion and Processing of Glass III

a value much higher than this – 100 mm/yr – has been reported[9]). The computed recession rates would be consistent with these observations if the NaOH(g) concentration is no greater than 125 ppm. Figure 4 compares computed and observed surface recession rates for air-fuel combustion and is qualitatively similar to the predictions for oxygen-fuel combustion. As before, computed recession rates are somewhat greater than those generally observed.

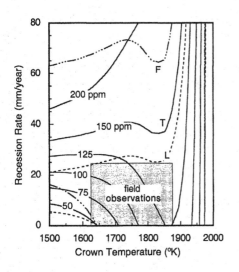

Figure 3. Computed recession rates for oxy-fuel combustion.

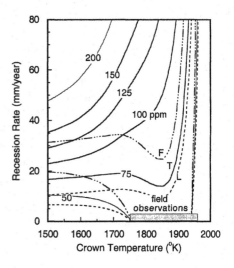

Figure 4. Computed recession rates for air-fuel combustion.

The recession rate may either increase or decrease with temperature, depending on the temperature range and the ambient NaOH(g) mole fraction. For NaOH(g) concentrations greater than ~ 150 ppm, the recession rate increases continuously with temperature, as apparent in Figures 3 and 4. The increase of recession with temperature is a consequence of the decreasing mole fraction of $Na_2O(g)$ in the melt. As discussed above, the equilibrium mole fraction of Na_2O in the corrosion product, x, decreases with temperature, from about 10% at 1473 K to less than 1%, at 1873 K. Thus, the amount of SiO_2 removed by each mole of NaOH(g) increases by a factor of ten, increasing the corrosion rate by that amount (Equation 2). As the temperature approaches the silica melting point, the equilibrium mole fraction (x) approaches zero, causing the corrosion rate predicted by Equation 2 to become infinite, a clearly unphysical result. Although this problem could be solved by including heat transfer to the crown in the model (to account for the heat required to melt the material), this addition is not warranted partly because heat transfer rates are highly dependent on the details of a particular furnace configuration and, more importantly, because silica crowns are not operated at temperatures much above 1873 K to remain within mechanical limits of the refractory. We thus believe that a reasonable upper limit for the crown temperature in the current model is 1873 K.

Although computed recession rates based on gas-phase transport generally exceed observations, the results are very sensitive to NaOH concentrations and crown surface temperature which are not accurately known. Moreover, it is very encouraging to see that computed recession rates decrease with crown temperature for sufficiently small values of the ambient NaOH(g) concentration. This result is consistent with industry experience showing that increasing crown temperatures by the addition of thermal insulation to the backside of the refractory greatly reduces corrosion. Thus, it is likely that the operative physics are correctly modeled, but that some of the inputs to the analysis may be inaccurate. Two realistic possibilities are inaccuracies in the NaOH(gas) concentration and the refractory temperature. For example, actual NaOH(gas) concentrations may be a factor of two less than the conventional estimates of ~ 200 ppm for oxy-fuel and 40 – 80 ppm for air-fuel. Walsh et al. measured concentrations of 150 ± 50 ppm in an oxy-fuel furnace.[4,10,11] Alternatively, their is probably sufficient uncertainty in the thermochemistry to assume that the predicted equilibrium NaOH(gas) concentrations could be low by 20-30%, or similarly, that the predicted Na_2O concentration in the melt is too low. Any of these explanations or some combination thereof could bring the predictions into better agreement with the data.

Liquid-phase diffusion

Since a liquid product layer quickly forms when NaOH(gas) reacts with silica, sodium from the gas phase must diffuse through this liquid in order to react with the refractory grains. It is conceivable that corrosion could be limited by the rate at which this transport occurs. The maximum thickness of the liquid melt layer on the brick face can be estimated by application of the Rayleigh-Taylor criterion for stability of a thin film. Droplet formation is likely to occur when the thickness of the film exceeds a critical thickness. We estimate a maximum stable layer thickness of ~2 mm, corresponding to a surface tension of 300 mN/m, which is typical of SiO_2 at 1773 K. This is only a few times greater than the surface tension of water at room temperature. It appears likely that the critical layer thickness will be quite similar for air- and oxygen-fired conditions and independent of layer composition.

Although the diffusivity of Na_2O in silicate melts has not been directly measured at the temperatures of interest, there is a great deal of information regarding the diffusivity of H_2O as well as Na, O, H, and OH ions. Much of this is found in the geophysical literature concerning silica rich magmas, and these studies appear consistent with recent measurements by Mesko and Shelby[12] for commercial float glass and TV panel glass. In general, sodium ions are known to diffuse very readily, having a diffusivity approaching 10^{-4} cm^2/s at 1600 C.[13,14] Oxygen species generally diffuse much slower because they bond to silicon in the glass and must diffuse from one bond to another in the silicate network. Thus, the diffusion of Na_2O is likely to be limited by diffusion of oxygen species, suggesting that estimates can be made on the basis of previous studies addressing H_2O, O, and OH.

An analytical model employing these diffusivities and the estimated layer thickness indicates that diffusion of Na_2O through the liquid product layer is not rate-controlling in full-scale furnaces. Computed recession rates limited only by diffusion of Na_2O through a 2-mm liquid film are shown in Figure 5 for ambient NaOH concentrations of 80, 200, and 600 ppm. The upper and lower sets of calculations utilize the diffusivity estimates of Mesko and Shelby[15] and Pfeifer,[13] respectively. The upper set, thought to be the best estimate, is in reasonable agreement with recession rates observed by ORNL[16,17] and Corning[8] in crucible tests. Based on the Mesko and Shelby diffusivity and an NaOH concentration of 100-200 ppm, maximum recession rates of 1 m/year would be attainable in the field. Since this greatly exceeds the earlier estimates based on gas phase transport, it follows that liquid phase diffusion is not the rate-controlling consideration in the corrosion process observed in the field.

Since low-density silica refractory contains about 20% open porosity, it is possible that NaOH(gas) could diffuse through the pores and react in the interior of the brick at lower temperatures than are present at the outer surface facing the glass melt. It appears that this can be ruled out on the basis of scanning electron micrograph (SEM) pictures and EDX analysis of corrosion products. We conducted a detailed SEM investigation of samples exposed to simulated furnace gases for short times (~75 hours) as well as post-mortem analysis of samples obtained from the crown of a oxy-fuel container-glass furnace during a rebuild. A typical photomicrograph of a sample obtained from the short-term tests at Monofrax Inc. is shown in Figure 6.[18]

Vapor-phase transport through pores

The micrographs show that the small pores fill first with an amorphous phase composed primarily of oxides of silicon, sodium, and calcium. Large pores are still open and do not show evidence of reaction within them (there are no halos of

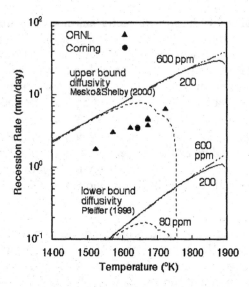

Figure 5. Computed recession rates limited by Na_2O diffusion through a 2 mm silicate liquid layer on the crown hot face

amorphous material evident). The outer surface of the brick (at the top of the photomicrograph) appears to be sealed by the amorphous phase, which presumably was liquid when the brick was at furnace temperature. Similar observations can be made in examining the postmortem samples from the container furnace. We thus conclude that NaOH(gas) does not penetrate the pores except during very early stages of reaction. Whether the filling of the small pores with the amorphous phase occurs when NaOH(gas) initially reacts with the silica or when reacted material formed at the front face is pulled into the refractory by capillary suction cannot be determined from the SEM analysis. However, an analysis of capillary suction rates indicates that this form of transport is consistent with the Na_2O profiles measured by EDS,[3] supporting the conclusion that corrosion occurs primarily on the front face of the refractory.

Pore condensation

In small pores, overlapping attractive fields of the pore walls can promote condensation of NaOH(gas). The partial pressure at which condensation can occur is thereby reduced, while the maximum temperature at which it can occur increases. This phenomenon is described by the Kelvin equation, in which the condensation pressure within a pore relative to the nominal condensation pressure outside the pore is a function of the surface tension σ, the pore diameter d, the liquid density ρ, and the temperature T:

$$P_{pore}/P_{nom} = \exp(-\sigma / d \rho RT) \qquad (3)$$

Application of this model to pore diameters typical of low-density silica indicates

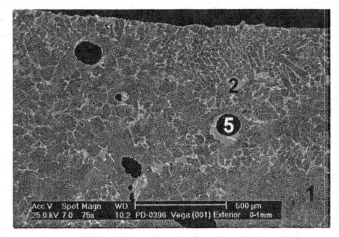

Figure 6. SEM photograph of a sample of Vega low-density silica (Harbison Walker Refractories) exposed to simulated furnace gases for 75 hours to 600 ppm NaOH(gas) in an oxy-fuel combustion atmosphere at 1883 K. (1) Unreacted silica grain. (2) Amorphous Si-Na-O-Ca phase. (5) Large open pore.

that condensation effects should only be present in pores smaller than 0.1 μm at temperatures between 1073 K and 1873 K (Figure 7). Only about 10% of the total pore volume is of this size (the pore size distribution shown in the figure is for the Vega refractory manufactured by Harbison Walker Refractories). The rate of corrosion occurring by this route is limited by the inward diffusion of NaOH(gas). Since this rate becomes rather slow for such small pores, it is likely that faster processes such as capillary suction of liquid formed near the surface will rapidly seal the smallest pores before condensation effects can produce significant corrosion.

COMPUTATIONAL FLUID DYNAMIC MODELING OF CORROSION

It is desirable to predict the effects of furnace operating conditions on corrosion as a function of location on the crown. Due to the complexity of the heat transfer and fluid dynamics within furnaces, it is desirable to couple the front-face corrosion model developed above with a model that uses computational fluid dynamics (CFD) to determine spatial distributions of NaOH(gas), refractory temperature, and NaOH(gas) flux to the crown. To accomplish this, we used the *Athena* glass-furnace model[19] developed by Air Liquide to obtain maps of NaOH(gas) adjacent to the crown and the crown temperature, then post-processed these results through the corrosion model to predict the recession rate. Preliminary results are displayed in Figure 8.

Predicted crown temperatures are in very good agreement with measurements made by optical pyrometry of breast wall temperatures. Initial results provided a profile whose shape was in good agreement with the measured profile,[4] but which was unrealistically hot (some temperatures were above the melting point of silica), even though the known thickness and thermal properties of the crown

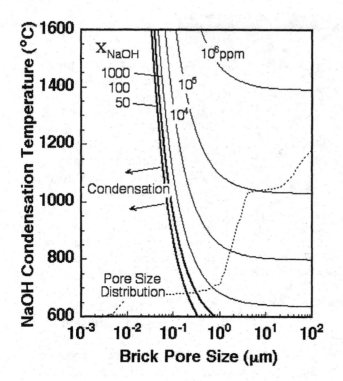

X_{NaOH}
1000
100
50

10^6ppm
10^5
10^4

Condensation

Pore Size
Distribution

Figure 7. Temperature at which NaOH(gas) is predicted to condense as a function of pore diameter and NaOH(gas) partial pressure. The dotted line indicates the pore size distribution for Vega low-density silica refractory.

refractory and insulation were used. Consequently, we reduced the thickness of the insulation to bring the values of the temperature more into line with those thought to be typical in glass furnaces. This yields a temperature profile at the junction between the crown and the breast wall that agrees very well with the measured temperatures, and a centerline profile that is somewhat hotter. As will be seen below, the centerline temperatures in the ~ 9 – 13 m region are probably still too high, with the peak temperature at 1630 °C.

The predicted trend in the NaOH(gas) concentration as a function of axial position in the furnace also agrees well with the measurements of Buckley et al.[10], and even agree quantitatively in the 8 – 15 m portion of the furnace. They appear to be underpredicted in 0 – 8 m region, although there is only one data point at ~ 5 m for comparison. The underprediction may be a consequence of the adjustment in the furnace temperature described above, which also had the effect of lowering the glass temperature somewhat. However, it may also be due to incomplete knowledge of the precise conditions in the furnace at the time of the measurements. Later measurements performed in this furnace indicate higher NaOH(gas) concentrations in this region (~ 125 ppm at 1.0 m);[4] however, the burner design was changed in the interim, so it is difficult to determine at this point whether the model or the measurements are at the root of the disagreement. Importantly, in most portions of the furnace, the NaOH(gas) predicted by *Athena* exceeds that at equilibrium, which is a requirement in our model for corrosion to occur.

Combining the NaOH(gas) concentrations with the refractory temperatures and an average gas velocity of 1.5 m s^{-1} yields recession rates that are within the range of field observations for oxygen-fuel furnaces (0 – 30 mm/y). However, disagreement exists in specific regions of the modeled furnace environment. In the 0 – 5 m region, measured recession rates are a factor of 3 – 3.5 higher than the model predicts, even though the predicted NaOH(gas) concentrations appear to exceed the measured ones. One source of this disagreement may be the average gas velocities used in the calculation. In this furnace zone, the direction of flow along the side walls is toward the exhaust, which may result in higher than average NaOH(gas) mass transfer rates. Alternatively, high corrosion rates at the batch end of a furnace are often attributed to the presence of batch dust, the effects of which are not included in our model.

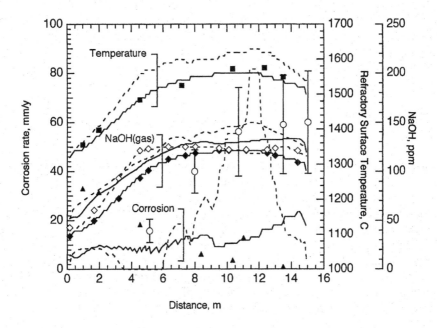

Figure 8.   Comparison of measured temperatures, NaOH(gas) concentration, and corrosion rate in the Gallo Tank 1 furnace with model predictions. Solid lines: furnace centerline. Dashed lines: 4.1 m from the centerline, at the junction between the breast wall and the crown. Squares: measured breast wall temperatures (Ref. 13); circles: measured NaOH(gas) concentrations (Ref. 11); diamonds: NaOH(gas) predicted by equilibrium at the crown temperature; triangles: measured corrosion rates, post-mortem analysis (Ref. 13).

In contrast, a large peak in the centerline recession rate (dashed line, Figure 8) appears in the 9 – 13 mm region. These corrosion rates substantially exceed the measured rates in the furnace. In this case, the anomalous predictions are largely the result of the predicted crown temperature, coupled with a large difference between the NaOH(gas) predicted by the CFD code and equilibrium. The high surface temperature decreases the mole fraction of Na_2O in the corrosion product (x in equation (2)), leading to increased corrosion rates, while increasing ΔX in equation (2) also increases the rate. We find that the recession rate decreases by almost a factor of 2 (to 45 mm/y) if the maximum crown temperature is decreased only 30 °C to 1600 °C, an amount that seems likely to be well within the measurement uncertainties. This recession rate is still somewhat high; the remaining difference may be a result of inaccuracies in the two models responsible for determining the NaOH(gas) concentration gradient (the sodium volatilization model in *Athena* and the thermodynamic model). Small changes in the NaOH(gas) predicted by each of these models (on the order of 5-10%) would be sufficient to reduce the corrosion rate to a value consistent with the measurements. Thus, predicted recession rates are extremely sensitive to the NaOH(gas) concentration gradient.

CONCLUSIONS

The combination of thermodynamic analysis with analytical models of transport phenomena presented here strongly supports the notion that corrosion of silica refractories in gas melting furnaces occurs primarily at the front (hot) face of the crown, rather than by motion of fluids (liquid or gas) into the porous interior. The rate-limiting process is most likely the transport of NaOH(gas) through the mass-transport boundary layer from the furnace atmosphere to the crown surface. Corrosion rates predicted on this basis are in better agreement with observation than those produced by any other mechanism, although the absolute values are highly sensitive to the crown temperature and the NaOH(gas) concentration at equilibrium and at the edge of the boundary layer. Model fidelity would definitely benefit from more accurate knowledge of these variables.

Substantial progress toward the goal of predicting detailed maps of crown corrosion has also been made by coupling the gas-transport model and corrosion thermochemistry with a sophisticated CFD furnace model. Clearly, a number of factors are at work in this system that affect the accuracy of the model and improvements in several areas are needed to increase prediction accuracy. In particular, the submodels predicting the crown temperature and NaOH(gas) concentrations are critical and could probably be improved by better knowledge of crown temperatures and actual NaOH(gas) concentrations. Use of more complete thermodynamic data to predict the equilibrium NaOH(gas) concentration would undoubtedly help as well. The present thermodynamic model accounts for only the SiO_2 and Na_2O components of the corrosion product, whereas it is known that the presence of calcium increases the corrosion rate. In general, however, the results are encouraging and suggest that models capable of predicting corrosion across an entire furnace crown are within reach.

ACKNOWLEDGEMENTS

The authors wish to thank R. D. Moore and J. Neufeld (Gallo Glass Co., Modesto, CA) for providing detailed information concerning operating conditions and the design of the Tank 1 oxy-fuel furnace, and N. Yang for SEM analysis and S. Griffiths for technical discussions (both of Sandia National Laboratories). This work was supported by the U.S. Dept. of Energy (DOE) Office of Industrial Technologies Glass Industry of the Future Program, with continued support from American Air Liquide, BOC Gases, and PPG Industries Inc.

REFERENCES

(1) K. E. Spear; M. D. Allendorf. "Thermodynamic analysis of alumina refractory corrosion by sodium or potassium hydroxide in glass melting furnaces" *J. Electrochem. Soc.*, 2002, **149**, B551, (2002).

(2) M. D. Allendorf; K. E. Spear. "Thermodynamic Analysis of Silica Refractory Corrosion in Glass-Melting Furnaces" *J. Electrochem. Soc.*, 2001, **148**, B59-B67, (2001).

(3) R. H. Nilson; S. K. Griffiths; N. Yang; P. M. Walsh; M. D. Allendorf; B. Bugeat; O. Marin; K. E. Spear; G. Pecoraro. "Analytical Models for High-Temperature Corrosion of Silica Refractories in Glass-Melting Furnaces" *Glass. Sci. Technol.*, 2003, **76**, 136, (2003).

(4) P. M. Walsh; R. D. Moore; J. Neufeld; L. Lemings; J. T. Brown; K. T. Wu. "Sodium volatilzation and silica refractory corrosion in an oxygen/natural-gaso-fired soda-lime-silica glass melting furnace"; XIX Int. Cong. Glass, 2001, Edinburgh, Scotland.

(5) D. W. Bonnell; J. W. Hastie. *High Temp. Sci.*, 1990, **26**, 313, (1990).

(6) K. E. Spear; T. M. Besmann; E. C. Beahm. "Thermochemical Modeling of glass: application to high-level nuclear waste glass" *MRS Bull.*, 1999, **24**, 37, (1999).

(7) "Phase Diagrams for Ceramists", in Vols. 1-12, The American Ceramic Society: Westerville, OH 1964-1996.

(8) J. T. Brown; R. F. Spaulding; D. S. Whittemore; H. E. Wolfe. "New Silica Refractiory of Oxy/Fuel Glass Melting" *Int. J. Glass., Proc. XV A.T.I.V. Conf. Parma*, 1999, 120-128, (1999).

(9) A. J. Faber; O. S. Verheijen. "Refractory corrosion under oxy-fuel firing conditions" *Ceram. Eng. Sci. Proc.*, 1997, **18**, 109-119, (1997).

(10) S. G. Buckley; P. M. Walsh; D. W. Hahn; R. J. Gallagher; M. K. Misra; J. T. Brown; S. S. C. Tong; F. Quan; K. Bhatia; K. K. Koram; V. I. Henry; R. D. Moore. "Measurements of sodium in an oxygen-natural gas-fired soda-lime glass melting furnace" *Ceram. Eng. Sci. Proc.*, 2000, **2**, 1, 183-205, (2000).

(11) P. M. Walsh; R. D. Moore. "Na vapor and gas temperature measured in glass container furnaces" *The Glass Researcher*, 2000, **10**, 1, 3-4, (2000).

(12) M. G. Mesko; J. E. Shelby. "Solubility and diffusion of water in melts of a TV panel glass" *Phys. Chem. Glasses*, 2001, **42**, SC2, 17-22, (2001).

(13) T. Pfeiffer. "Viscosities and Electrical Conductivities of Oxidic Glass-Forming Melts" *Sol. State Ionics*, 1998, **105**, 277-298, (1998).

(14) H. A. Schaeffer. "Diffusion controlled processes in glass forming melts" *J. Non-Cryst. Sol.*, 1984, **67**, 19-33, (1984).

(15) M. G. Mesko; J. E. Shelby. "Water solubility and diffusion in melts of commercial silicate glasses" *Glass Sci. Technol.*, 2000, **73**, SC2, 13-22, (2000).

(16) C. D. Roberts; B. A. Pint, personal communication, 1998.

(17) A. A. Wereszczak; M. Karakus; K. C. Liu; B. A. Pint; R. E. Moore; T. P. Kirkland "Compressive Creep Performance and High Temperature Dimensional Stability of Conventional Silica Refractories," Oak Ridge National Laboratories, 1999.

(18) A. Gupta, Monofrax Inc., personal communication, 2000.

(19) C. Schnepper; O. Marin; C. Champinot; J.-F. Simon. "A modeling study comparing an air- and an oxy-fuel fired float glass melting tank"; Proc. XVIII Int. Cong. Glass, 1998, San Francisco.

HOW THE PROPERTIES OF GLASS MELTS INFLUENCE THE DISSOLUTION OF REFRACTORY MATERIALS

George A. Pecoraro
439 Dakota Drive
Lower Burrell, PA 15068

ABSTRACT

This presentation reviews the equations that describe refractory dissolution by molten glass and gives some practical examples of how corrosion rates can be minimized in glass furnaces.

INTRODUCTION

The high wear areas of glass contact materials that influence the service life of glass furnaces are usually the up-tank basin walls, the roof and walls of the throat, the areas of the bottom adjacent to stirrers, bubblers and electrodes. The metal line region in the batch melting area of the furnace wears very rapidly and usually requires the addition of water coolers or overcoat blocks to extend furnace life. High wear areas in throats, electrodes, and bubblers can result in premature furnace repairs and glass leaks.

To find optimum furnace wall designs and operational parameters, corrosion mechanisms of refractories used for electrodes and furnace hardware need to be understood.

Because of the important economic and technical consequences of minimum glass furnace life (glass defects, glass leaks, thermal efficiency loss, etc.), there has been considerable effort in recent years to:
• Measure corrosion rates of the materials of furnace construction in commercial glass melts,
• Formulate predictive equations for the thermodynamics and kinetics of corrosion, and

- Better understand the principles involved.

The principles of corrosion of glass contact refractories and refractory metals are found in the theories of mass-transport kinetics of heterogeneous systems and phase-equilibria thermodynamics. Mass transport equations have been developed for dissolution kinetics of a variety of applications. These are reviewed. Some suggestions for designing longer-lasting and better-performing refractory material systems are presented.

THERMODYNAMICS OF GLASS MELT CORROSIVITY

The driving force for dissolution of a solute in a solvent is the reduction of free energy of the system. The driving force for dissolution of a refractory (solute) in molten glass (solvent) can be thought of simply as the difference between the concentration of the refractory in the saturated glass adjacent to the refractory and the unsaturated (unaltered) glass. The liquidus curve of the binary equilibrium phase diagram of the solute and the solvent represent the saturation composition at which the free energy of the solute equals its free energy of solution. The liquidus determines the amount of a solid that can be dissolved into the liquid (saturation composition) at a given temperature. The simplest model for dissolution of a pure refractory A dissolving in glass melt A-B at temperature T_1 is shown in the phase diagram in Figure 1. The free energy difference is depicted by points B and L. As temperature increases, the free energy difference increases. Hence, the thermodynamic driving force increases.[1]

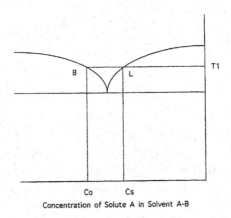

Concentration of Solute A in Solvent A-B

Figure 1. Binary phase diagram of glass A-B, where the refractory is component A

REFRACTORY DISSOLUTION – GENERAL CONSIDERATIONS

For most common systems, the dissolution rates of oxide solids in glass melts depend on three phenomena:[2]

1. The solubility of the oxide in the melt
2. The mobility of the reacting species in the melt (Li_2O, Na_2O, and K_2O).
3. The mobility of the dissolved oxide in the melt (examples are Al_2O_3 and ZrO_2).

The greater the mobility of reacting species in the melt, the greater the solubility of the refractory in the melt, the greater the removal rate of the solute from the interface, and the greater the dissolution rate. Any natural or forced convection process that removes saturated liquid from the solid-liquid interface accelerates the dissolution rate.[2-4] The movement disturbs the chemical equilibrium of the region in the glass adjacent to the solid that is saturated with refractory.

A heterogeneous corrosion process is said to be 'diffusion-controlled' if its slow step is either the introduction of reacting species in the glass melt toward the refractory-glass interface or the removal of the products of dissolution from the refractory-glass interface. In either of these cases, the dissolution process is said to be controlled by the laws of diffusion kinetics.

The process is said to be reaction-rate controlled if the rate of reaction at the phase boundary is the slow step in the process.

The time and temperature characteristics of the overall corrosion process are defined by which of these two mechanisms is controlling. For example, if the diffusion of reaction products away from the surface is limiting, then the reaction order is zero or reciprocal first order.[5]

DISSOLUTION KINETICS EQUATIONS
The kinetics of four types of solute/solvent transport-controlled systems are considered in this section. They include most of the dissolution mechanisms of glass furnace refractories.
- Stagnant liquid or molecular diffusion (furnace paving)
- Natural or free convection (furnace side walls)
- Forced convection (throat erosion)
- Surface tension gradient driven (metal line, upward drilling)

Stagnant Liquid
The Frick equation has been presented in the following form and applied to refractory dissolution by glass melts.[6] *(See Table II for designation of symbols).*

$$j/A = (C_s - C_o)D/\delta_{eff} \tag{1}$$

This expression is the product of two terms: a thermodynamic term (Cs-Co), and a mobility term, D/δ_{eff}. The 'δ_{eff}' term is defined as the tangent to the concentration profile at the interface. The mass transport of refractory per unit area is therefore proportional to the product of the concentration gradient and the effective diffusion coefficient.

An example of this type of dissolution mechanism is the glass furnace bottom near the sidewalls away from bubblers and electrodes where there is little, if any, flow of molten glass. Over time, a thick, viscous layer of refractory-rich glass builds up on the furnace bottom refractory. This layer protects the refractory from dissolution, because the distance of diffusion of the reacting species becomes increasingly larger with time. The velocity of propagation of a species from an infinite media is proportional to $t^{1/2}$ [7]. The parabolic time dependency is the reason that furnace bottoms can last for 20 years.

TEMPERATURE DEPENDENCY OF DISSOLUTION KINETICS
In most refractory dissolution processes, the concern is with heterogeneous reactions involving an interface between two reacting phases, e.g., solid-liquid. Arrhenius showed that, for some processes, for example, diffusion of ions in glass melts, the rate may be related to temperature by the relation:

$$D = D_o \exp^{-E/RT} \qquad (2)$$

The activation energy "E" is the height of the potential barrier that must be overcome by an atom or ion in the diffusion process. Generally, there is an 'E' for each step of the process. Transport of ions to and from the interface is usually an activated processes.

For viscosity, the effect of temperature is defined by Fulcher's equation.

$$\mu = \mu_o \, e^{\, E/R(T-To)} \qquad (3)$$

Ionic diffusion usually is faster as the viscosity is decreased, since bonds between atoms are more easily broken as the diffusing species moves through the glass melt.

The diffusion boundary layer thickness, δ_D, is not constant with temperature. The δ_D for laminar flow is defined by the following equation:[8]

$$\delta_D = \delta_V (D/\mu)^{1/3} \qquad (4)$$

If diffusion is the rate-limiting step, then the temperature dependency of 'D' may approximate the temperature dependency of the mobility term. However, the second term in equation (1), (C_s-C_o) is not exponential, as can be seen from a phase diagram.

REFRACTORY DISSOLUTION – NATURAL CONVECTION CONDITIONS

Natural convection is the transport of a substance in a naturally sinking (or rising) fluid medium. As the refractory dissolves in the glass melt and the boundary layer becomes thicker with time, the system becomes fluid-dynamically unstable. The denser melt containing Al_2O_3 and ZrO_2 begins to sink. The boundary layer thickness becomes constant with time. This situation describes dissolution kinetics with natural convection. An example of glass furnace sidewall block dissolution in free convection is depicted in Figure 2.

The phenomenon of molecular and free convection in a wide range of refractory-glass melts has been demonstrated by many investigators. [9-17] From some of these studies, the following equation was derived for free-convection-driven dissolution from a vertical refractory.

$$j = 1.8(\Delta C)*\{(g\Delta\rho D^3) / \mu x\}^{1/4} \qquad (5)$$

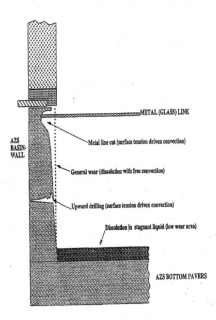

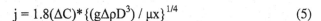

Figure 2. Schematic cross sections of glass-furnace refractory dissolution

The ΔC term is the thermodynamic term, and the remaining expression is the mobility term. The mass transport per unit is influenced by the refractory concentration difference of the glass melt, the change in density as the glass near the wall enriches in refractory, the kinematic viscosity of the enriched melt, and the vertical distance from the glass surface. As shown in Figure 2, a typical wall wear profile shows less dissolution as the distance from the glass surface increases.

In highly dense lead glasses, the alumina and silica refractory enrichment can actually lower the density. This would result in upward movement and a profile that is thicker at the metal line than below it.

REFRACTORY DISSOLUTION – SURFACE TENSION DRIVEN

Surface tension gradients at the glass melt-refractory interface, like melt density differences, can generate motion at the interface because of the Marangoni effect.[8-9] Metal-line cut has been explained in terms of surface tension forces by several authors.[6,10] Refractory dissolution controlled by surface-free convection occurs in the meniscus of the glass melt under two conditions where:

1. The surface tension of the melt increases with the concentration of the dissolved refractory (i.e. Al_2O_3 and ZrO_2), and
2. The change in surface tension is large in proportion to the increase in density.

Examples of surface-tension-driven convection at the metal line are shown in Figure 2, where there are relatively rapid dissolution processes occurring at the refractory-glass-atmosphere interface. Thermal gradient from top-to-bottom is not the cause of this dissolution profile, since this effect is seen in iso-thermal lab melts. The surface-tension effect is not included in equation (5).

If the fresh glass being convected toward the basin wall is miscible with the refractory-enriched glass at the refractory-glass interface, then eddy currents result. The turbulence in the flow system accelerates the dissolution process, because it removes refractory-enriched glass from the interface.

An equation from which the dissolution rate of the metal line can be calculated is as follows:

$$J = B(C_i-C_o)*(D_o 2\Delta\sigma g/\upsilon)^{1/3} \qquad (6)$$

REFRACTORY DISSOLUTION – FORCED CONVECTION DRIVEN

The influence of forced-convection-driven dissolution of refractories by glass melts has been studied by several investigators.[10-15] Glass furnace refractory dissolution is enhanced by forced convection in areas near mechanical stirring and bubblers, and where a geometric restriction such as a throat is present. It is shown by equation (1) that the dissolution rate of refractories in molten glass is proportional to the concentration gradient $(C_i-C_o)/\delta$ of the dissolving material at the glass-refractory interface. Forced convection decreases the dimension of the boundary layer, and therefore increases the rate of dissolution.

The velocity boundary layer thickness "δ_v" is given by[15]

$$\delta_v = 3y / (v^* y_s/n)^{1/2} \tag{7}$$

when "n" is the zero coordinate starting point of the flow (where y=0) and increases as the distance from the starting point increases. The mass flow density of the material dissolved away becomes:

$$k = 1/3(C_i-C_o)^*(v/y)^{1/2}(D^2 v^{-1/2})^{1/3} \tag{8}$$

KINETICS OF DISSOLUTION OF A ROTATING DISC.

Cold modeling and dissolution studies of refractories at 1400°C using a rotating disc or the bottom surface of a rotating rod were conducted by Busby and Turner to:

1. Empirically distinguish between natural and forced convection
2. Increase the understanding of metal line and subsurface corrosion, and
3. Obtain dissolution results in a shorter time.[14]

The conditions of flow adjacent to a rotating cylinder under laminar flow conditions are well understood. The expression for hydrodynamic boundary layer thickness for a rotating disc is[15]

$$\delta = 3.6(v/\Omega)^{1/2} \tag{9}$$

An equation for calculating the mass flux from disc surfaces of refractory-glass melt systems is given by the following equation.[15]

$$j = 0.62 \, (Ci-Co)^* D^{2/3} v^{-1/6} \Omega^{1/2} \tag{10}$$

Equation (10) is of the same general form of the other mass transport equations presented above. There is a thermodynamic term and a kinetic term. Using the rotating disc allows a distinction to be made between the dissolution rates of different refractories. At rpms of 10 to 40, the flux line virtually disappears.

In Table I, the influence of the properties of glass melts cited in the equations above is summarized. Notice that a similar pattern exists for most properties.
- Concentration differences are to the first power
- Diffusivities have exponents from the 3/4 to 1st power
- Kinematic viscosities have an exponential range from $-1/6^{th}$ to $-1/3^{rd}$.

Table I. Glass Melt Properties That Influence Its Corrosivity to Refractories

Glass Property	Stagnant Melt Convection	Free Convection	Metal Line Corrosion	Forced Convection	Rotating Disc
Solubility	Ci-Co	Ci-Co	Ci-Co	Ci-Co	Ci-Co
Effective Binary Diff. Coeff.	D	$D^{3/4}$	$D^{2/3}$	$D^{2/3}$	$D^{2/3}$
Kinematic Viscosity		$v^{-1/4}$	$v^{-1/3}$	$v^{-1/6}$	$v^{-1/6}$
Surface Tension			$\Delta\sigma$		
Density Difference			$\Delta\rho^{1/4}$		

In Table II, the description of the symbols used in the equations is listed.

Table II. Description of Symbols

English Symbols
Cs = saturated concentration of solute dissolved in solvent
Co = concentration of solute in unaltered solvent
D_{eff}= effective binary diffusion coefficient
E = activation energy
g = gravity constant
j = dN/dT =mass of solute corroded per unit time
k = first-order rate constant for heterogeneous reaction at the interface
 = reaction constant
t = time
T = absolute temperature
v = velocity
x = distance from top or from leading edge
y = coordinate in a forced convection system

Greek Symbols
δ_{eff} = effective boundary layer thickness for diffusion
δ_V = velocity boundary layer
δ_D = diffusion boundary layer
μ = viscosity
v = kinematic viscosity

Advances in Fusion and Processing of Glass III

Ω = angular velocity of rotating disc or rod
σ = surface tension
ρ = density

APPLICATION OF DISSOLUTION KINETIC PRINCIPLES TO REFRACTORY DESIGN-REFRACTORY CHEMISTRY

This review of refractory dissolution kinetic equations has provided a foundation for application of the principles to actual furnace wall design and material chemical characterization.

Refractory Chemistry

For glass contact applications, refractories should be used that have the lowest solubility in the molten glass. ΔC has been shown to be the thermodynamic driving force for dissolution. The common oxides that have the lowest solubility at glass melting temperatures are ZrO_2 and Cr_2O_3. Figure 3 plots dissolution versus time for refractories of three distinct chemistries in S-L-S glass.

In stabilized form, ZrO_2 is used to make small crucibles for glass melting. Fusion-cast refractories having a ZrO_2 concentration as great as 95% are commercially available. There are very corrosion resistant in borosilicate glasses. However, in soda-lime glass, the 95% ZrO_2 products are, at best, equivalent to fusion-cast refractories having 41% ZrO_2 concentration. Schlotzhauer and Hutchins[16] offer a model of the AZS-glass interface that explains how the combination of Al_2O_3 and ZrO_2 work together to provide excellent resistance to dissolution by glass melts. This explains why ZrO_2 dissolves at a faster rate than 41% ZrO2 in the AZS refractory in Figure 3 (from Schlotzhauer et al.[16]).

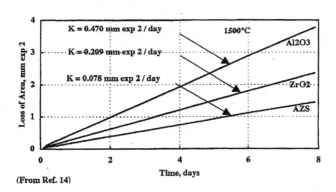

Figure 3. Dissolution Rates of Refractories of Various Chemical Make-up.

The effect of adding solute oxides Al_2O_3 and ZrO_2 to the glass to reduce ΔC is shown in Figure 4.[17] Note that the dissolution rate is depressed by as much as a factor of 5 at constant temperature. However, the possible need to raise temperature to process the glass may negate this effect.

Cr_2O_3 is less soluble in most glass melts than ZrO_2. However, since Cr_2O_3 imparts a green color to most commercial glasses, its application as a glass contact refractory is limited to small, high-wear areas such as bonded Cr_2O_3 over coat blocks of basin walls, entrance blocks of throats in container furnaces, and the liner of fiberglass furnaces. Fused-cast Cr_2O_3-Al_2O_3-SiO_2 refractories are used in special applications where the highest degree of corrosion resistance is required, but some color potential can be tolerated.

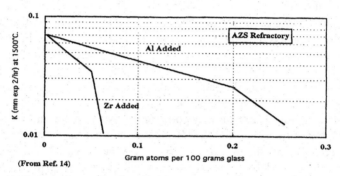

(From Ref. 14)

Figure 4. Depression of dissolution rates by adding solute to the glass composition

Al_2O_3 fused-cast products can be used in soda-lime furnaces at or below 2550°F, because their solubility is low below this temperature. These products are less costly and have very low defect potential. This temperature range includes the conditioner and glass-forming applications such as canals, channels, and float lips.

Effect of Temperature

Many of the properties of glass melts that influence refractory dissolution are strongly temperature dependent. Examples are:

- Solute solubility
- Diffusivity
- Density
- Kinematic viscosity

Therefore, the refractory-glass melt interface is usually force cooled to reduce dissolution rates. Kucheryavyi, et-al.[17] claim that the glass-refractory interface temperature can be reduced by as much as 400K by application of cooling, and that furnace life can be increased from 5 to 15 years if the refractory-glass interface temperature is reduced from 1320 to 1100°C for a 41% ZrO_2 fusion-cast basin wall. Air velocity ranged from 3.5 to 6.5 m/s.

Four designs for slowing the rate of metal-line dissolution are depicted in Figure 5. Figure 5A shows a cross section design for enhancing the cooling effect of metal-line cooling air applied to the cold face of the block. The block's cross-section is the thinnest at the metal line (top) where the glass is the hottest and where surface tension effects are the greatest. This design also causes a thicker boundary layer thickness, since the downward flow rate is reduced because of its horizontal component. As shown in Figure 6, dissolution rates were reduced by a much as a factor of three in crucible-scale tests because of a slower transport rate of solute from the interface.[19] This effect has been verified in actual float furnace operation.

Figure 5B shows a water-cooled rectangular pipe placed against the hot face of the block just below the metal line. The cooler has two functions.
- The first is to reduce the hot-face temperature to reduce both solute solubility and to physically block the downward flow of glass driven by density and temperature differences.
- The second is to reduce the downward flow that increases the boundary layer thickness for diffusion.

Figure 5C shows a molybdenum plate embedded in the internal vertical wall just under the metal line. The plate retards the density- and temperature-driven downward flow in the same manner as depicted by the water cooler in Figure 5B.

A basin wall design that has a notch to accommodate a refractory patch block or a water-cooled bow is shown in Figure 5D.

Combinations of refractory chemistry, tapered shape and either the internal water cooler or the embedded molybdenum plate may provide the optimum performance.

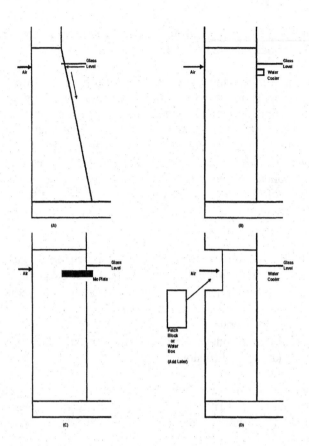

Figure 5. Four schemes for increasing melter basin wall refractory life

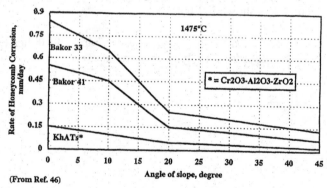

(From Ref. 46)

Figure 6. Depression of refractory dissolution by sloping the basin wall shape.

CONCLUSIONS

Equations describing how glass chemistry and properties influence the dissolution of refractories can be helpful in designing glass contact refractory walls, bottoms and throats. Four examples are given that use forced cooling, block chemistry, and geometry to minimize the effects of dissolution.

REFERENCES

[1] A.R. Cooper,Jr, *Advances in Glass Technology,* 6th Ed. Plenum Press, New York, NY,1962.

[2] G.V. McCauley, Glass and Refractory Symposium, *Bull. Am. Ceram. Soc.,* **4, [11]** 605-610 (1925).

[3] W.K. Brownlee, "Insulation of Glass Tanks," *J. Am. Ceram. Soc.,* **7,** 457 – 464 (1924).

[4] F.C. Flint and A.R. Payne, "Tank Block Corrosion by Shelving," *J. Am. Ceram. Soc,* **9,** 613-7 (1926).

[5] G.M. Schwab et al., *Catalysis,* Van Nostrand, 1937.

[6] W.D. Kingery, *Kinetics of High-Temperature Processes*, The Technology Press of MIT , Cambridge, MA, 1959.

[7] A.R. Cooper, Jr., "Kinetics of Refractory Corrosion," *Ceramic Engineering and Science Proceedings,* **2, [11-12]**1063-1089 (1981).

[8] T.S. Busby *Glass,* "Flux Line Corrosion," *Glass* **39** [4]182-189 (1962); "Refractory Wear," Glass **44** [7]311 (1967).

[9] L. Reed, *Trans. Br. Ceram, Soc,.* **53,** 180-202 (1954).

[10] M. Dunkl and R. Bruckner, "Bestimmung der Korrosionsrate von feuerfesten Baustoffen durch Glasschmelzen bei freier laminarer Dichte und Grenzflachenkovektion," *Glastech. Ber.,* **53,** 321 (1980).

[11] L. Reed and L. R. Barrett, "The Slagging of Refractories," *Trans. Brit. Ceram. Soc.,.* **54,** 671-676 (1955).

[12] H.R. Moore and R. Heeley, "An Experimental Investigation of Alumino-Silicate Refractories," *J. Soc. Glass Tech.,* **34,** 274 (1950).

[13] W. Trier and P.J. Koros, *Glass Furnace Design, Construction and Operation*, Society of Glass Technology, Charlesworth and Co., Huddersfield, England, 1987.

[14] T.A. Busby and R.J. Turner, "The Development of a Rotating Corrosion Test for Glass Contact Refractories," *Glass Technology,* **22 [1]** 15-23 (1981).

[15] V.G. Levich, *Physicohydrodynamics*, Prentice Hall, Engelwood Cliffs, NJ, 1962.

[16] L.R. Schlotzhauer and J.R. Hutchins III, "Corrosion Resistance of Alumina-Zirconia-Silica Refractories," *The Glass Industry,* **47,** 26-29,48 (1966).

[17] M.N. Kucheryavyi, O.N. Popov, A.S. Astakhov, "New Channel Design for a Glass Furnace," *Glass and Ceramics,* **45,** [7-8] 251-253 (1988).

EVALUATION OF CROWN REFRACTORIES UNDER OXYFUEL ENVIRONMENT

M. Velez, M. Karakus, X. Liang, W. L. Headrick, R. E. Moore
University of Missouri-Rolla
Ceramic Engineering Dept.,
Rolla, MO 65409-0330

J. G. Hemrick
Metals and Ceramic Division
Oak Ridge National Laboratory
Oak Ridge, TN 37831-6069

J. M. Almanza
CINVESTAV
Carr. Saltillo-Mty. Km 13,
Apdo. Postal No. 663
25000, Saltillo, Coah., Mexico

ABSTRACT

The use of refractory materials for glass melters under oxyfuel combustion, regarding corrosion resistance, the potential to generate glass defects under various conditions, and thermomechanical properties is summarized in this work. Silica brick will continue to serve in crown construction in many melters due to the excellent creep resistance, low conductivity, and low cost. The fused products used for glass contact appear to be functional alternatives for crowns but with additional structural costs. The next-in-line products appear to be bonded and fused spinel-based and mullite-based compositions. A fused version of spinel has demonstrated high creep resistance at crown temperatures and high corrosion resistance. Some monolithic products were also evaluated under oxyfuel simulation, as they have never been evaluated in full-scale furnaces.

INTRODUCTION

Oxyfuel technology in the glass industry has begun to be adopted, but opportunity for wider acceptance is still available[1]. Our own survey (2003) indicates that approximately 10% of container glass furnaces, 35% of fiber glass furnaces, 5% of float glass furnaces and 23% of specialty glass furnaces have been converted to oxyfuel firing. Benefits include reduced gas and particulate emissions, and decrease in capital investment, in comparison with air-fuel combustion. However, oxyfuel firing has also increased the rate of degradation of crown silica refractories with an increase in NaOH vapor, compared to air-gas combustion, due to reduction of nitrogen and a higher water concentration in the combustion chamber. In traditional crown silica refractories, the silica rapidly reacts with NaOH during oxyfuel firing forming a glassy phase that may quickly consume the refractory. To address this, different experiments have been reported to obtain information about corrosion mechanisms and process parameters; some on a laboratory scale[2-7] and some in industrial furnaces[8]. Thermodynamic calculations indicate that NaOH vapor concentrations below 15 ppm do not react with silica under either air- or oxy-fired conditions, since this is the smallest p_{NaOH} in a system containing crystalline SiO_2 in equilibrium with a variable composition sodium-silicate liquid phase in the range 1400-1700°C[9-10].

Refractories for both oxy- and air-fuel-fired furnace superstructures are subjected to high temperatures during service and may appreciably creep or subside if the refractory material is not creep resistant or if it is subjected to high stress[11]. Published engineering creep data are scarce for almost all commercially available refractories used for glass furnace superstructures. Additionally, the various refractory suppliers typically conduct their mechanical testing differently, and interpret and report their obtained data differently, making it difficult for furnace designers to compare competing grades of candidate refractories in an equitable manner. Furthermore, the refractory supplier's data is often not in an available form that can be used for furnace design or the prediction of long-term structural integrity of furnace superstructures.

Work at the University of Missouri–Rolla (UMR) and Oak Ridge National Laboratory (ORNL) directly addresses these issues, generating a database of the high-temperature mechanical, thermal and chemical behaviors of crown refractories, and realizing the energy savings and pollutant abatement oxy-fuel firing may yield. The resulting database could then be used in three ways. First, it will be made available to the glass manufacturers for assisting them in the selection of the appropriate refractories. Second, the data will be used to refine existing models describing the stress, temperature, and microstructural dependencies of the creep rate. Third, the data will be used in conjunction with finite element analysis (FEA) to predict the time-dependent deformation of glass-tank crowns. Previous studies have resulted in the generation of data for several refractory systems including mullite[12], silica[13], AZS[14], fusion-cast alumina[15], and bonded and fusion-cast spinel[16].

MICROSTRUCTURAL CHARACTERIZATION

Optical microscopy (reflected light and cathodoluminescence imaging, RL/CL) was the main technique used to characterize samples. A cold cathode CL microscopy system (model Mk4, Cambridge Imaging Technologies, Cambridge, UK) mounted on a standard Nikon petrographic microscope (model Labophot-Pol with an electronic image acquisition system) was used for these microstructural analyses. SEM, XRD, and chemical analysis were used to corroborate composition and phase analyses in some cases. Most of the information is being stored in a web site under a National Science Foundation program[17]. The CL technique is unique and provides information that often cannot be obtained by any other technique[18].

CORROSION OF CROWN REFRACTORIES

A number of brick compositions, bonded or fused, have been promoted as substitutes for silica crown bricks in oxyfuel furnaces[2,3]. Yet, the cost effectiveness of silica is very difficult to match. In addition, there are higher costs associated with the installation of heavier refractories (Table I).

Table I. Refractories for Crown Applications[based on 19]

Property	Refractory				
	Silica	Mullite	Fused-cast Spinel[20]	Fused-cast AZS	Fused-cast alpha/beta-Alumina
Relative cost	1	6	7	10	11
Density g/cm^3	1.79	2.5-2.7	2.9-3.0	3.7	3.4
Resistance to NaOH vapor	Poor	Good	Excellent	Medium	Excellent
Construction Type	Type I	Bonded: Type I / Fused: Type II	Type II	Type II	Type II
Thermal Conductivity W/mK @ 1200°C	2.0-2.3	2.5-3.5	3.0-4.0	5.0	5.19
Thermal Shock Resistance	Poor (<500°C)	Good	Medium	Limited due to low porosity	Limited due to low porosity
Thermal Expansion	Non-linear	Linear	Linear 0.85% at 1000°C	Non-linear	linear
Defect Potential	Very low	Low	Low	High	Medium

Type I: Standard bonded end arch
Type II: Ground block, investment in supporting steel work

The selection of refractories for oxyfuel require that they must be physically stable when exposed to chemical species in any combination within operating conditions; and/or that the reaction kinetics of any degradation must be minimized. Given these constraints, the standard isothermal tests developed for silica and silicate refractories are not applicable to most fused oxides. Long exposure times are needed, for instance, to assure that alterations to the most resistant crown substitutes do not preclude long-term service. To address this, a pilot-size oxyfuel furnace has been built and used to simulate continuous melting of glasses under oxyfuel combustion[21], to assess crown refractories at high temperatures[22-25], and to test sensors[26,27].

Silica Refractories: Table II shows a summary of properties and parameters of unused silica refractories. Those refractories were tested in the oxyfuel furnace simulator (OFS) under different parameters [1400 to 1550°C, 20 to 600 ppm NaOH (by volume), and up to 2-weeks duration of tests].

Table II. Summary of Properties for Unused Silica Bricks[based on 13,24]

Data\Sample	Gen-Sil	Vega 1	Vega 2	Vega ZED
Apparent Porosity, %	20-23	23.3	19	21
Bulk Density, g/cm^3	1.79-1.83	1.80	1.91	1.83
Permeability, cD	170	95-160	20	--
Median Pore Size, microns	20	16	8	--
Cold Crushing Strength, MPa (psi)	20.7-34.5 (3,000-5,000)	27.6 (4,000)	27.6 (4,000)	31.5 (4,570)
Modulus of Rupture, MPa (psi)	6.9-9.7 (1,000-1,400)	5.5 (800)	4.8 (700)	6.0 (870)
Reheat Change in Dimension	-0.1% @ 1600°C	+0.2 @ 1500°C	+0.1 @ 1500°C	+0.2%
SiO_2, %	95.69	96.4	98.0	96.3
Fe_2O_3, %	0.9	0.5	0.4	0.4
TiO_2, %	0.1	Trace	Trace	0.1
CaO, %	3.0	2.9	0.8	2.7
Al_2O_3, %	0.3	0.2	0.6	0.3
MgO, %	--	Trace	0.1	0.1
Alkalis, %	0.1	0.02	0.1	0.1
Residual Quartz, %	<1	<1	<1	<1
Cristobalite-to-Tridymite Ratio	0.75	1.42	2.12	2.05

The corrosion findings from the oxyfuel furnace simulator (OFS)[22-25] can be summarized as: (1) the ratio of cristobalite/tridymite changes with distance from the hot face to the cold face, being highest at the hot face; (2) sodium concentration close to the hot face is high and decreases with distance, and with porosity of the silica brick; (3) the hot face is sealed with a glassy Na-silicate phase which drips to the glass bath; and (4) corrosion is more severe at higher NaOH concentration (400-600 ppm). The use of the OFS allows more detailed information than the use of the standard procedure ASTM C987[13]. Fused-silica castable (i.e., Metsilcast of Magneco Metrel) was also tested and seemed to perform better to high NaOH concentrations, than commercial silica bricks, as the CaO content is below 0.1%.

Corrosion of Alumina-Zirconia-Silica-Containing Castables: Several sintered AZS-based castable developmental materials (pre-form wedges, Table III) were tested as crown materials in the OFS (200-400 ppm NaOH during 2 weeks exposure time) and characterized by optical microscopy (reflected light and cathodoluminescence, RL/CL), XRD and ICP chemical analysis.

Table III. Chemistry of Materials from Emhart Glass Manufacturing, wt%

Sample	Mineralogy	Al_2O_3	SiO_2	ZrO_2
A	Alumina + (Silica)	~ 94	~ 2.5	
B	Alumina + Zirconia	~ 83	~ 1.5	~ 15
C	Alumina + Zircon + Mullite + Zirconia	~ 74	~ 9.5	~ 15
D	Alumina + Zircon + Mullite	~ 75	~ 15	~ 9
E	Mullite + (Alumina)	~ 72	~ 26	
F	Mullite + Alumina	~ 80	~ 18	
G	Zircon		~ 37	~ 57

Alumina refractories (sample A) revealed β-alumina formation at the hot face with a sharp boundary between β-alumina and α-alumina material. Small amounts of glassy phase and nepheline in the matrix between α-alumina aggregates and fine grains were formed. This sample showed that the ultimate reaction product was β-alumina at the exposed refractory surface:

$$11Al_2O_3 + 2NaOH = Na_2O.11Al_2O_3 + H_2O$$

Zirconia grains, added to the alumina matrix as fine zirconia particles (sample B), virtually remained unreacted. The alteration layer showed α-alumina grains in the matrix as well as large tabular and fused alumina aggregates that were totally converted to β-alumina. Only the alumina component interacts with NaOH.

Mullite-based refractories and mullite-based castable with tabular alumina grains (samples C-F) showed interaction indicating a sequence of reaction zones, from hot face to cold face: (1) β-alumina + nepheline zone, with relatively few pores or voids; (2) mullite + nepheline zone, in which mullite aggregates are dissociated to form α-alumina from mullite aggregates and prismatic recrystallized α-alumina from mullite matrix; (3) a dense zone formed as a result of large quantities of liquid-phase development; (4) a dense mullite matrix zone and no recrystallized prismatic α-alumina is observed; and (5) unaltered refractory.

Zircon refractories (sample G) produced zirconia and large amounts of glassy phase. Three basic zones were observed: (1) totally altered zirconia + glass zone (hot face), (2) a transition zone, and (3) an unaltered zone. Zircon normally dissociates at relatively high temperatures; however, this temperature is lowered below 1500°C in the presence of NaOH and other impurities.

Corrosion of Fusion-Cast α/β Alumina Refractories: Corrosion testing on both a lab and a pilot-size furnace have shown little recession of these materials due to interaction with normal glass compositions and the associated vapor species[28]. Due to the nature of the fused-casting process, these refractories have a heterogeneous microstructure with a columnar zone of elongated or column-like grains oriented along the direction of cooling on the exterior of the block and a more homogeneous equiaxed structure on the interior. Fusion-cast alumina materials have high refractoriness, spall resistance, and low levels of porosity (<5%) and possess low levels of glassy phase in their structure (<2%), making them less susceptible to deformation by the mechanisms of creep[15].

Due to their higher thermal conductivity, heat flow through a furnace crown composed of fusion-cast alumina is higher than through traditional silica crowns, affecting back-up insulation designs and cooling methods employed on the exterior of the furnace. Also, the reversion of β-alumina to α-alumina during the life of the refractory brick occurs in the presence of batch dust or silica. A small amount of silica destroys the β-alumina matrix by reacting with sodium. The result of the Na evolution is the formation of a pure α-alumina layer/section on the exposed surfaces of the refractory.

Corrosion of Fusion-Cast Spinel Refractories: Cored samples were subjected to NaOH levels of 200-400 ppm for 10 days at 1500°C. No reaction of the spinel was observed using RL/CL microscopy, qualifying these refractories as excellent regarding chemical resistance towards NaOH vapor (Table II). However, forsterite formation was observed due to reaction with silica from contact with underlying AZS refractories. This was confirmed with XRD; magnesia peaks were reduced as the glass content increased slightly[16].

THERMOMECHANICAL PROPERTIES

Thermal Conductivity: Thermal conductivity of crown materials was measured using the laser flash technique, which measures the thermal diffusivity of a material using a laser and the heat capacity, measured by calorimetry. In conjunction with the bulk density, the thermal conductivity can then be calculated. Figure 1 shows the thermal conductivity calculated for both bonded and fusion-cast spinel from 100 to 1425°C. The thermal conductivity of fusion-cast alumina is shown in Fig. 2 from room temperature to 1425°C.

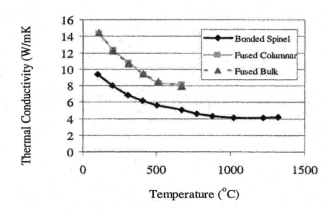

Figure 1. Thermal conductivity of spinel[16].

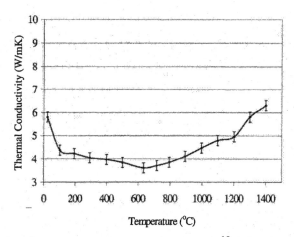

Figure 2. Thermal conductivity of fusion-cast alumina[15].

Creep: Creep of refractories is normally measured over 25, 50, or 100 hours and loads of 0.2 MPa[11]. Bonded spinel materials showed significant creep at 0.3 MPa and 1550°C, while the fusion-cast spinel showed very little creep at this stress level at 1650°C. Therefore, testing has been performed at an increased stress level of 3.0 MPa, as shown in Fig. 3. Following the first 100 hours, the creep rate slightly increased for fusion-cast spinel to 1.44 x 10^{-6}/hr along the columnar microstructure in the chilled zone and 5.33 x 10^{-6}/hr for the bulk material away from the chilled zone at 1650°C. Creep of fusion-cast alumina was measured at multiple temperatures and stresses, as shown in Fig. 4. Significant creep was not induced in this material until testing the highest temperature/stress regime of 1650°C/1.0 MPa.

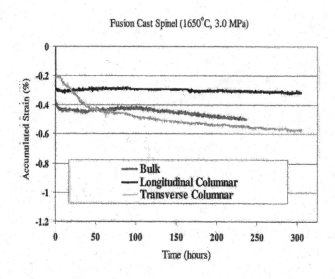

Figure 3. Creep of fusion-cast spinel at 1650°C and 3.0 MPa[16].

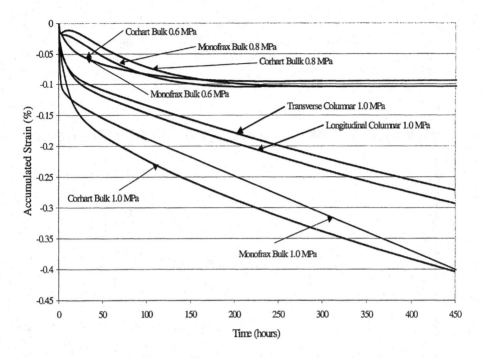

Figure 4. Creep of fusion-cast alumina refractories at 1650°C [15].

MODELING

Limited work exists on the modeling of crown superstructure refractories. FEA (finite element analysis) was performed at ORNL and UMR using ANSYS 5.4 and 5.6 to simulate stress concentrations in fused-cast alumina crowns due to thermal expansion at 1600°C and 0.2 MPa[15]. Analysis was based on the physical characterization data (density, elastic moduli, thermal properties), along with the creep data and the microstructure. Attention was given to changes in physical properties with exposure to temperature and load.

Modeling of a mullite glass-tank crown is being created using ABAQUS FEA. Data for silica and mullite crowns previously developed at ORNL are used for the initial model[12]. The model predicts the initial opening of cold-face joints, as seen on commercial crowns. Creep and corrosion slowly causes the cold face joints to close. Currently, creep data are not available at the high compressive stress found at the hot face for any refractory materials, hindering completion of the models. The primary load-carrying part of crown is shown as brick components on Fig. 5. Temperature gradients in the crown cause two primary changes: an expanded size due to increase in temperature, and rotation due to thickness temperature differences. The brick joints separate under this condition, affecting the stress distribution and heat transfer. Transient heat transfer was used for the establishment of temperature field, with 1600°C at the hot-face and a steady state was defined as when the cold face reached 1400°C. Figure 6 shows the deformed arch after initial heat up compared with the original arch, illustrating that all the bricks contact surfaces are separated during heat up.

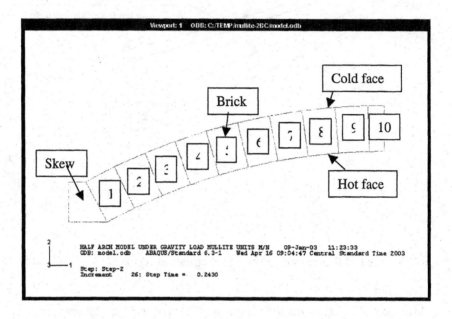

Figure 5. Glass tank crown model bricks definition.

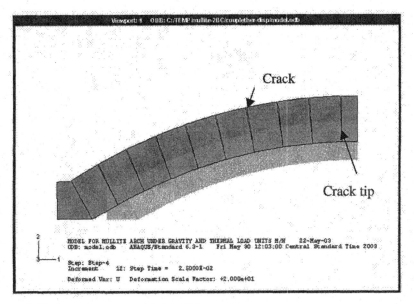

Figure 6. Deformed crown after initial heat up.

CONCLUDING REMARKS

The University of Missouri-Rolla and ORNL Refractories Groups have worked over the last decade on models and testing related to refractories for the glass industry. The facilities include two digitally-controlled compressive creep frames that can test to 1700°C with high-resolution extensometry, and thermal conductivity (ORNL), an oxyfuel simulator furnace, and dynamic corrosion testing (UMR). Our vision of the glass industry indicates that the following tasks would be of value to pursue: (1) creep testing on heat-treated samples to compare behavior to current results; (2) creep testing and physical characterization of salvaged fusion-cast refractory bricks from oxy-fuel furnaces; (3) theoretical modeling of composite creep behavior exhibited by fusion-cast alumina[15], for instance; and (4) non-linear finite element analysis of furnace crowns.

ACKNOWLEDGEMENTS

We wish to express our appreciation to the NSF Industry-University Center for Glass Research (NSF Grant No. 0128040), BOC Gases, Emhart Glass Manufacturing, Magneco/Metrel, Corhart Refractories, and the NSF/University of Dayton Digital Library Program DUE-0121540.

REFERENCES

[1]R.W. Schroeder, "Development of Oxyfuel Technology in the Glass Industry," *The Glass Researcher*, 8[1] 1; 3-4 (1998).

[2]L.H. Kotacska, T.J. Cooper, "Testing of Superstructure Refractories in a Gas-Oxy Atmosphere Against High-Alkali Glasses," *Ceram. Eng. Sci. Proc.*, **18**[1] 136-145 (1997).

[3]A. Gupta, S.M. Winder, "Ongoing Investigation of Oxy-Fuel Firing Impact on Corrosion of Nonglass Contact Refractories," *Ceram. Eng. Sci. Proc.*, **17**[2] 112-121 (1996).

[4]G. Duvierre, A. Zanoli, Y. Boussant-Roux, M. Nelson, "Selection of Optimum Refractories for the Superstructure of Oxy-Fuel Glass Melting Furnaces," *Ceram. Eng. Sci. Proc.*, **18**[1] 146-163 (1997).

[5]A.J. Faber, R.G.C. Beerkens, "Reduction of Refractory Corrosion in Oxy-Fuel Glass Furnace," 18th International Congress on Glass. Section A9. Refractory Materials, pp. 13-18 (1998).

[6]J. Boillet, W. Kobayashi, W.J. Snyder, C.A. Paskocimas, E.R. Leite, E. Longo, J.A. Varela, "Corrosion of Silica and Mullite Refractories used in Glass Furnaces under 100% Oxy-Firing Process," *Ceram. Eng. Sci. Proc.*, **17**[2] 180-188 (1996).

[7]A.J. Faber, O.S. Verheijen, "Refractory Corrosion under Oxy-Fuel Firing Conditions," *Ceram. Eng. Sci. Proc.*, **18**[1] 108-119 (1997).

[8]C.A. Paskocimas, E.R. Leite, E. Longo, W. Kobayashi, M. Zorrozua, J.A. Varela, "Determination of Corrosion Factors in a Glass Furnace," *Ceram. Eng. Sci. Proc.*, **19**[1] 75-89 (1998).

[9]K.E. Spear, M.D. Allendorf, "Mechanisms of Silica Refractory Corrosion in Glass-Melting Furnaces," pp. 439-448 in *High Temperature Corrosion and Materials Chemistry, Proc. Per Kofstad Memorial Symposium.* Edited by M. McNallan, E. Opila, T. Marayama, and T. Marita, The Electrochemical Society Proceedings Series, Pennington, NJ, 1999.

[10]M.D. Allendorf, K.E. Spear, "Thermodynamic Analysis of Silica Refractory Corrosion in Glass-Melting Furnaces," *J. Electrochem. Soc.*, **148**[2] B59-B67 (2001).

[11]J.G. Hemrick, A.A. Wereszczak, "Creep Measurement and Analysis of Refractories," in Fundamentals of Refractory Technology, pp. 171-193 in *Ceram. Trans., Vol. 125.* Edited by J. P. Bennett and J. D. Smith, The American Ceramic Society, Westerville, Ohio, 2001.

[12]J.G. Hemrick, A.A. Wereszczak, M. Karakus, K.C. Liu, H. Wang, B.A. Pint, T.P. Kirkland, R.E. Moore, Compressive Creep and Thermophysical Performance of Mullite Refractories, Technical Report ORNL/TM-2002/84, 2002.

[13]A.A. Wereszczak, M. Karakus, K.C. Liu, B.A. Pint, R.E. Moore, T.P. Kirkland, Compressive Creep Performance and High Temperature Dimensional Stability of Silica Refractories, Technical Report ORNL/TM-13757, 1999.

[14]A.A. Wereszczak, J. Heide, T.P. Kirkland, G.V. Srinivasan, and S.M. Winder, "High Temperature Deformation of an AZS Refractory," pp. 379-384 in *Advances in Fusion and Processing of Glass II, Ceramic Transactions, Vol. 82,* Edited by A.G. Clare and L.E. Jones, The American Ceramic Society, Westerville, Ohio, 1998.

[15] J.G. Hemrick, Creep Behavior and Physical Characterization of Fusion-Cast Alumina Refractories, PhD Thesis, University of Missouri-Rolla, Ceramic Engineering Dept., 2001.

[16] W.L. Headrick, R.E. Moore, J.G. Hemrick, M. Ferber, "Characterization and Modeling of Glass Tank Refractories at UMR and ORNL," *Refractories Applications & News,* **8**[4] 13-15 (2003).

[17] M. Velez, M. Karakus, R.E. Moore, "UMR Digital Library: High Zirconia AZS Refractories," *Refractories Applications & News,* **7**[6] 36-37 (2002).

[18] M. Karakus, R.E. Moore, "CLM – A New Technique for Refractories," *Ceram. Bull.,* **77**, 55-61 (1998).

[19] C. Windle, "Rebonded magnesia-alumina spinel the oxy-fuel solution," *Int. Ceram.,* **1**, 41-43 (2000).

[20] G. Boymanns, F. Gebhardt, M. Dunkl, H.D. Schlacht, "Spinel Bricks for Highly Stressed Roofs in Glass Melting Furnaces," *Glastechn. Ber. Glass Sci. Technol.,* **73**[10] 293-298 (2000).

[21] M. Velez, L. Carroll, C. Carmody, W.L. Headrick, R.E. Moore, "Oxy-fuel Simulator Glass Tank Melter," pp. 47-54 in *Environmental Issues and Waste Management Technologies in the Ceramic and Nuclear Industries VI, Ceramic Transactions, Vol. 119.* Edited by J. P. Bennett and J. D. Smith, The American Ceramic Society, Westerville, OH, 2001.

[22] M. Velez, J.M. Almanza, R.E. Moore, M. Karakus, W.L. Headrick, "Degradation of Crown Silica Bricks in Glass Tank Melters," pp. 1277-1286, in *Proc. UNITCER'01* (Unified International Technical Committee on Refractories), Vol. 3, Cancun, Mexico, Nov. 4-8, 2001.

[23] M. Velez, M. Karakus, W.L. Headrick, "Oxyfuel Firing Effects on Refractories," in Fundamentals of Refractory Technology, pp. 223-233, in *Ceramic Transactions, Vol. 125.* Edited by J. P. Bennett and J. D. Smith, The American Ceramic Society, Westerville, Ohio, 2001.

[24] M. Velez, J.M. Almanza, M. Karakus, "Microstructure and Corrosion Performance of Crown Silica Bricks under Oxyfuel Combustion," *Am. Ceram. Soc. Bull.,* **80**[9], 33-39 (2001).

[25] R.E. Moore, M. Velez, M. Karakus, J.M. Almanza, P. Sun, W.L. Headrick, "Silica Corrosion Studies Using the UMR Oxy-fuel Simulator Furnace," *Ceram. Eng. Sci. Proc.,* **22**[1] 79-90 (2001).

[26] J.M. Almanza, The Use of β"-Alumina Electrolyte as a Sodium Vapor Sensor for the Determination of Sodium Hydroxide in a Glass Tank. PhD Thesis, University of Missouri-Rolla, Ceramic Engineering Dept., 2003.

[27] B. Varghese, R. Zoughi, M. Velez, W.L. Headrick, R.E. Moore, "Refractory Wall Thickness Measurement Using Time Domain Reflectometry at Microwave Frequencies," pp. 223-234 in *Proc. 39th Annual Symposium on Refractories,* The American Ceramic Society, St. Louis Section, April 9-10, 2003.

[28] H.T. Goddard, L.H. Kotacska, J.F. Wosinski, S.M. Winder, A. Gupta, K.R. Selkregg, S. Gould, "Refractory Corrosion under Air-Fuel and Oxy-Fuel Environments," *Ceram. Eng. Sci. Proc.,* **18**[1] 180-207 (1997).

KINETICS AND MECHANISMS OF NIOBIUM CORROSION IN MOLTEN GLASSES

Renaud Podor, Nicolas David, Christophe Rapin and Patrice Berthod
Laboratoire de Chimie du Solide Minéral, UMR7555
Université Henri Poincaré - Nancy I
54 506 Vandoeuvre-Les-Nancy, France

ABSTRACT

The kinetics and mechanisms of corrosion of pure niobium in three molten borosilicate glasses (Zn or Fe-enriched) were investigated in the 1050 – 1400°C temperature range. The corrosion layers formed at the metal/glass interface, the elemental concentration profiles in the glass, and the corrosion rates were determined. Corrosion rates are reported on an Arrhenius plot. They lead to the determination of the activation energies of corrosion processes that depend on the glass composition. The rate-limiting process is probably the diffusion of oxygen through one of the corrosion layers. The corrosion mechanisms are described in the form of successive redox reactions between the glass and the substrate. They are characteristic of a metal-glass couple. From these results, the Nb^{II}/Nb^0 standard potential is estimated to be below $-1.4V/YSZ$, while the Nb^{IV}/Nb^{II} and Nb^V/Nb^{IV} standard potentials are estimated to be in the $-0.96V$ to $-0.76V$ range. The formation of an intermetallic compound containing boron (NbB) as a corrosion layer is seen for the first time.

INTRODUCTION

The development of new materials that are resistant to the corrosion by molten glass is an essential challenge for the glass-making industry. In particular, the use of adapted metallic materials, whose manufacturing processes are easy to condition, should allow the development of metal parts that are not easily realizable with ceramic materials. Accordingly, Nb-based alloys could be good candidates. The objective of this study is to determine corrosion rates and associated mechanisms of corrosion of metallic niobium in three borosilicate glasses.

MATERIALS AND METHODS

Metal plates - Nb, 99.8% purity (Strem Chemicals) – initial thickness = 1050 ($\pm$ 5) μm. Glasses - three different borosilicate glasses were provided by industrial partners in the form of 5-mm-diameter pellets. Their compositions are reported in Table I.

Table I. Glass compositions expressed in weight percent

	SiO$_2$	B$_2$O$_3$	Al$_2$O$_3$	Li$_2$O	Na$_2$O	K$_2$O	CaO	ZnO	Fe$_2$O$_3$	ZrO$_2$
G-Zn	58.8	18.2	4.3	2.6	7.0	-	5.2	3.2	-	0.7
G-Fe	46.5	18.5	10.0	-	20.0	-	-	-	5.0	-
G-S	64.5	4.5	3.40	-	16.0	1.20	7.2	-	0.15	-

Immersion tests – Niobium is an easily oxidizable metal at temperatures exceeding 500°C in air, and the oxide scale formed is not protective. In order to limit the metal oxidation in air before immersing it into the molten glass, the metal coupon was quickly introduced in the molten glass directly from room temperature to the working temperature. This immersion test procedure was established in order to observe corrosion of the metallic pieces and to preserve the corrosion-layer sequence at the metal/glass interface.

Sample observations - After immersion, the sample was quenched in air, embedded in epoxy resin, and polished in order to obtain metallographic cross-sections. Thickness measurement of the layers, determination of phase compositions, and systematic element concentration profiles in glass are obtained from the sections' observation by optical microscopy, scanning electron microscopy (SEM), and electron-probe micro-analyses (EPMA).

RESULTS

The corrosion of Nb in the three glasses at T=1200°C, 1300°C, and 1400°C was systematically studied for different immersion durations. The results obtained are presented for each metal-glass couple. Only the most representative SEM photomicrographs are reported. The compositions of the main phases formed at the metal-glass interface are reported directly on the figures.

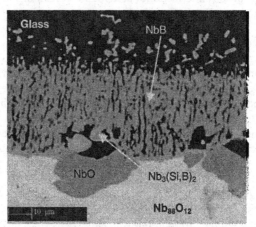

Figure 1. SEM photomicrography of a niobium plate immersed in the G-S glass at T=1400°C for 8 hours (Back-scattered electron mode)

Metal/glass interface morphologies

G-S glass: The cross-section morphologies corresponding to the niobium samples immersed in the G-S glass exhibit a complex metal/glass interface characterized by the phase succession NbB, $Nb_3(Si,B)_2$ and NbO from the glass to the metal (Figure 1). The metallic substrate was also oxidized into NbO_x, with x varying with the immersion temperature according to the Nb-O phase diagram. The presence of boron in the NbB compound was confirmed by acquiring wavelength dispersion spectra from an SX100 electron microprobe (Figure 2).

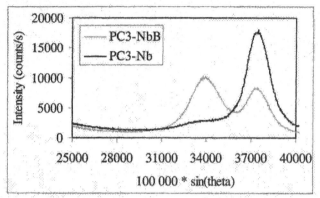

Figure 2. Wavelength dispersion spectra of pure Nb and NbB recorded with a PC2 spectrometer (SX100 electron microprobe) at the characteristic position of B.

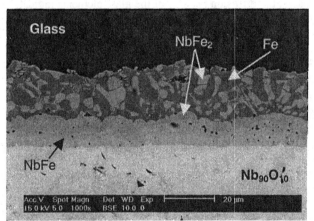

Figure 3. SEM photomicrograph of a niobium plate immersed in the G-Fe glass at T=1400°C for 0.5 hour (Back-scattered electron mode)

G-Fe glass: The metallic substrate was also oxidized into NbO_x. The corrosion products form compact-like layers at the metal – glass interface (Figure 3). From the metallic substrate to the glass, the layer succession consists of a two-phase domain constituted by an assemblage of NbO_x and NbFe, compact layers of NbFe and $NbFe_2$, and a $NbFe_2$ - Fe two-phase domain. A regular decrease of the substrate thickness with increasing immersion time and temperature was measured on these samples. It is correlated with the corrosion layer's thickness increase. Furthermore, for the longer immersion times, NbO precipitates in the NbO_x substrate.

G-Zn glass: The corrosion layers formed were NbO_x, NbO_2 and Nb_2O_5, from the metallic substrate to the glass (Figure 4). Large bubbles were also formed at the metal/glass interface.

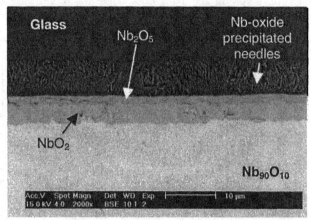

Figure 4. SEM photomicrograph of a niobium plate immersed in the G-Zn glass at T=1400°C for 1 hour (Back-scattered electron mode)

In all cases, the diffusion profiles of the glass elements indicate an enrichment of the glass with Nb close to the metal – glass interface. It corresponds to the dissolution of a part of the corrosion products into the melt. The glass surrounding the metal – glass interface was impoverished in iron or zinc species in the G-Zn and G-Fe glasses respectively (Figure 5). The main constitutive element concentrations in the glass (Al_2O_3, SiO_2, CaO, Na_2O) do not vary more than is caused by the Nb, Fe, and Zn variations.

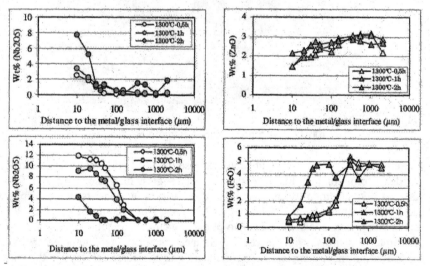

Figure 5. Diffusion profiles of Nb-Zn and Nb-Fe, in the G-Zn and G-Fe glasses respectively as a function of immersion time at T=1300°C.

DISCUSSION

Corrosion kinetics: Assessments of corrosion kinetics of niobium in molten glass were based on niobium substrate thickness measurements. The corrosion rates are calculated assuming that the $Nb_{0.9}O_{0.1}$ density is equivalent to that of pure Nb:

$V = \Delta e/2 * \Delta t$, where Δt is the run duration, and Δe is the thickness variation of the substrate during Δt. The corrosion rates are time independent. They were determined for each glass and temperature and reported on an Arrhenius diagram (Figure 6). The Ln(V) variations *versus* 1/T are linear, indicating that the corrosion mechanisms of niobium in glass are thermally activated in the explored temperature domain. The activation energies derived from these variations depend on the glass composition (Table II).

Table II. Corrosion rates and activation energies determined in three glasses.

Glass type	T=1200°C	T=1300°C	T=1400°C	E_a (kJ.mol^{-1})
G-Zn (mm.yr^{-1})	163	733	2341	280
G-Fe (mm.yr^{-1})	121	210	194	41
G-S (mm.yr^{-1})	5	17	36	202

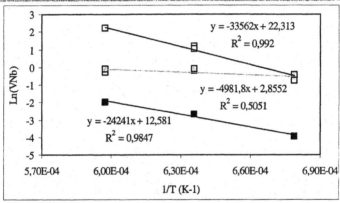

Figure 6. Arrhenius data for corrosion of niobium in G-Zn (open squares), G-Fe (filled gray squares), and G-S glasses (filled black squares).

Corrosion mechanisms: Even if each niobium/glass couple corresponds to a unique corrosion mechanism (characterized by a corrosion morphology and an activation energy), a general corrosion process can be proposed. In the very beginning of metal/glass contact, the niobium substrate oxidizes into NbO_x, inducing a fast and local redox decrease in the glass surrounding the metal plate. As a consequence, the oxidized metal species of the glass were reduced to the metallic state following the redox standard potential series of the considered metallic couples in the glass. Then, the metallic nodules formed during this process can react with the niobium coupon and form specific corrosion layers.

- In the G-Fe glass, metallic iron reacts with niobium. The succession of intermetallic phases that were formed during this process corresponds to the diffusion of Fe in the niobium substrate. It is in good agreement with the phases reported in the Nb-Fe phase diagram[1]. It is to be noted that the outermost layer (NbFe$_2$ + Fe) observed on the sample quenched after immersion in glass at T=1400°C is molten at high temperature. Nevertheless, the corrosion mechanism is not changed.

- In the G-S glass, boron (and probably a limited quantity of silicon) was reduced at the niobium-plate contact. The corrosion layers that are formed correspond to the Nb-B (Si)-O phase diagram.
- In the G-Zn glass, Zn^0 is formed. The boiling point of this element is lower than the immersion test temperatures. This causes the formation of bubbles in the glass. In this case, no metal can react directly with the niobium substrate. The niobium plate still oxidizes, and both NbO$_2$ and Nb$_2$O$_5$ are formed. Surprisingly, NbO was not observed in these samples.

The thickness and number of corrosion phases that develop in the corrosion layers depend on immersion time: the longer the time is, the higher the phase number and/or corrosion layer thickness are. Furthermore, the niobium substrate oxidizes into NbO. Continuing the immersion of the plates for longer times would have probably caused the complete oxidation of niobium into NbO$_2$ and Nb$_2$O$_5$ (as observed in the case of zirconium[2], where the initial zirconium substrate is finally oxidized into ZrO$_2$).

Limiting processes: The exact determination of the rate-limiting processes of niobium corrosion in molten glasses is difficult to attempt. However, it can be seen from the results reported herein that the metallic substrate oxidizes simultaneously with the corrosion layers' development. This indicates that the diffusion of oxygen through the metal/glass interface layers or in the molten glass is probably the phenomenon that limits the niobium corrosion.

Electrochemical implications: The relative positions of the Nb standard potentials can be derived from the corrosion-layer formation mechanisms determined previously. No data are available in the literature concerning the behavior of this metallic element in molten glass. In the G-S glass, the NbO layer forms between the NbB and NbO$_x$ layers. As a consequence, the Nb^0/Nb^{II} standard potential is lower than –1.4V. In a similar way, NbO$_2$ does not form in the presence of iron, while it forms in the G-Zn glass (as Nb$_2$O$_5$ does). The Nb^{II}/Nb^{IV} and Nb^{IV}/Nb^V standard potentials are included between the Fe^0/Fe^{II} and Zn^0/Zn^{II} standard potentials, that is to say between –0.95V and –0.76V.

The study of the niobium corrosion by molten glasses points out the necessity of considering at least three parameters to predict or understand the formation of complex corrosion layers. The knowledge of the relative positions of the metallic elements present in the glass standard potentials allows the determination of the potential of the metal piece introduced into the melt, and which species (that is to say the oxidation state of metallic elements present in the glass) are stable in the glass surrounding this piece. Then, the formation of the corrosion layers can be determined by looking at the phase diagram between the different metallic species present at the metal/glass interface. This latter step can become complicated by additional reactions with oxygen or oxides present in the glass. For example, in the case of niobium immersed in the G-Zn glass, the formation of $LiNbO_3$ or $NaNbO_3$ phases is probable (as observed with tantalum[3]), even if these phases were not finally observed.

For example, a niobium plate was immersed in a $Na_2O-3.5SiO_2$ melt for two hours at T=1400°C. The corrosion layers that can be predicted as proposed above are $Nb_{0.9}O_{0.1}$ + NbO, Nb_5Si_3 and probably $NbSi_2$. The results obtained are summarized on Figure 7 (metal/glass interface) and are consistent with the Nb-Si phase diagram[1]. They are in perfect agreement with the expected corrosion layers.

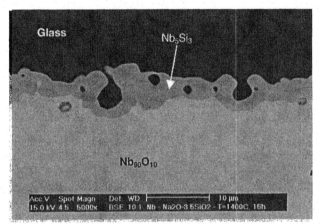

Figure 7. SEM photomicrograph of a niobium plate immersed in a $Na_2O-3.5SiO_2$ glass at T=1400°C for 2 hours (Back-scattered electron mode)

CONCLUSION

The niobium corrosion rates in molten glasses were high in glasses containing additional metallic elements susceptible to reaction with the niobium piece, while they remain relatively low in simple glasses (not containing multi-valent species). This could have important consequences on the high-temperature behavior of the new generation of niobium- and molybdenum-base composites (such as Mo-Si-B[4], Nb-Cr-Si alloys[5] and coatings[6]) designed for aeronautic applications, for which the oxidation layer compositions are close to that of a simple glass melt. In this case, the introduction of a foreign metallic element in the oxidation layer could drastically increase the oxidation rate of the inner piece.

REFERENCES

[1]T. B. Massalski, "Binary alloy Phase Diagrams," American Society for Metals, Metals Park, OH (1990)

[2]R. Podor, C. Rapin, N. David and S. Mathieu, "Kinetics and mechanisms of Tantalum corrosion in glass melts," submitted to *Glass Science and Technology*

[3]R. Podor, N. David and C. Rapin, "Experimental study of zirconium corrosion by molten glass. Part 1: Iron-bearing borosilicate glass," submitted to *Glass Science and Technology*

[4]J. S.Park, R. Sakidja and J. H. Perepezko, "Coatings designs for oxidation of Mo-Si-B alloys," *Scripta Materialia*, 46, (2002), 765-770

[5]J. C. Zhao, M. R. Jackson and B. P. Bewlay, "Oxidation resistant coatings for niobium-based silicide composites," European patent application EP 1 229 146 A2, 15pp., 07/08/2002.

[6]M. Vilasi, H. Brequel, R. Podor and J. Steinmetz, "Silicide coatings for niobium alloys," *Elevated Temperature Coatings: Science and technology II*, Edited by N. B. Dahotre and J. M. Hampikian. The Minerals, Metals and Materials Society, 1996

GLASS TANK REINFORCEMENTS

Wolfgang Simader and Hermann Walser
PLANSEE AG
A-6600 Reutte/Tyrol
Austria

ABSTRACT

The lifetime of different glass tanks can vary greatly between a few months and several years. It depends on many factors such as the glass composition and temperature, but also daily production quantities.

The duration of a tank campaign is determined by the rate of wear within the tank and the subsequent failure of important functions. The following areas of the glass tank are critical:
- dog house
- throat channel
- bubble maker
- wall

Cladding with molybdenum sheet protects these critical areas of the glass tank against wear, maintaining their form and reliability for longer periods. This of course helps to optimize the manufacturing process and glass quality and significantly improves the service life of the glass tank.

INTRODUCTION

The production of glass began hundreds of years ago. Despite this long history, the manufacturing processes for glass are still being constantly evaluated and optimised. Various topics dominate current discussions. Key words such as higher strength, low weight, less impurities and greater cost effectiveness are frequently heard at glass conferences and in discussions between glass producers. Competition from alternative materials, such as plastics, for bottles is increasing rapidly.

Every glass producer is forced to improve product quality, but at the same time to reduce the costs for consumers. Possible ways to increase cost effectiveness, for example, are to reduce the production of reject parts and prolong the campaign length of a glass tank.

Great effort has already been made to increase the lifetime of a glass tank. The quality and corrosion resistance of the refractory bricks have been improved year by year. Sintered refractory bricks have been replaced by cast refractories to protect the most heavily worn sections of a furnace in particular. But even the cast quality exhibits corrosion rates that cannot be ignored. The only way to drastically improve the corrosion resistance of glass tank parts that are exposed to heavy wear is to protect them with metal. Only a few metals can withstand the high temperatures required for the production of glass. Figure 1 shows a comparison of the corrosion resistance of different metals and AZS material to the most commonly used glass melts. This diagram shows how limited the possibilities are.

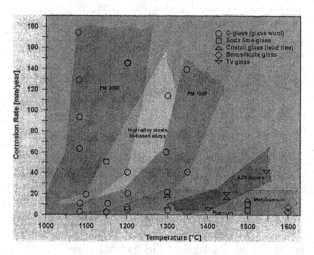

Figure 1. Corrosion resistance of different metals and AZS material

Iron & iron-based alloys and nickel & nickel-based alloys are not able to fulfill the high requirements of glass production. The melting point of these metals or alloys is much too low and even when used below this temperature, they show high corrosion rates and pollute the glass heavily when immersed in the melt. Two metals have already been proven to exhibit good corrosion resistance without pollution of the glass melt - platinum and molybdenum. The corrosion resistance of platinum is unrivalled, but closely followed by molybdenum. Two main differences exist between these metals:

>   the oxidation resistance
>   and the price

Platinum is the only metal currently used which can withstand corrosion and oxidation. The oxidation resistance (Figure 2) of molybdenum is poor and, therefore, it requires special protection from oxidizing atmospheres until it is immersed completely in the glass melt. Nowadays this can be achieved with a coating, called SIBOR®, which will be described in the following paragraph. The big difference between molybdenum and platinum is the price. Platinum is a precious metal and must be priced at a market rate, which is determined day-to-day by the stock market. 10 to 15 grams of platinum cost approximately the same as 1 kg of molybdenum sheet already coated with SIBOR®. The quantity of platinum required is much too high for most uses and it is therefore limited to very special applications such as platinum feeders for special glasses with very high purity levels.

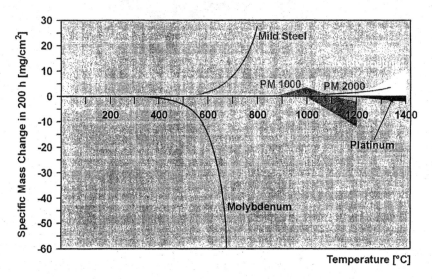

Figure 2. Oxidation resistance of several materials

SIBOR® COATING

As mentioned above and shown in Figure 2, molybdenum has a poor oxidation resistance at temperatures above 600 °C (1112 °F). In principle, various solutions exist to protect refractory metals like molybdenum against oxidation: alloying, packing with ceramic powders, glassification, cooling, application of protective gases (e.g. hydrogen, argon, ...), cladding with platinum or - last but not least – coating with an impervious layer.

The only coating which can guarantee an "oxidation free" period is the so-called SIBOR® coating. This patented coating consists of silicon and 10 % boron by weight and it is applied to sand blasted molybdenum surfaces using a plasma spray process. The coated molybdenum parts are then annealed to ensure outstanding oxidation resistance. Figure 3 shows three cross sections of a molybdenum sample with SIBOR® coating after each production step.

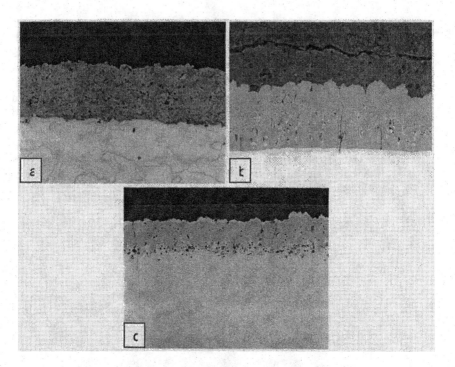

Figure 3. Cross sections of molybdenum with SIBOR® coating
3a ... SIBOR® coating as sprayed
3b ... SIBOR® coating after annealing,
3c ... SIBOR® coating after 400 hours in use in air at a temperature of
1450 °C (2642 °F)

The SIBOR® coated molybdenum parts (glass melting electrodes or glass tank reinforcements) can be installed in a cold glass tank before the up-tempering process starts. The parts will remain in tact without any oxidation throughout heat-up, even with a slow up-tempering rate of 5 to 10 °C per hour. It is guaranteed to last as follows: 5000 hours at 1200 °C (2192 °F), 500 hours at 1450 °C (2641 °F) and 50 hours at 1600 °C (2912 °F). This time/temperature performance enables glass producers to install the molybdenum parts easily and safely in a cold tank. Further advantages of the SIBOR® coating are the properties of the layer. The SIBOR® coating is not as brittle as ceramic coatings like the SiCrFe coating and silicide coatings (e.g. MoSi) and can withstand normal handling during the installation process without chipping. The SIBOR® coating will be dissolved by the glass within a few days. At the beginning, bubbling can occur, but it will decrease rapidly after 24 hours. Due to the composition of this coating (Si, B) it will not cause any discoloration or contamination of the glass melt.

GLASS TANK REINFORCEMENTS

The lifetime of different glass tanks can vary greatly between a few months and several years. It depends on many factors such as the glass composition and temperature, but also daily production quantities. Opal glass, for example, is a very aggressive glass and a tank campaign lasts only a few months. Glass tanks for container glass (soda-lime glasses) have a service life of up to 10 years (6-8 years on average). The duration of a tank campaign is determined by the rate of wear within the tank and the subsequent failure of important functions.

The performance of the SIBOR® coating mentioned above now makes it possible to use the good corrosion resistance of molybdenum to protect the areas of a glass tank that are exposed to heavy wear in most glass melts (Figure 1). Some of these areas are critical for the lifetime of the glass tank, others are critical for the performance of the tank and the glass quality.

The critical sections are marked in Figure 4, which shows a schematic diagram of a typical glass tank:

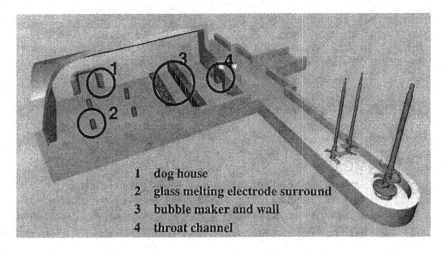

1   dog house
2   glass melting electrode surround
3   bubble maker and wall
4   throat channel

Figure 4. Schematic diagram of a glass tank showing critical areas

- The "doghouse" – where the raw materials for glass production are introduced into the glass melt – heavily stressed due to the oxygen content of the batch and the formation of foam
- Areas surrounding the glass melting electrodes, which are normally heavily stressed due the high temperature and corrosive and erosive convection streams.
- The "bubble maker" – where large, defined bubbles are added to the molten glass to agglomerate the little bubbles – heavily stressed by corrosion and erosion processes

- The "crosswall" – controls the convection streams in the melting area and the transition time of the glass melt – heavily stressed by corrosion and erosion processes
- The "throat channel" – the transition between the melting area and the working end and the feeders – heavily stressed by corrosion and erosion processes

The appearance of the crosswall and the throat channel at the end of the tank campaign as shown in Figures 5 and 6 are very familiar to all glass producers.

Figure 5 and 6. Corroded crosswall and throat channel at the end of a tank campaign

Cladding with molybdenum sheet protects these critical areas of the glass tank against wear, maintaining their form and reliability for longer periods. This of course helps to optimize the manufacturing process and glass quality and significantly improves the service life of the glass tank.

EXAMPLES OF GLASS TANK REINFORCEMENTS:
Generally speaking, glass tank reinforcements are molybdenum sheets (thickness 6-10 mm, 0.25-0.4"), which are made into various shapes and forms using bending and machining processes. 100 % of the surface is then coated with SIBOR® to achieve complete oxidation resistance of the whole assembly.

Tank components made of molybdenum can easily be fixed to the tank using different methods:
1) Clamping between the refractory bricks
2) Fixing with bolts that are inserted through the sheet into the refractory brick

Advances in Fusion and Processing of Glass III

3) Simple covering of the parts requiring protection such as the wall or the bubble maker

Figure 7 shows a doghouse reinforcement. Simply formed molybdenum sheets are fixed with pins onto the corner bricks of the doghouse. This helps to prevent corrosion of the bricks as the batch enters and often produces foam in this area.

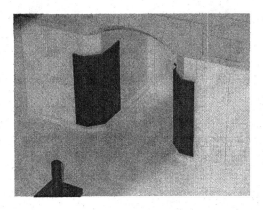

Figure 7. Doghouse reinforcement

Figure 8 shows how the area surrounding a glass melting electrode can be protected. This protection prevents corrosion and erosion, which can be caused by the convection streams produced by the temperature differences in the glass bath. Such corrosion leads to problems when the electrode holder and the glass melt come into direct contact. Corrosion of the electrode holder will result in impurities and discoloration of the glass melt and local overheating at the end of the electrode holder. This local overheating will also damage the glass melting electrode by alloying the molybdenum with iron and nickel. This alloying causes low melting eutectics to form leading to a reduction in the electrode diameter and eventual breakage of the glass melting electrode. Direct contact between the electrode, its holder and the protection plate must of course be avoided (otherwise it would function as a plate electrode).

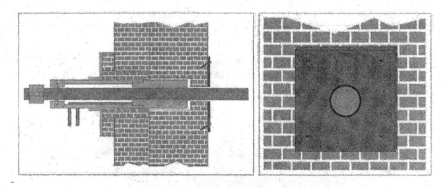

Figure 8. Plate protection for glass melting electrode surrounds

Figure 9 shows a simple bubble maker cover. The same principal can also be used to protect walls. The "U – channel" is simply placed over the refractory bricks and clamped between the bottom bricks. It ensures that the form and function of the bubble maker (or wall) are maintained for a long period.

Figure 9. Bubble maker / wall protection

The most critical area in a glass tank is of course the throat channel. This section is located between the glass tank itself and the working end and controls transition time and glass flow. The throat channel construction consists of several refractory bricks (two side bricks, one top brick and several bricks for the channel). To guarantee good corrosion protection it is also necessary to cover the joins in the brick. This is only possible if the reinforcement plates used have a greater width and height than the individual bricks and they have to be fixed to the outside of the bricks (cladding). Figure 10 shows how a throat channel can be effectively protected.

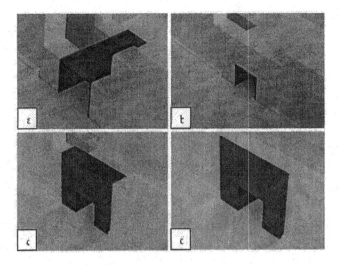

Figure 10. Schematic diagram of a throat channel reinforcement

The throat channel protection assembly consists of two parts: the front plate and a "U"-channel (Figure 11). This is necessary for ease of handling and installation. The connection between the two parts is specially designed and produced to prevent any penetration of the glass.

Figure 11. Throat channel protection - front and back views

Figure 12 shows an installed throat channel protection from the front.

Figure 12. Installed throat channel protection

Similarly other components such as flow pipes inserted in a crosswall (used to generate additional convection currents) and delivery pipes can be produced in molybdenum and protected against oxidation. In such cases, SIBOR® coating is required on all surfaces.

The design of each glass tank reinforcement is adapted to suit individual customer requirements after discussion between the glass producer and/or the furnace constructor and PLANSEE.

Such parts can be installed easily by either the glass producer, the furnace constructor or by PLANSEE.

GLASS COMPOSITION DEPENDENCE OF METAL CORROSION BY MOLTEN GLASSES

Renaud Podor, Nicolas David, Christophe Rapin,
Patrice Berthod and Pierre Steinmetz
Laboratoire de Chimie du Solide Minéral, UMR7555
Université Henri Poincaré - Nancy I
54 506 Vandoeuvre-Les-Nancy, France

ABSTRACT

Ta, Nb, Ti, and Zr corrosion for six different molten glasses have been experimentally studied at 1400°C. Both corrosion-layer compositions and corrosion rates have been determined. Mechanisms of corrosion-layer formation, mainly based upon both redox reactions between metal and glass and solid-state diffusion, are proposed. The consequences of metallic element additions to the glass on metal corrosion are discussed in comparison with a metal-free reference glass. The corrosion rates of niobium are specifically determined in a Na_2O-$3.5SiO_2$ glass modified by additions of metallic elements. A correlation is established between the corrosion rate and the formal redox potential of the M^{II}/M^0 couple.

INTRODUCTION

The knowledge of the corrosion behavior of metals in molten glass is an important challenge for the glass industry. If the corrosion reactions can be predicted from both immersed material and glass composition, this would be a great advance in this field of interest. A series of four metals have been immersed in six different glasses. The results reported herein point out the general mechanism of metal corrosion by a molten glass and the effect of the addition of an oxidant to the glass.

MATERIALS AND METHODS

Metal plates – pure metals were respectively provided by Strem Chemicals (Nb, 99.8% purity and Ta 99.95%), Chempur (Ti, 99.6%), and Interchim (Zr, 99.8%). Glasses - Complex borosilicate glasses were both provided by industrial partners in the form of 5-mm-diameter pellets. The Na_2O-$xSiO_2$ binary glasses were prepared in the laboratory from oxides and carbonates. Their compositions are reported in Table I. Five weight percent of metallic oxides were added to the G-A glass to pursue specific studies.

Table I. Glass compositions expressed in weight percent (* industrial glasses)

	SiO_2	B_2O_3	Al_2O_3	Li_2O	Na_2O	K_2O	CaO	ZnO	Fe_2O_3	ZrO_2
G-Zn*	58.8	18.2	4.3	2.6	7.0	-	5.2	3.2	-	0.7
G-Fe*	46.5	18.5	10.0	-	20.0	-	-	-	5.0	-
G-S*	64.5	4.5	3.40	-	16.0	1.20	7.2	-	0.15	-
G-A	70.5				29.5					
G-B	66				34					
G-C	59.2				40.8					

Immersion tests – The metallic plates were directly immersed in the molten glass from room temperature to the working temperature in order to limit their oxidation in air.

Sample observations - After immersion, the samples are quenched in air, embedded in epoxy resin, and polished in order to obtain metallographic cross-sections. Thickness measurement of the layer, determination of phase compositions, and systematic element concentration profiles in the glass are obtained from the sections' observation by optical microscopy, scanning electron microscopy (SEM), and electron probe micro-analyses (EPMA). This procedure was defined in order to exhibit glass corrosion processes of the metallic pieces and to preserve the corrosion-layer sequence at the metal/glass interface.

RESULTS

The corrosion of the four metals was systematically studied in the six glasses at 1400°C, for different immersion times. The metal-glass interface morphologies are reported for each metal-glass couple. Only the most representative SEM photomicrographs are reported hereafter. The compositions of the main phases formed at the metal-glass interface are reported directly on the figures. For each case, a plot of the corrosion rates determined at 1400°C is presented in the form of a histogram.

Metal/glass interface morphologies

In a general way, the metal-glass morphologies are specific to one metal-glass couple. However, some general remarks can be pointed out. The corrosion reaction of the M metal in a molten glass is mainly controlled by redox reactions occurring between the metal and one element present in the glass. These reactions yield to the oxidation of the M metal and to the reduction of species present in the glass. The redox reaction occurs according to the scale of the redox couple potentials (Figure 1). The M metals chosen in the present study are characterised by redox couple potentials lower than –1.2V/YSZE, that is to say lower than the Si^{IV}/Si^0 and/or B^{III}/B^0 couple potentials. These metals can react with several redox couples present in the glass, depending on the glass composition. Nevertheless, according to the corrosion layers formed at the metal-glass interfaces during the immersion of the metal in the glass, it appears that the redox couple that first reacts with the M metal is the one whose redox potential is the higher in the redox potential scale. In this case, the O_2/O^{2-} redox couple must be "activated" prior to any other redox couple. It is probably the case, but the concentration of O_2

dissolved in the glass is low, and the oxygen diffusion is sufficiently low, to explain the fast consuming of the total O_2 present in the glass near the metal plate. This yields a local decrease of the glass redox near the metal plate. A glass potential measure was realized thanks to electrochemical methods before and after immersion of a zirconium plate in molten glass. The glass potential was decreased from +0.05V (glass equilibrated with air) to –0.5V (after the metal immersion).

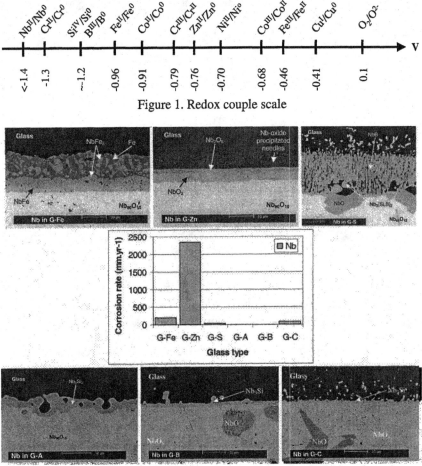

Figure 1. Redox couple scale

Figure 2. Metal-glass morphologies and corrosion rates of niobium in molten glass determined at T=1400°C (Back-scattered electron mode)

Then, when the glass potential is sufficiently low, a change in the oxidant species is observed. The second redox couple involved in the corrosion of the M metal is the one that is directly inferior to the O_2/O^{2-} redox couple, according to the scale of the redox couple potentials. This depends on the glass composition. In the G-Fe case, metallic iron nodules are formed in the glass and/or directly at the surface of the M metal, while the metallic substrate still oxidizes. Then, this element combines with the M metal, according to the M-Fe phase diagram[1]. In the case of the G-Zn glass, metallic Zn forms when reducing ZnO is present in the glass. However, the low boiling point of this element ($T_{boiling} = 904°C$) induces an intense bubbling near the metal plate that corresponds to the volatilization of Zn. No Zn is observed in the corrosion layers. The observed bubbling is characteristic of the gaseous Zn and of the local redox decrease near the metal coupon.

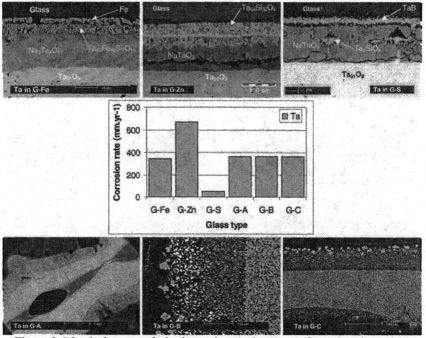

Figure 3. Metal-glass morphologies and corrosion rates of tantalum in molten glass determined at T=1400°C (Back-scattered electron mode)

When the concentration of the oxidant present in the glass (Fe^{II}, Fe^{III} in G-Fe) near the metallic plate is too low, one can observe another change in the oxidant involved in the metal corrosion. As an example, the case of Zr immersed in G-Fe is characteristic: the iron content in glass becomes equal to zero, then zirconium borides and silicides are formed.

In the case of the G-S glass, both Si^{IV}/Si^{0} and/or B^{III}/B^{0} redox couples can react with the M metal (because the potential values of both couples are very close). The metallic species formed during the glass elemental reduction react with the M substrate to form corrosion layers composed of intermetallic compounds.

Advances in Fusion and Processing of Glass III

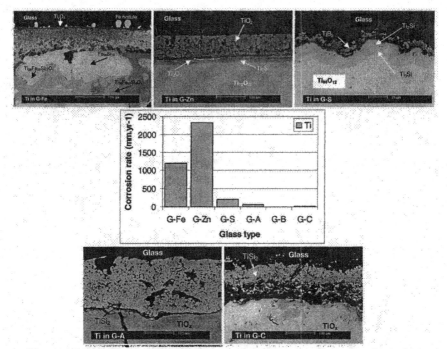

Figure 4. Metal-glass morphologies and corrosion rates of titanium in molten glass determined at T=1400°C (Back-scattered electron mode)

At the same time, the oxidation of the M metal causes the formation of the different oxides that are stable in the redox conditions imposed by the glass. When the substrate is completely oxidized into MO_x, then the local glass redox re-equilibrates with the surrounding atmosphere, and the oxidation of the metal is pursued up to its complete transformation into the oxide corresponding to the higher oxidation degree of the metal (Ta^V, Nb^V, Zr^{IV} or Ti^{IV}). The formation of the oxides is in competition with their dissolution in the glass. Both phenomena are simultaneous and generally cause the complete degradation of the metal plate. In some specific cases, when the solubility of the oxide in glass is very low, the oxide layer formed can protect the metal from further corrosion. This is the case of some superalloys (Fe-base, Ni-base, and Co-base) and pure chromium, which behavior in molten glasses has been described elsewhere[2]. The oxides formed during the oxidation of the M metal can also react with the glass and cause the formation of other complex phases such as $NaTaO_3$ or $LiNbO_3$.

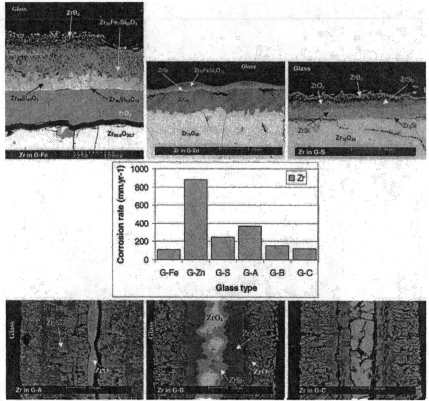

Figure 5. Metal-glass morphologies and corrosion rates of zirconium in molten glass determined at T=1400°C (Back-scattered electron mode)

Another series of experiments corresponding to the immersion of Nb plates in G-A glass (Na$_2$O - 3.5SiO$_2$) modified by the addition of 5 wt% of metal oxide was conducted at 1400°C for 12 hours. Eight glasses were tested: no addition, additions of Cr$_2$O$_3$, Fe$_2$O$_3$, CoO, NiO, CuO, ZnO, and a mixture of Fe$_2$O$_3$ and NiO. The metal-glass interface morphologies are reported in Figure 6. The corrosion layers formed are in complete agreement with the explanations given before and with the redox potential scale. The impossibility of depositing chromium at the niobium surface is correlated to the too-low redox potential of the Cr^{II}/Cr^0 couple, in comparison with both Si^{IV}/Si^0 and B^{III}/B^0 redox potential values. These experiments show that several elements can be deposited at the same time at the surface of a metallic substrate.

The corrosion of metals by molten glasses is a complex phenomenon that is based upon redox equilibrium between the metal and the elements present in the glass. The corrosion phases that are formed can be predicted from the initial glass composition.

Advances in Fusion and Processing of Glass III

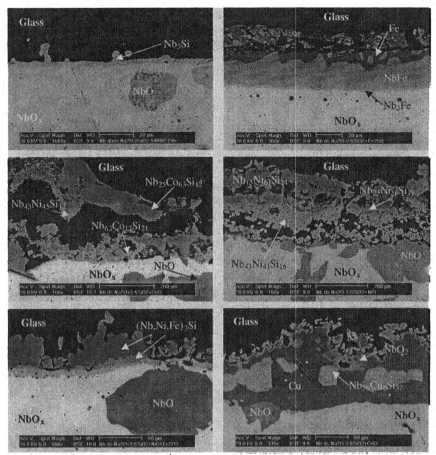

Figure 6. Metal-glass morphologies of niobium in glass (Na$_2$O - 3.5SiO$_2$) modified by the addition of 5 wt% of metal oxide, determined at T=1400°C (Back-scattered electron mode)

Corrosion rates

The corrosion rates determined for each metal-glass couple cannot be easily correlated to simple parameters. It seems that the corrosion rates determined in the G-Zn glass are higher than those determined in the G-Fe glass, the ones that are higher than those determined in the G-S glass, for all the metallic elements tested in this study. However, the differences between the phases present in the corrosion layers, and the formation of an important bubbling in the G-Zn glass does not allow the proposal of further interpretations.

The corrosion rates determined in the case of niobium plates' immersion in Na$_2$O –3.5SiO$_2$ modified by metal oxide additions are correlated with the redox potentials of the M^{II}/M^0 couples (Figure 7). The higher the difference between the Nb^{II}/Nb^0 potential and the M^{II}/M^0 potential is, the higher the corrosion rate is. In the particular cases of zinc- and copper-enriched glasses, the relatively high or low corrosion rates associated with these elements are probably due to the fact that the corrosion mechanisms are different from the other cases. (The formation of bubbles in the ZnO-enriched glass induces a continuous renewal of the glass at the metal-glass interface, while only one electron is exchanged in the case of the copper-enriched glass.)

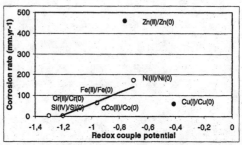

Figure 7. Niobium corrosion-rate variations in $Na_2O - 3.5SiO_2$ modified by the addition of metal oxide, reported as a function of the metal-couple potential.

The first results trying to correlate the corrosion rates with the glass basicity are not self-consistent. Indeed, they are perturbed by the alumina dissolution in the $Na_2O-xSiO_2$ binary glasses used for these experiments.

REFERENCES

[1]R. Podor, D. David, C. Rapin, P. Berthod and P. Steinmetz, "Experimental study of zirconium corrosion by molten glass," GRC on High Temperature Corrosion. New London, NH, USA, July 20-25, 2003.

[2]J. Di Martino, C. Rapin, P. Berthod, R. Podor and P. Steinmetz, "Corrosion of metals and alloys in molten glasses. Part 2: Nickel and cobalt high chromium superalloys behaviour and protection," *Corrosion Science*, to be published.

[3]R. Podor, C. Rapin, N. David and S. Mathieu, "Kinetics and mechanisms of Tantalum corrosion in glass melts," submitted to *Glass Science and Technology*

CORROSION OF SUPERALLOYS IN MOLTEN GLASS - ELECTROCHEMICAL CHARACTERIZATION OF THE PASSIVE STATE

Jean Di Martino, Christophe Rapin, Patrice Berthod
Renaud Podor, Pierre Steinmetz

Laboratoire de Chimie du Solide Minéral (UMR 7555)
Université Henri Poincaré Nancy 1
BP 239, F54506 Vandoeuvre-les-Nancy – France

ABSTRACT

Superalloys were studied in a molten glass at 1050°C. The corrosion rates were estimated by classical electrochemical techniques as polarization resistances determinations. Linear polarizations were used to characterize the passive behavior of alloys. All tested alloys exhibit a low polarization resistance value of about $20\Omega.cm^2$ corresponding to a corrosion rate of a few cm/yr. Moreover, all alloys clearly evidence a polarization curve of passivable alloys. The active state is characterized by a rapid dissolution of the constitutive elements of the alloy in the glass melt. The passive state is characterized by a lower corrosion rate (Rp = 200 - 800 $\Omega.cm^2$) and can be obtained by a prior air oxidation or with a temporary anodic polarization. The obtained passive state is due to the presence of a thin protective chromia (Cr_2O_3) scale. The experimental apparatus allows electrochemical measurements on rotating electrodes. This is useful to approach the consequences of relative movement of glass versus alloy. Several speeds were tested from 1 to 5 m/min. In the case of an erosion phenomenon, a favorable galvanic coupling allows a restoration of the protective layer. This microgalvanic coupling explains the healing and stability of the passive layer in the glass.

INTRODUCTION

In the glass industry, the lifetime of many metallic parts in contact with molten glass is mainly controlled by high-temperature corrosion. The degradation of the alloys immersed in the molten glass is governed by oxidation-reduction reactions. Then electrochemical techniques were developed for the characterization of alloy corrosion [1, 2]. The aim of this paper is to describe the corrosion behavior of superalloys containing about 30% of chromium in a soda-lime molten glass at 1050°C and to determine the mechanism of passivation.

MATERIALS AND METHODS

This study was performed in an industrial borosilicate glass ($64.5\%SiO_2$ - $0.15\%Fe_2O_3$ -$3.40\%Al_2O_3$ -$0.1\%SO_3$ -$7.2\%CaO$ -$16.0\%Na_2O$ –$3\%MgO$ - $1.20\%K_2O$ -$4.5\%B_2O_3$ in weight %) at T = 1050°C. The studied alloys are high-chromium Ni-base (base Ni, 28wt% Cr, 0.7wt%C, 7.5wt%W, 7.5wt%Fe) and Co-base commercial superalloys, containing Fe, Ta, W and C as additional elements (base Co, 30wt%Cr, 0.4wt%C, 3wt%Ta, 6wt%W, 8.5wt%Ni). These alloys are classical high-temperature superalloys, mechanically strengthened by carbide precipitation (MC or $M_{23}C_6$ type). The behavior of a few selected pure metals (Co, Ni, Cr and Fe) is also characterized in molten glass. These metals were chosen because of their importance either in glass compositions (Fe) or in high-chromium nickel-base, cobalt-base superalloys compositions.

The apparatus used [3, 4] is in a furnace in which a Pt-10%Rh crucible is placed in air atmosphere. The electrodes' tips are immersed in 1.5 kg of glass melt. The working temperature is 1050°C. The working electrode for glass study is a Pt wire (Φ = 1 mm), whereas the working electrode for metal study is a 5.5-mm-diameter alloy rod polished with 1200 grit SiC paper. The counter electrode is a platinum plate (25*2*10 mm). A reference electrode, a stabilized zirconia electrode, was used. It was constructed using a stabilized zirconia rod (⊔⊔⊔mm) cemented with zirconia inside a mullite tube. The electrode was flushed with air as the reference gas introduced with a syringe needle. All potentials given in this paper are referred to the potential of the stabilized zirconia electrode flushed by air. These high-temperature adapted electrodes are extensively detailed in reference [5-7].

Parc M263A potentiostats were used to perform the electrochemical measurements. Free potential recording, polarization resistance measurement (Stern-Geary's method), linear polarization, cyclic and square wave voltammetry methods were used. Long-lasting free-potential recordings are always coupled with a polarization resistance determination every two or four hours. Polarization resistance (R_p) measurements are performed by polarizing the working electrode from E_{corr} -15mV to E_{corr} + 15mV, with a 600 mV/h scan rate.

Cross sections are made on resin-coated samples (epoxy type). Sample preparation begins with SiC polishing (from grit 400 to 2400) before diamond finishing ($3\mu m$ then $1\mu m$) and ultrasonic-alcohol final wash. Optical observations were performed on an Olympus Vanox AHMT, and SEM observations on a Philips XL30.

RESULTS

The solvent's electro -activity domain extends approximately from -1.2 to +0.1 V. It is limited in the negative potentials by the reduction of the silicate network in accordance with $SiO_4^{4-} + 4 e^- \rightarrow Si + 4 O^{2-}$ [2]. The oxidation anodic limit of the solvent corresponds to oxygen gas formation, due to the oxidation of the O^{2-} ions ($O^{2-} \rightarrow \frac{1}{2} O_{2 (g)} + 2 e^-$), then of the silicate network ($SiO_4^{4-} \rightarrow SiO_3^{2-} + \frac{1}{2} O_{2 (g)} + 2$

e⁻). Even if numerous species are present in this glass, only Si^{IV}/Si^0, Fe^{III}/Fe^{II}, Fe^{II}/Fe^0 and O_2/O^{2-} couples are electro active.

Corrosion of pure metals in C-glass

The corrosion potentials of Cr, Ni, Fe and Co in the C-glass were measured (Table I). The comparison of these values with redox-potential [2, 4, 8-10] values allows the assignment of anodic and cathodic reactions (Table I). Polarization resistances (noted Rp) recorded each 4 hours during 50 hours are stable (Fig. 1). The values are in the 9-35 $\Omega.cm^2$ range. They are low and characteristic of both material in the active state and high corrosion rates [11].

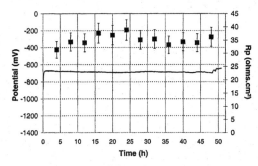

Figure 1. E_{corr} and R_p (■) of nickel in glass at 1050°C

Table I : Electrochemical parameters of metals and alloys in molten glass

Alloys /metals	Active state				Passive state			
	Ecorr (V)	Rp ($\Omega.cm^2$)	Oxidant	Corrosion rate (cm.yr⁻¹)	Ecorr (V)	Rp ($\Omega.cm^2$)	Oxidant	Corrosion rate (cm.yr⁻¹)
Fe	-1.10	14	Si^{IV}	3.2	-	-	-	-
Co	-0.85	15	Fe^{III}, O_2	3.0	-	-	-	-
Ni	-0.65	35	Fe^{III}, O_2	1.3	-	-	-	-
Cr	-1.30	9		5.0	-0.1	80		0.4
Co base 30Cr-6W	-1.10	25		1.8	-0.55	200	$Fe^{III},$	0.2
Co base 30Cr	-	-	Si^{IV}	-	-0.60	400		0.1
Co base 30Cr-2Al	-1.20	12		3.9	-0.25	250	O_2	0.2
Ni base 25 Cr	-1.20	20		2.3	-0.65	800		0.05

Cr, Ni and Co polarization curves exhibit a passivation plateau (Figure 2). Only the chromium passivation current is low enough ($\approx$ 0.2 mA.cm⁻²) to offer a possible protection of an alloy by the development of a protective layer.

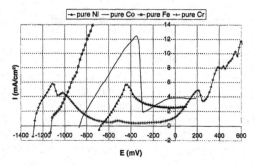

Figure 2. Polarization curves of pure metals, C glass

Corrosion of alloys in the C-glass

In a preliminary study, $Ni_{1-x}-Cr_x$ alloys have been synthesized, and their corrosion currents (I_{corr}) were determined in the soda-lime glass at T=1100°C [2]. The minimum I_{corr} is observed for a 30 wt% Cr content in the alloy. All the tested alloys (containing about 30wt% Cr) exhibit a corrosion potential of –1.2V, which is characteristic of the active state. The active state is characterized by a rapid dissolution of constitutive elements of the alloy in the glass (Figure 3) and a glass penetration at the grain boundaries. The associated corrosion rates are equal to 1-5 cm.yr^{-1}(Table I). The polarization curves recorded on both 30 wt% chromium Co and Ni-base commercial alloys are similar to the one recorded on pure chromium. An example of a recording on a Ni-base alloy is reported in Figure 3. This typical curve exhibits ❶ an anodic peak representative of the alloy oxidation mainly into Cr^{II}, ❷ a passivation plateau with a corrosion current equal to ~0.2 mA.cm^{-2}, and ❸ a transpassivation domain for potentials higher than –0.1V. The passive state, which corresponds to a limited corrosion of the alloy in the glass, can be reached by different methods [4, 11], such as pre-oxidation and anodic polarization of the alloy. It is characterized by the presence of a 3-5μm protective chromia scale (Figure 3) at the metal-glass interface. This layer is responsible for the passivation of the alloy and prevents glass penetration.

The polarization curve recorded on an alloy that is in the passive state (Figure 3) exhibits no anodic peak and a high corrosion potential (E_{corr} = -800/-100 mV) located in the passivation plateau (oxidants are Fe^{III} and dissolved O_2). Rp values are then usually from 200 to 800 Ω.cm^2, corresponding to an alloy corrosion rate of nearly 2 mm.yr^{-1}, that is to say one order of magnitude lower than for the active alloy.

Advances in Fusion and Processing of Glass III

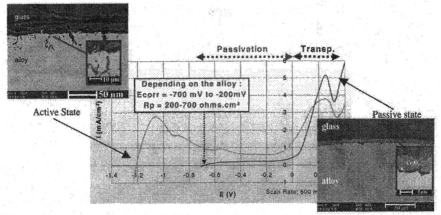

Figure 3. Polarization curve of cobalt base alloy in active and passive state; corresponding cross sections.

Stabilization of passive state

Passivity for alloys implies the formation of a protective scale of chromia (Cr_2O_3). In our case of high-chromium alloys, chromia should play this role, but as-polished alloys (with a E_{corr} value about –1.2V) release Cr^{II} ions instead of spontaneously forming chromia. Three efficient methods have been developed to reach a passive state.

(1) The addition of oxidizing species to the glass (Mn^{IV}, Fe^{III}...) leads to an increase of corrosion potential, resulting in Cr_2O_3 formation at the alloy/glass interface. If modification of glass composition is not suitable, two other ways are left.

(2) A temporary anodic protection is possible, by imposing a potential located in the passivation plateau (-0.7<E<-0.1V) for one hour. It also leads to chromia formation at the alloy/glass interface. Yet this method can be difficult to use in dynamic industrial processes. The last one is easier [11].

(3) The formation of a chromia protective layer before glass contact can be obtained by hot-air oxidation of the alloy. Classically, one hour in the furnace just above the crucible (called "preoxidation") is sufficient. The duration and the temperature of this treatment have been determined for different alloys) [11].

These methods have been applied on numerous cobalt-base and nickel-base alloys (Table 1), having in common a high chromium content (25-30%). 100 hours testing have shown that the obtained passive state is reproducible and long lasting. The thickness-loss measurements on a cobalt-base industrial component (1000 hrs lifetime, same temperature, same glass) yielded a corresponding Rp value of about 400 $\Omega.cm^2$. It is in very good accordance with the electrochemically measured values in the laboratory.

Quantification of glass erosion

The experimental apparatus allows electrochemical measures on rotating working electrodes. This is useful to approach the consequences of relative movement of glass versus alloy. Several speeds were tested from 1 to 5 m/min (60 to 300 rpm with 5.5-mm-diameter electrode) on a Co-base alloy. The electrodes were always preoxidized one hour at 1150°C. The results of corrosion potential and polarization resistance measurements are shown in Figure 4.

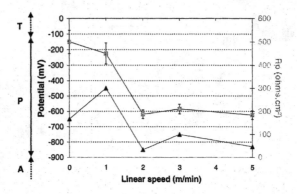

Figure 4. E_{corr} (▲) and R_p (■) of Co-base alloy after 20 h in glass, for different linear speeds; T: transpassivation P: passivation. A: activation domain

In every case, the passive state is still effective after 20 hrs in glass, with Rp values between 180 $\Omega.cm^2$ and 500 $\Omega.cm^2$. For a higher speed, the E_{corr} and Rp values become lower but still representative of a passive state ($E_{corr} \approx$ -800mV and Rp $\approx$ 200 $\Omega.cm^2$). SEM observations show on the one hand a dense and continuous chromia scale, and on the other hand a cracking and fragmentation of chromia, due to erosion phenomena. The thinning of the Cr_2O_3 layer seems responsible for the Rp decrease.

Determination of passivation mechanisms

In order to determine the stabilization mechanisms of the passive state, galvanic couplings were realized between an active electrode and a passive one. These experiments were performed on the Ni-base alloy. First, an electrode was preoxidized. After immersion, it stayed in a passive state (E_{corr} = -430 mV and Rp = 800 $\Omega.cm^2$). Thus, this electrode was coupled with a half-surface as-polished immersed electrode (active state). Figure 5 shows in (a) the coupling potential and in (b) the coupling current. The current quickly decreases from 5.2 mA/cm^2 to 0 mA/cm^2 in 4 hours. Correspondingly, the potential of the active electrode increases from the activation domain (-900 mV) to the passivation domain (- 300 mV).

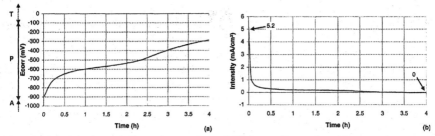

Figure 5. Galvanic coupling of a passivated Ni-base electrode with an active one. $S_c/S_a = 2$; (a) potential (b) current. T=1050°C.

In conclusion, these results allow some assumptions about the stabilization mechanisms of the passive state. If a part of the Cr_2O_3 protective layer is damaged, a galvanic coupling can lead to the restoration of the layer. In this study, a ratio of cathodic surface over anodic surface, Sc/Sa of 2 is favorable, while Sc/Sa = 1 is not. These experiments allow partial understanding of the role of the preoxidation (Figure 6):

(a) After air oxidation, the sample is covered with a Cr_2O_3 layer.

(b) Some parts of the alloy may become active because of unavoidable solubility of Cr_2O_3.

(c) With a favorable Sc/Sa ratio, the galvanic coupling leads to the restoration of the passive state.

(d) Finally, after a long immersion time, the whole surface of the alloy is covered with Cr_2O_3 generated in the glass.

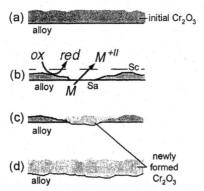

Figure 6 : Schematic representation of the stabilization of a passive scale of Cr_2O_3 after preoxidation

CONCLUSION

Classical electrochemical methods with an adapted experimental apparatus are suitable for the characterization of alloy corrosion in molten glasses. The electrochemically measured corrosion rates are in good agreement with the thickness losses of an industrial component.

All the tested high-chromium alloys are spontaneously in an active state, corresponding to a corrosion rate of a few cm/year. In the passive state, these alloys can be protected by the presence of a thin chromia layer on their surface. Chromia protection allows a decrease of at least one order of magnitude in corrosion rates. The three following methods are efficient to reach a passive state: addition of oxidizing agents to the glass, temporary anodic protection, and preoxidation. The erosion phenomena were shown on a rotating electrode. In the case of passive scale damage, a favorable galvanic coupling allows a restoration of the protective layer.

REFERENCES

[1] A. Parent, "Etude électrochimiquede la corrosion des métaux et alliages par le verre fondu," *Verres et Réfractaires*, **23** 300-311 (1969).

[2] D.Lizarazu, P. Steinmetz, J.L. Bernard, "Corrosion of nickel-chromium alloys by molten glass at 1100°C: An electrochemical study," *Materials Sciences Forum*, **251 & 254** 709-720 (1997).

[3] J. Di Martino, C. Rapin, P. Berthod, R. Podor, P. Steinmetz, "Use of electrochemical techniques for the characterization of alloys corrosion in molten glasses," Proceedings of the 6th ESG Conference, Glass Odyssey, June 2002, Montpellier, France.

[4] J. Di Martino, C. Rapin, P. Berthod, R. Podor, P. Steinmetz, "Electrochemical study of metals and alloys corrosion by molten glasses," Proceedings, 15th ICC, Granada, September 2002.

[5] B. Tremillon, "Electrochimie Analytique et Réactions en Solution," Tome 1, Ed Masson, Paris (1993).

[6] F. Baucke, "High-Temperature Sensor for Oxidic Glass-Forming Melts" in " Sensors A comprehensive Survey," Volume 3 part II, p 1155-1198 Edited by W. Göpel, VCH Pub. (1992).

[7] F. Baucke, "Electrochemical cells for the oneline measurements of oxigen fugacities in glass forming melts," *Glastech. Ber.*, **61** [4] 87-90 (1988).

[8] E. Freude and C. Rüssel, "Voltammetric methods for determinig polyvalent ions in glass melts,",*Glasstech. Ber.*, *International Journal of Glass Science and Technology* **6**, p. 202-204 (1987).

[9] C. Rüssel, "The electrochemical behavior of some polyvalent elements in a soda-lime-silica glass melt," *J. of Non Crystalline Solids*, **119**, 303-309 (1990).

[10] J. Di Martino, C. Rapin, P. Berthod, R. Podor, P. Steinmetz, "Corrosion of metals and alloys in molten glasses: Part 1: Glass electrochemical properties and pure metals (Fe, Co, Ni, Cr) behaviours," *Corrosion science*, in press.

[11] J. Di Martino, C. Rapin, P. Berthod, R. Podor, P. Steinmetz, " Corrosion of metals and alloys in molten glasses: Part 2: Nickel and Cobalt high chromium superalloys behaviour and protection ," *Corrosion science*, in press.

ELECTROCHEMICAL STUDY OF COBALT-BASE SUPERALLOY CORROSION BY A MOLTEN GLASS: INFLUENCE OF ALLOY MICROSTRUCTURE AND CHEMICAL COMPOSITION OF THE GLASS

Jean Di Martino, Sylvain Michon, Patrice Berthod, Christophe Rapin,
Renaud Podor and Pierre Steinmetz
Laboratoire de Chimie du Solide Minéral,
Université Henri Poincaré Nancy 1
Boulevard des Aiguillettes, BP239
54506 Vandoeuvre-lès-Nancy, Cedex, France

ABSTRACT

The corrosion of three high chromium cobalt base alloys was studied in different molten glasses: soda-lime, silicoaluminate, and borosilicate. Their behaviors were electrochemically characterized (polarization curves, corrosion potential and polarization resistance). Generally alloys are active, but they are able to become passive. High-temperature oxidation in air before immersion allows them to be directly passive in the soda-lime glass. Temperature and duration of this preoxidation must be sufficient to achieve passivation, depending on the alloy's microstructure. Carbides appeared important for a good stability of the passive state over long time. No preoxidation is required for the second alloy in the silicoaluminate glass in which it is spontaneously passive, because of the presence of strong oxidant species in this glass (Fe^{III}). On the contrary, the same alloy cannot be passive in the borosilicate glass, maybe because of a higher solubility of chromia in this glass.

INTRODUCTION

Superalloys figure among the materials used in the hottest parts of glass-making machines. These metallic alloys bring their excellent mechanical properties where tensile resistance and fracture toughness are of great importance. Unlike many refractory oxide materials, superalloys are affected by both high-temperature oxidation and corrosion by molten glass, which limit their life times. On the second point, it is possible for superalloys to better resist against glass corrosion, for example by reaching and keeping a passive state [1]. As was seen in this study for cobalt-base superalloys, this is more or less difficult, depending on the alloy and glass characteristics, and also on some process parameters, such as hot-air exposure before contact with molten glass.

EXPERIMENTALS

Synthesis of the alloys

Three cobalt alloys were synthesized by high-frequency induction melting and solidification in a ceramic crucible. Their chemical compositions are reported in Table I. Two are cobalt-base superalloys, containing eutectic carbides in grain boundaries, which are the main strengthening phases for this superalloy's family [2, 3]. The Co1 alloy only contains TaC carbides, while Co2 contains both TaC and $(Cr,W)_{23}C_6$ carbides. In order to identify the specific role of the latter, a third cobalt alloy, named Co0, and representing Co2 without its carbides, was also prepared by HF induction melting from the chemical composition of the Co2 matrix measured by microanalysis. The microstructures of these cobalt alloys are shown in Figure 1. In addition, an industrial high-chromium Ni-base superalloy (Ni-30Cr-0.7C-7Fe-7W in %wt) and pure chromium (99.5%) were also studied.

Samples of the alloys were metallographically prepared. Initial microstructures and corroded surfaces were examined using a Philips XL30 Scanning Electron Microscope equipped with an Energy Dispersive Spectrometry apparatus.

Table I. Chemical compositions of the studied cobalt alloys

Alloy	Co	Ni	C	Cr	Ta	W	Microstructure
Co1	bal.	9.0	0.4	29	3	6	Matrix + TaC
Co2	bal.	9.0	0.4	28	6	/	Matrix + TaC + $M_{23}C_6$
Co0	bal.	9.0	/	27.9	0.81	5.1	Matrix (=matrix of Co2)

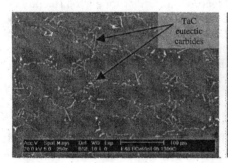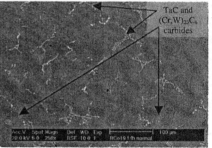

Figure 1. Microstructures of the Co1 (left) and Co2 (right) superalloys

Electrochemical measurements of corrosion

The corrosion behavior of the alloys immersed in molten glass was characterized using a three-electrode apparatus adapted from aqueous corrosion (Figure 2) already described by Di Martino et al. [1]. It can be used in this new field both for pure metals and metallic alloys in contact with molten glass [4,-6]. Glass is contained in a Pt-10%Rh crucible placed in air atmosphere inside a furnace. The working electrode, which can be put in rotation, is a 5.5-mm-diameter alloy rod polished with 1200 grit SiC paper. A platinum plate (25 x 10 x 2 mm³) functions as counter electrode. All potentials were measured versus an yttria-stabilized-zirconia electrode (YSZE) based on the O_2/O^{2-} redox couple [7]. The electrodes' tips are immersed in about 1.5 kg of molten glass. The chemical compositions of the three glasses are given in Table II. Experiments were performed at a temperature depending on the glass nature: 1050°C (SL glass), 1200°C (SA glass), and 1250°C (BS glass).

Using Parc M273 and M263A potentiostats, several electrochemical measures were determined: linear polarization curve (from E_{corr} to YSZE + 0.1V with a 600mV.h^{-1} scan rate), evolution of free potential with time (mainly on 20 hours), and polarization resistance measurements (every 2 hours) by Stern-Geary's method (from E_{corr}−15mV to E_{corr}+15mV at 600mV.h^{-1}).

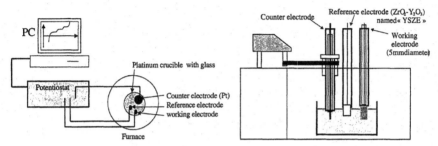

Figure 2. The electrochemical apparatus for glass corrosion measurements

Table II. Chemical compositions and working temperatures of the three glasses

Glass	SiO_2	Na_2O	CaO	Al_2O_3	B_2O_3	Fe_2O_3	MgO	T
SL	bal.	16	7.2	3.4	4.5	0.15	3.0	1050°C
SA	bal.	6.7	15	22	/	4.9	0.77	1200°C
BS	bal.	0.6	22	14	7.4	0.31	0.35	1250°C

RESULTS AND DISCUSSION

Behavior of the alloys in the SL glass

When directly immersed into the SL molten glass without any exposure in hot air before, all the superalloys studied are in an active state. Corrosion potential (E_{corr}) remains near −1.2V / EMZS. According to the standard potential scale [6], the anodic reaction is assigned to Cr -> Cr^{II} + 2e, and the cathodic one is Si^{IV} + 4e -> Si^0. The polarization resistance (Rp) value is about 20Ω.cm^2, which corresponds to a corrosion rate of the alloy in the glass of a few cm/year. This active state is characterized by the absence of any oxide layer, glass being directly in contact with the alloy, as is illustrated by the first micrograph of Figure 3, obtained for the Ni-base alloy.

Nevertheless, all alloys are able to reach a passive state, as shown by the polarization curves obtained (Figure 4). Indeed, with the imposed potential increasing from E_{corr}, corrosion current jumps over an anodic peak. Then the alloy becomes passive and develops an external protective chromia scale, as is also illustrated by the second micrograph of Figure 3. The anodic reaction then becomes Cr -> Cr^{III} + 3e. The cathodic one can be O_2 + 4e -> $2O^{2-}$ or Fe^{III} + 1e -> Fe^{II}.

This passive state can be obtained with an exposure to air at high temperature before immersion into the molten glass (treatment called "preoxidation"). Then, alloys are initially in a passive state, which is characterized by a polarization curve without any anodic peak, and the anodic part of which begins at a higher potential (Figure 5). Thus corrosion potential and polarization resistance are higher than for an alloy in an active state (more than −0.7 V / EMZS and $200\Omega.cm^2$ respectively). Then the corrosion rate, which is directly related to the inverse of the polarization resistance, is greatly lowered (to a few mm/year).

Efficiency of the preoxidation treatment

Obtaining the passive state using hot-air oxidation prior to immersion depends on both preoxidation parameters and microstructure of the alloys. Indeed, each alloy needs specific values of both temperature and duration of preoxidation. For example, a preoxidation at 1050°C during 1 hour is enough for Co2 to be passive in the SL glass as early as immersion. But this state is not necessarily stable, and the alloy can fall again into an active state after a few hours of immersion. On the other hand, a longer preoxidation, such as 4 hours at 1050°C, leads

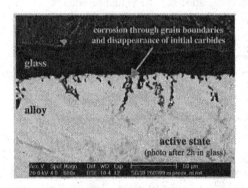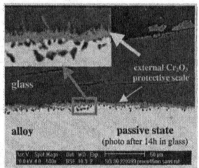

Figure 3. The Ni-base superalloy surface state after glass corrosion

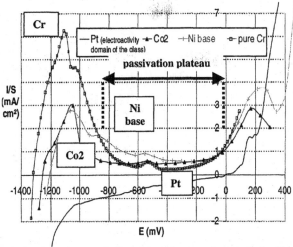

Figure 4. Polarization curves obtained for alloys and pure Cr in SL glass (1050°C)

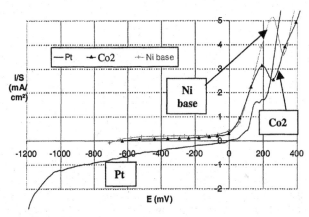

Figure 5. Polarization curves in SL glass of preoxidized alloys (1050°C)

to a real stable passive state for this alloy. The same result can be reached using a shorter preoxidation done at higher temperature. A preoxidation at 1150°C for 1 hour leads to a good stability of the passive state and a corrosion resistance similar to the one obtained at 1050°C for 4 hours. Another example, 15 minutes at 1150°C, followed by a cooling down to 1050°C over 20 minutes, leads to about the same polarization resistance as 1 hour at 1050°C. Thus different temperature/duration combinations can provide a similar result. All these observations are summarized in Figure 6.

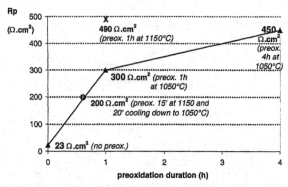

Figure 6. Evolution of the corrosion resistance of Co2 in SL glass, with temperature and time of preoxidation (values obtained after 4h immersion)

Role of the alloy microstructure

The ease of obtaining the passive state and the long-time stability of the latter also depend on the alloy's characteristics. Thus, a 1-hour preoxidation at 1050°C is not sufficient for Co1 to get and keep a passive state. This alloy needs to be longer preoxidized than Co2 to be really passive in molten SL glass. It is to be attributed to the nature of the carbides: Co1 only contains TaC, while Co2 has both TaC and chromium carbides, for the same final density, due to 0.4%C.

The specific role of the carbides was highlighted by electrochemical experiments performed on the Co0 alloy, which represents the Co2 matrix, which is to say Co2 without carbides. This alloy is also active and capable of being passivated in the molten SL glass, and has nearly the same passivation current as Co2 (Figure 7 left). It is more on the stability of the passive state (obtained by preoxidation) that the absence of carbides has an important effect. Indeed, after only a few hours, Co0 falls in the active state (Figure 7 right). To explain that, two hypotheses can be made: $M_{23}C_6$ can be considered as grain-boundary Cr reserves, and all carbides act as specific diffusion paths for Cr. Then, when such carbides are present, the supply of Cr is easier, and the chromia layer can remain stable during a longer time.

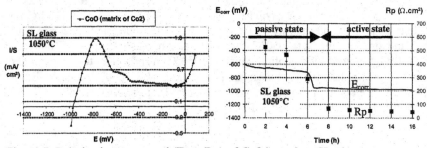

Figure 7. Polarization curve and (E_{corr}, Rp) of Co0 in molten SL glass

Role of the glass composition

Glass composition can also have an influence of the ability of the alloys to reach the passive state and the stability of this latter state, and that in two opposite ways. In the case of the SA glass, it appeared that Co1 is passive as early as immersion, although no preoxidation has been applied. In addition, this passive state remains for at least 80 hours. The polarization curve is characteristic of an alloy already passive: no anodic peak and relatively low passivation current for this higher temperature (Figure 8). However, Rp of Co1 is divided by 4, and its passivation current (then its corrosion rate) is four times higher compared to the results obtained in the SL glass, because of the higher temperature. Over this long period, E_{corr} progressively raises from its initially high value, the oxidant action of the Fe^{III} specie being replaced by the O_2 one. Although Rp remains at a high level, there is a slow decrease, caused by this evolution of the oxidant nature (number of electrons involved in the Stern relation). This type of favorable behavior is directly related to the presence of strong Fe^{III} oxidants in a

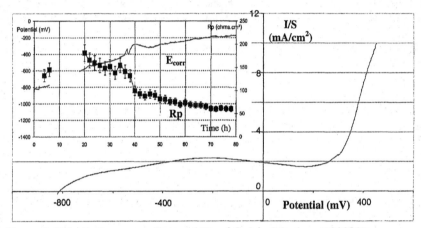

Figure 8. Polarization curve, E_{corr} and Rp of Co1 in SA glass (1200°C)

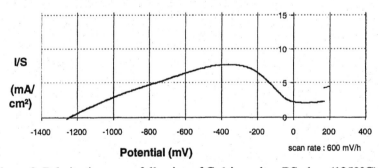

Figure 9. Polarization curve following of Co1 in molten BS glass (1250°C)

relatively high amount. On the other hand, in the BS glass, no passivation seems to be possible. The polarization curve obtained on the Co1 alloy in this molten glass at 1250°C does not present any passivation plateau. Corrosion current is very high everywhere on the recorded curve (Figure 9). In such a glass, Co1 is not able to become passive, even after a preoxidation. Possible reasons are a too-high solubility of chromia in this glass or a too-high run temperature.

CONCLUSION

Generally, high-chromium cobalt alloys are able to be passive in molten soda-lime glasses, as it can be also observed on high-chromium nickel alloys and pure chromium. A possibility for alloys to reach this passive state is an oxidation in hot air before contact with molten glass. Two main parameters of this preoxidation, temperature and duration, have to be high enough. Their minimal values depend on the alloy's microstructure, for example, on the type of carbides it contains. Carbides also seem to stabilize this passive state during long-time experiments. For each alloy, (T,d) combinations can be found to achieve passivation, and a choice is possible to allow the realization of an efficient preoxidation in an industrial process. Thus, oxidant species present in the glass, such as Fe^{III}, can lead to a passivation as early as immersion. On the contrary, other glasses do not allow chromia-forming alloys to become passive, maybe because of a high chromia solubility in the glass. Then, other means have to be found in order to protect the alloys.

REFERENCES

[1]J. Di Martino, C. Rapin, P. Berthod, R. Podor and P. Steinmetz, "Use of Electrochemical Techniques for the Characterization of Alloys Corrosion in Molten Glasses," Proceedings of the 6th ESG Conference, Glass Odessey, June 2-6, 2002, Montpellier, France

[2]E.H. Bradley, "Chapter 3: Microstructure," 44-47, Superalloys: A Technical Guide, ASM International, Metals Park, OH 44073, 1988

[3]C.T. Sims and W.C. Hagel, "Chapter 5: Cobalt-Base Alloys," 150-154, The Superalloys, John Wiley & Sons, New York, 1972

[4]J. Di Martino, C. Rapin, P. Berthod, R. Podor and P. Steinmetz, "Electrochemical Study of Metals and Alloys Corrosion by Molten Glasses," Proceedings of the 15th International Corrosion Congress, Sept. 22-27, 2002, Granada, Spain

[5]J. Di Martino, C. Rapin, P. Berthod, R. Podor and P. Steinmetz, "Corrosion of Metal and Alloys in a Molten Borosilicate Glass, Part 1: Glass Electrochemical Properties and Pure Metals (Fe,Co,Ni,Cr) Behaviors," Corrosion Science, in press

[6]J. Di Martino, C. Rapin, P. Berthod, R. Podor and P. Steinmetz, "Corrosion of Metal and Alloys in a Molten Borosilicate Glass, Part 2: Nickel and Cobalt High Chromium Superalloys Behavior and Protection," Corrosion Science, in press

[7]F.G.K. Baucke, "Electrochemical cells for the online measurement of oxygen fugacities in glass forming melts," Glastech. Ber. , 61 [4] 87-90 (1988)

GLASS-SILICIDE COVERINGS

A.G. Hambardzumyan,
Scientific Production Enterprise
of Material Science (SPEMS)
of Republic of Armenia
17 Charents Street
Yerevan, 375025, Armenia

G.A. Kraveckiy, and V.V. Rodionova
Research Institute of Graphite,
2 Electrodnaya,
Moscow, 111524, Russia

ABSTRACT

For reliability and thermal stability of carbonic silicated materials which in manufacturing do not generate dense and unbroken carbide films, coverings are developed that assure the healing of discrete parts of silicated surface at temperature 1500^0C. Coverings based on borosilicate high siliceous glasses with additions of molybdenum disilicide were proposed as an antioxidant protection of silicated carbonic materials at the temperature 1500^0C. Developed coverings have high speed of stress relaxation at high temperature, their coefficient of thermal linear expansion (TCLE) is comparable to TCLE of undercoats, and they assure self-healing of defects appearing in operation. These features and simplicity of coatication allow to apply them to parts of large dimension and complicated configuration.

INTRODUCTION

The unique properties of graphite materials (thermal resistance, due to their low coefficient of thermal expansion, chemical inertia in corrosive medium, low density, increase of resistance with increasing temperature) enable the use of ceramic ware in different branches of industry (ferrous and non-ferrous metallurgy, engineering, aircraft and space technology). Oxidation during heating in oxygen-containing gas medium (starting at 800^0C) is the central failure mechanism of carbonic materials. Volumetric and surface methods are used to protect carbonic materials from oxidation.

Currently applied methods of surface protection are either unecological (noise, dust content) or do not ensure gas-proof cover (plasma spraying), or need complex facilities, using toxic chlorine-containing materials and demand strict adherence to the technological parameters of the process (temperature, uniform supply of gas components to the surface of coated parts). The newest methods of covering (electron-beam, ion-plasmous) are suitable for small-size parts, in which case the thickness of covering is not more than 10-20 μm.

Volumetric protection of carbonic materials by siliconizing or borosiliconizing provides the formation of a surface oxycarbide film, which ensures protection of parts at temperature up to 1200^0C. But due to the non-uniform porous structure of carbonic materials, in particular carbon-carbonic materials, the oxycarbide film does not cover the entire surface. As a result the heating of silicicated parts in the air causes formation of local unprotected areas, which destroys coated parts. We should mention that the oxidation of silicicated and boro-silicicated parts during heating is three to five times less than the oxidation of non-silicicated carbonic materials when heat is applied.

GOALS AND TASKS

Development of covering

The Research Institute of Graphite and The Institute of Chemistry of the Silicates of Russian Academy of Sciences jointly with the Scientific Production Enterprise of Material Science (SPEMS) of The Republic of Armenia developed a glass-silicide covering aimed to protect carbonic as well as silicicated and boro-silicicated materials including composition carbon-carbonic materials. The covering is used to protect carbonic parts while heat up to 1500-1600^0C is applied in an air atmosphere at small rarefaction in the current of dissociated air.

In order to produce this covering the power-calcining technology in an argon or air atmosphere is used. The covering thickness is 0.2-0.3 mm. The original components are finely divided powders of boro-siliceous glass, alloyed by aluminium oxide and molybdenum disilicide ($MoSi_2$).

Molybdenum disilicide

Among all oxygen-free refractory compounds, molybdenum disilicide is known to be the most heatproof in an air atmosphere up to 1700^0C. Gas-proof films, forming in an air atmosphere at 1100-1200^0 C, consisting of silica, and holding on the surface of molybdenum disilicide, cause the high heat resistance of molybdenum disilicide. The most favourable temperature for using it is within 1300-1700^0 C. The film thickness of silica after 2000 hours of exposition in an air atmosphere at the 1300^0 C is 0.02-0.03 mm, and after 3000 hours exposition at 1700^0C it is 0.05-0.1 mm [1,2]. Thin films on the molybdenum disilicide, in contrast to films 0.05 mm thick are resistant to crystallization.

There are other compounds with infusibility similar to that of $MoSi_2$ (Mo_3Si, Mo_3Si_2, Mo_5Si_3, $MoSi_{0.65}$) in the Mo-Si system. This especially applies to Mo_5Si_3 [1]. The perfect co-existence of molybdenum disilicide with siliceous films at high temperature confirms the advisability of the choice of silica (SiO_2) and disilicide of molybdenum for the synthesis of heatproof coverings.

The choice of the quantitative ratio $MoSi_2$: SiO_2 is dependent upon the following factors. A higher content of molybdenum disilicide in covering insures higher infusibility and refractoriness, thus the inclusion of the maximum possible amount of molybdenum disilicide in the covering is very advisable. But when its content is more than 50%, vitrification of the covering is difficult, which reduces refractoriness and air-tightness of the covering due to the formation of a porous "cake" instead of refractory glass. Reduction of molybdenum disilicide improves the formation of covering and decreases its infusibility. A content of molybdenum disilicide in coverings of ≤ 50 mass% enables them to be used at temperatures up to $1500\text{-}1600^0C$.

Siliceous glasses

The usage of pure silicon dioxide for the synthesis of coverings is undesirable, as it has many crystalline modifications (quartz, cristobalite, tridymite, coesite, keatite, stysovite). Stiskovite siliceous glasses (noncrystalline silica) can be obtained in different states, which have very different density. This allows to consider them as a single modification. Such transformations of silica arising during heat treatment are adverse for the structure and properties of covering material (especially for thermal expansion).

It is necessary to use alloying of siliceous glasses for the covering synthesis to prevent these transformations.

Studies of a great number of different types of glasses showed alkali free borosiliceous glasses, alloyed by a small amount of aluminium oxide, to be the most effective as glass-matrix. The boron oxide content should not be higher than 20 mass%. The higher the exploitation temperature of the covering, the less the amount of boron oxide it should contain. When the content of boron oxide is ≥ 5 mass% the covering formation is good in an air atmosphere. In an argon atmosphere the covering is formed when the minimum content of B_2O_3 is ≥ 10 mass%. One of the ways to decrease the content of boron oxide below < 10 mass% is to increase the covering treatment temperature, while narrowing the change of treatment temperature interval.

The elements of the first and second groups of the periodic chart as well as the elements of the iron group interact with molybdenum disilicide very effectively, so their presence sharply reduces refractoriness of covering and can cause its destruction.

Boron and aluminium oxides have several functions as components of glass. First, they reduce the crystallization ability of silicon dioxide, decrease the structuredness of formed glass, make it more stable and insure convenience in glass production. To form 50-60 mass% of amorphous silica more than 10 mass% of boron oxide is needed. When the content of silica is close to 80 mass%, the refractoriness of the covering in the region 1500-1600^0C is assured by 15-20 mass% of boron oxide. Secondly, refractory molybdenum borides are formed as a result of interaction between molybdenum dicilicide and boron oxide. Refractory molybdenum borides increase the covering infusibility.

Thirdly, alloying by boron and aluminium oxides promotes the decrease of heat expansion of covering to values comparable to silicated carbonic materials, which increase heat shock resistance of the covering. The addition of aluminium oxide is favorable for the decrease of viscosity of the covering at high temperature (1500^0C) and this prevents erosion of the covering in high-speed and high-temperature air flows. But when the content of aluminium oxide is > 3 mass%, the fusible $Al_2O_3 - SiO_2$ eutectic is formed and that leads to the decrease of infusibility of covering.

Construction of furnace

The technological process of glass melting was developed in the Scientific-Production Enterprise of Material Science (SPEMS) of the Republic of Armenia. The technology is modified for industrial production in electrical furnaces (garnissage fusion). Electrical garnissage furnace of direct heating (EGFDH) consists of a cylindrical or conical metallic water-cooled case, three molybdenic electrodes which transverse the furnace case at angle of 120^0C to the centre and the central (null) electrode whose lower part gets out from bottom plate (exhaust tube) for discharge of melted glass. For additional heating of exhaust tube is aimed a high-frequency inductor. The power source for induction heating of exhaust tube is a thyristor transformer.

In EGFDH the glass melting process occurs under the layer of charge, which permits avoidance of volatilization of volatile components and to get homogeneous glasses of stable composition.

As initial components, nondeficit powders produced by local industry (quartz sand, boric acid, aluminum hydroxide, molybdenum disilicide) are used. Heat treatment is carried out in an argon atmosphere (sort A).

The process of covering formation

In inert atmosphere at high temperature the oxidizing process in covering is realized only at the expense of the oxygen of the melt. When heating up to $1300-1500^0$C partial oxidizing of molybdenum disilicide occurs and films of silicon dioxide are formed on the surface of molybdenum disilicide grains. The glassy components have low viscosity during covering formation and they form a heatproof glassy matrix. This matrix includes parts of molybdenum disilicide in silicon or boride shells which are evenly distributed in the glassy mass. In addition alloying of silicon dioxide occurs as a result of the partial dissolution of molybdenum disilicide in it.

The formation of a transition layer with a monotonically changing composition on the boundary between covering and base leads to increase of the bond strength between covering and carbonic base.

In order to investigate the properties of covering material, samples were made from a powder mixture of borosilicate glass and molybdenum disilicide by the method of hot pressing.

1. The density of covering material is 2.9 g/cm^3
2. The open porosity of covering, defined by the method of hydrostatic weighing of 10 samples, is 1.21.
3. The thermal coefficient of linear expansion was α x 10^6 K^{-1}.

Table I. Dependence of TCLE on temperature

Temperature intervals (°C)	TCLE (α x 10^6 K^{-1})
50 – 650	4.03
20 - 1000	3.75
20 – 1100	4.05
20 – 1200	4.0
20 – 1300	4.0

At temperature 1375^{0}C fusion of the covering and deformation of the sample occur. The measurements were carried out by Ulbrecht dilatometer.

4. Effective viscosity lgη x10^3 at different temperatures (^{0}C) was:

1016 – 8.29	1451 – 5.69
1106 – 7.76	1507 – 5.40
1310 – 6.35	1570 – 5.38
1387 – 5.82	

Viscosity of the covering material is twice the glass viscosity.

5. Heat resistance: at temperatures from 20^{0}C up to 1500^{0}C, number of cycles is > 10 (without cracks and without mass changing of sample)
6. The flexural strength at a temperature 20^{0}C, is– 9.33 ± 0.25μPa
7. The electrical resistivity ρ, is 10^4 Ohm.m.
The measurements were carried out on a sample 100 mm in length with cross-section area of 0.4 cm^2.
8. Heat-conductivity, λ, 1.72 ± 0.2 W/m•K
9. Chemical stability – it is stable in air, hydrogen, superheated vapor of sulfur at high temperature, in boiling solutions of hydrochloric, suphuric and nitric acids (according to measurements at Institute of Chemistry of Silicates of Russian Academy of Sciences).
10. Compatibility with other materials. It is compatible with quartz, corundum, beryl oxide, zirconium dioxide and heatproof alloys at 1200^{0}C, with molybdenum - in inert medium up to 1375^{0}C. It is unwettable by melt of indium, thallium, Kovar (according to measurements at the Institute of Chemistry of Silicates of the Russian Academy of Sciences).

Oxidation resistance of graphite silicated samples with glass-silicide covering in quiet air.

During thermocyclings at temperature interval of 20-1500^{0}C and a total test time of 20 hours (the total number of cycles was 50) losses of sample mass were not observed but an overweight of 0.14-0.6 % was measured. When the temperature of sample was 1500^{0}C in continuous regime losses of sample mass were not observed during 1000 hours (according to measurements at the Institute of Chemistry of Silicates of the Russian Academy of Sciences).

The maximum operating temperature of silicated samples with M-46 covering in the quiet air is 1600^{0}C.

Investigation of the oxidation resistance showed that the limiting stage determining working resources of covering at temperature up to 1500^{0}C, is evaporation of boron oxide. At present borosilicate glass which contains 18-20% of boron oxide is used, but the use of low-boron glasses appreciably increases the working resource of coated parts and increases the maximum operation temperature by 50-80^{0}C.

The main advantage of such covering is the high heat shock resistance. This can be explained by several factors:
1. Good agreement between the temperature coefficient of linear expansion (TCLE) of covering and undercoat
2. The covering material is in visco-plastic state at temperatures over 1200^{0}C, which leads to the relaxation of thermal stress
3. The visco-plastic state of covering to some extent promotes self-healing of small defects which arise in covering during operation

The developed covering has low heat-conductivity (1.7 W/m x K at 20^{0}C) and good insulating properties ($\rho = 10^4$ Ohm x m). After 3 weeks exploitation of the samples with covering, the water absorption was 10 times lower than that of samples without covering (0.043 and 0.33 respectively) though they do not differ in strength.

The developed covering has a high oxidation resistance both in statical test in air (1000 hours in air at the 1500^{0}C) and in dissocated airflow at 0.1 atm.

CONCLUSION
Thus, the main advantages of glass-silicide coverings are:
*   Exploitation temperature is of 1500^{0}C
*   Relaxation of thermal stress owing to visco-plastic state at high temperature
*   Agreement of TCLE of covering and undercoat
*   Self-healing of small defects arising in covering during operation
*   Reduction of water absorption by more than a factor 10 in sample with covering
*   Simplicity of technology
*   Possibility of applying covering to parts of large dimension (up to 2000mm), including ones with complicated configuration

REFERENCES

1. G.V. Samsonov, L.D. Dvorina, B.M. Rud, «Silicides», Metalurgiya, Moscow, p.p.177-190 (1979).

2. G.V. Samsonov, «Silicides and its using in technique», Kiev, Ukrainian Academy of Sciences, p.p.81-92, p.p.182-189 (1989).

Advances in Forming

MECHANICAL STRENGTH INCREASE DURING THE FORMING PROCESS OF GLASS

H. Hessenkemper
TU Bergakademie Freiberg – (Germany)

ABSTRACT

The strength of glass is a decisive attribute. During the forming process of glass this property is strongly influenced by the time dependant thermal situation due to the visco-elastic character of the glass melt. It is possible to increase the original strength of the glass following some physical topics. The complex influence parameters are shown and underlined by industrial tests.

INTRODUCTION

The strength of glass is a decisive property. Economic objectives demand a decrease of weight in combination with an increase of mechanical properties. More than 30% of the cost in the container glass production is directly bonded to the weight as energy and batch cost. Therefore the weight reduction is a strategic demand to the glass industry and this aspect is connected to the mechanical properties of glass. An increase of the strength of glass is therefore the most important challenge to the industry in the future.

BASIC CONSIDERATIONS

The strength of glass is directly combined with the surface damages and only in an indirect way to a small amount due to the chemical composition. Figure 1 shows this surface related strength. Therefore different strategies do exist to increase the strength: To avoid surface damages and to heal surface damages. For both possibilities different options are available and could be discussed in Fig. 2. Another distinction is to differentiate between the situation during the glass production and the cold glass. Although the use of the glass, the transport and other effects are important to the quality of the glass surface during their lifetime, an important part to the mechanical strength of glass is combined with the production process. This is especially true for the surfaces without direct mechanical contact to the surrounding like the inner part of glass tubes and bottles. Here the mechanical strength like impact strength is dominantly defined by the quality of this surface which again is defined by the production process. It is a common experience that a blow and blow process created better mechanical strength than a press and blow process. An explanation for this has been given in small particles resulting from the forming material being in contact with the melt. Extensive examinations proved that this is only true for a smaller part of the mechanical failures. Actually in most of the cracks origin no particle could be found. Therefore the question come up, what is the physical background and what possibilities do exist to improve the situation.

The basic considerations for the forming of cracks in glass melts are given in Fig. 3: The forming of cracks is due to stresses exceeding the strength of the melt. These stresses result from mechanical aspects like forming processes in addition to thermal induced stresses due to the cooling of the melt and the temperature dependant thermal expansion coefficient. In opposite to this there are relaxation processes. The relaxation time is proportional to the viscosity. These increase and decrease mechanisms define the time and place dependant stresses. If they exceed the strength of the melt a crack can occur and can create surface damages.

These considerations give an answer to the fundamental question of the strength increase: Because of the simplified correlation

$$\eta \sim \tau \qquad (1) \qquad \qquad \text{where } \eta : \text{Viscosity and}$$
$$\tau : \text{Relaxation time}$$

$$S(t) = S_0 \exp(-t/\tau) \qquad (2) \qquad \text{where } S(t): \text{Time dependant stress}$$
$$S_0 : \text{Stress at time } t = 0$$

it is fundamental to shift the temperature situation in the forming process to the highest possible values to avoid crack formation. Even some machine developments like the Heye 1,2 machine and the RIS machine are based to this principles.

TRANSFER INTO THE INDUSTRIAL CONTEXT

Figure 4 demonstrated more detailed the situation during the forming process. The forming process is basically a process of heat exchange from the melt to the surrounding. A strong heat exchange occurs during the contact of the melt surface to the metal. From the volume only 1-10% of these heat losses could be replaced. Therefore a cold glass surface is created with a high viscosity, high relaxation time and a strong creation of thermal induced stresses. By the way, this is the reason for the two step forming process: To give the surface the chance for reheating and decreasing the surface viscosity to enable the next forming step.

This thermal situation increased the probability for a crack creation in a more than exponential way. This crack could not propagate into the melt because of the low viscosity, low relaxation time beneath the surface. Partly this crack is healed in the following heat history, but a surface damage is existing after the cooling of the glass.

If this thermal problem is the main aspect for surface damages, Fig. 4 demonstrated all possibilities for influencing the mechanical strength created by the forming process of glass.

There should be a connection to the parameters involved to the reheating of the surface, like spectral behaviour of the melt. Actually it could be shown that there is a connection between the brittleness of the melt and the spectral behaviour. The same is true to the cooling situation between the melt and the forming material, the surface roughness, the heat transfer coefficient and the cooling situation of the forming material. Every activity influencing the heat transfer mechanism is influencing the mechanical strength: Swabbing, temperature instabilities resulting from the glass conditioning, different heat capacities and heat conductivities by using different forming or plunger materials and so on.

The different experiences in the industry proved that normally a stable strength increase of 20% and more is possible optimising just the classical parameters. Using specialised parameters like different plunger materials, even better results a possible. Figure 5 shows a SIMS analysis for the surface of a new plunger material. It has been the same bulk material. One plunger has suffered a different surface treatment in the area of about 20-30 nm. The chemical differences are shown and they resulted in quite different impact strength and improved standard deviation. This has been due to a heat transfer coefficient in Fig. 4 and in addition to different surface energies. This is influencing the stress situation as demonstrated in Fig. 3 and in introducing this material as plunger material the first tests proved the expected results for the impact test in the container glass production.

Another possibility for increasing the mechanical strength is the option to change the surface chemistry during the production process with simple means. This can create a smoother surface as it is shown in Fig. 6 by treating the hot glass surface with water steam. The reason could be an incorporation of OH-Groups into the glass surface and a resulting viscosity decrease with a better thermal healing of cracks. But new experiments gave a hint to more complex chemical interactions. Nevertheless the smoother surface is reported from different sides.

CONCLUSIONS

Summary and outlook

The increase of the mechanical strength of the glass can be reached by simple means in the production process just following the basic physics. It is necessary to understand the forming process first as a heat exchange process. From this different possibilities are created to optimise the mechanical strength of glass. In addition new materials and special treatments of the hot glass surfaces give more possibilities to optimise and conserve the obtained strength.

This enables the industry in combination with the already known techniques to improve the strength and reduce the weight.

If this could be combined with an online non-destructive test for every bottle to ensure a certain strength level, this could give a strategic change to the glass industry. Right now the first attempts are made to develop such a possibility.

Acknowledgement: This article was previously published in *The International Glass Journal*. It is published here with permission of *The International Glass Journal* and with the agreement of The American Ceramic Society.

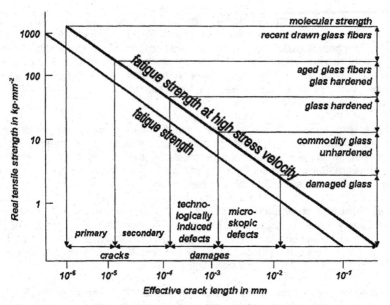

Fig. 1. Influence of surface damages on the tensile strength.

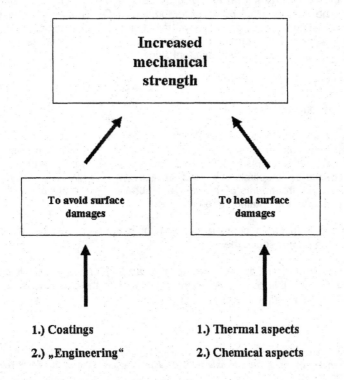

Fig. 2. Strategies to improve the strength.

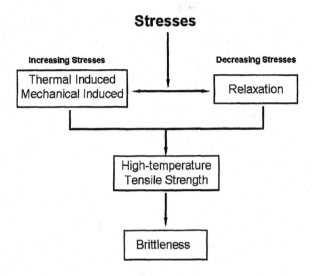

Fig. 3. Forming of cracks in glass melt.

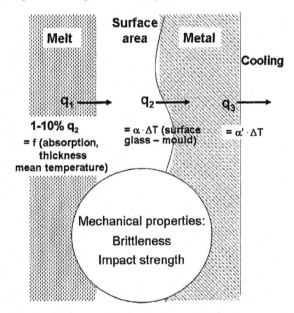

Different situations
1. **Free cooling**
2. **Contact melt – metal**

Fig. 4. Strength created by the forming process of glass.

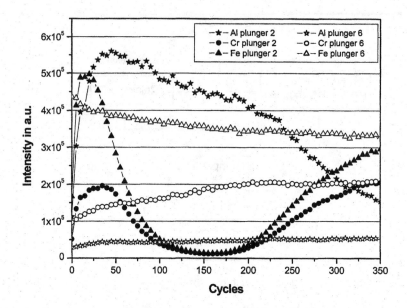

Fig. 5. SIMS analysis for the surface of plunger.

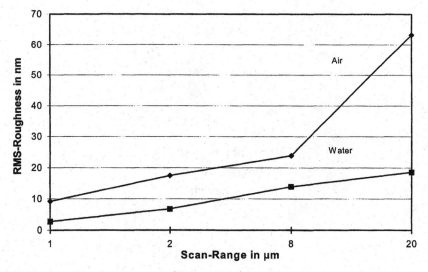

Fig. 6. Surface Roughness.

OPTIMISATION OF THE HEAT TRANSFER DURING FORMING OF GLASS

Andreas Kropp
Steam & Glass Technology GmbH
Dresdner Strasse 136
D – 01705 Freital
Germany

Stephen Follis,
Steam & Glass Technology GmbH
7 glebe Road, East Gillibrands
Skelmersdale-Lancashire WN8JP
Great Britain

ABSTRACT

The application of intelligent temperature and pressure measurement tools provides important information for process stabilization and improvement for glass container production. Applying appropriate blow-head design changes and adopting the optimum forming pressure significantly maximises the heat-removal process gained through the mould and from the final blowing medium.

INTRODUCTION

During the forming process of glass containers, the final blowing medium provides various attributes. Firstly, the blowing pressure forces the glass melt onto the internal mould surface. This induces an improved heat transition between the glass and mould whilst the shape of the finished article is produced. Secondly, the final blowing air convectively removes heat from the internal glass surface. Both attributes contribute to the heat removal process, thus influencing the quality and potential productivity rates. The purpose of removing heat during this forming stage is to reduce the integral glass temperature to a level at which the final product has sufficient stability to maintain shape after leaving the mould. The heat quantity removed by radiation and conduction through the mould has been mathematically calculated and practically measured at a value of ~90% of the total heat removed. The remaining ~10% can be removed through convective cooling provided by the final blowing medium. To achieve the optimal utilisation (maximum heat removal through conduction and convection) of the final blowing air, the correct internal pressure has to be applied together with a maximum flow within the container. Both parameters can be determined using suitable measurement equipment (available from Steam & Glass Technology) together with the implementation of methodical procedures.

MEASUREMENT EQUIPMENT

IPMS - Intelligent Pyrometrical Measurement System consists of a rapid detection, non-contact, temperature measurement device in conjunction with specifically designed software providing "real time" analysis. The pyrometer is aligned by means of an integrated laser and can be used at distances varying between 1 - 4 meters from the target area. Receiving continuous measurements (190/sec) a cyclic temperature curve is generated in accordance with the viewing field (Fig. 1). The temperatures captured include the parison, container and the internal / external mould surfaces.

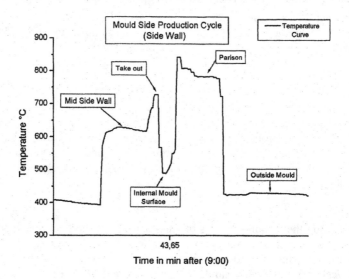

Fig. 1: Temperature signal for one machine cycle (mid side-wall viewing area).

The IPMS software recognises the machine cycle using a dedicated intelligent algorithmic filter (no machine signal required). The user can identify desired measurement locations (up to 4) within the cyclic signal. The extracted temperatures are simultaneously plotted onto individual graphs. The results instantly display the effect of implemented changes, optimising procedures and machine fluctuations. The same software incorporates multiple print and save facilities for full in-depth analysis.

POD100 – Pressure Optimisation Device is a mobile measuring instrument, intended to measure online forming pressures. The unit has multiple display options, data analysis and signal treatment facilities for the measurement of pressure.

Two pressure inputs enable the simultaneous measurement of two different locations, for example; the finish cooling air and the final blow. The data can be numerically or graphically displayed for an individual pressure or both at the same time. The Pod 100 uses a simple menu structure to organise its features and functions. The menu structure also consists of sub menus which are used to create alternative set-ups.

The supplied cable, for connection between the sensor pod and main processing unit, allows the user to analyse the data from a comfortable working distance from the somewhat hot process. A heat resistant hose completes the connection between the sensor pod and measurement location. This flexible link enables safe measurement of pressures on moving machinery, for example, the blow-head arm.

Research by Steam & Glass Technology conclusively confirms that the actual forming pressure, derived from the POD 100, differs substantially from the conventional measurement point in the manifold. Using the conventional manifold pressure gauge as a reference to the forming pressure is very misleading. This is mainly because the pressure shown on the manifold gauge is influenced by control valves, pipe work and blow head design etc. On the other hand, the POD 100 precisely measures the internal bottle pressure and indicates process variants.

METHODOLOGY
The forming pressure required to achieve the optimal heat transfer between glass and mould can be determined by monitoring the temperature of the internal mould surface. Over a given period a incremental pressure increase is applied to the final blowing pressure (internal bottle pressure).

From the resulting temperature curve (Fig.2) it becomes evident as to what level of internal bottle pressure is required (glass-to-mould pressure) to reach the heat saturation point of the moulds. It is worth noting that exceeding this saturation point does not enable a greater heat removal process through the moulds. If the saturation pressure level is lower than that of the standard machine setting, an increase in flow can be created to provide a far greater heat removal process to the internal glass surface.

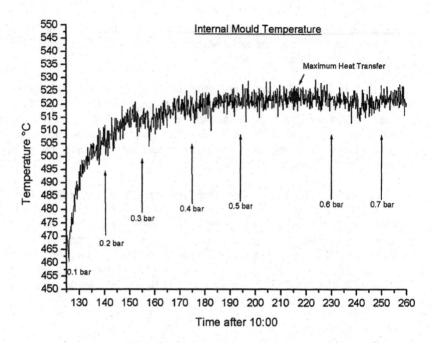

Fig. 2: Typical temperature curve from the internal mould temperature during a pressure incrimination exercise.

Whilst maintaining the optimal forming pressure (as derived from the pressure incremental exercise) the next goal is to maximise the flow within the container. To achieve this with a study of the cross-sectional surface area of all air passageways associated to the final blowing process is performed. Based on these results, changes in inlet and exhaust holes are calculated followed by modification to the blow-head design.

Using both the POD 100 (to confirm internal forming pressure) and the IPMS (to confirm temperature changes) the new blow heads are fitted to the machine. The online temperature received from the IPMS will clearly show the effect of an improved convective cooling or perhaps identify that the changes hindered the process (Fig. 3). Further optimisation is achieved by applying the same measurement techniques whilst varying the length, diameter and shape of the blow tube (always maintaining the optimum internal forming pressure). Small changes to blow-tube design (e.g., 3mm length) can result in 2% productivity increases: Measurement techniques provide accurate results, confirming which design provides the maximum cooling effect.

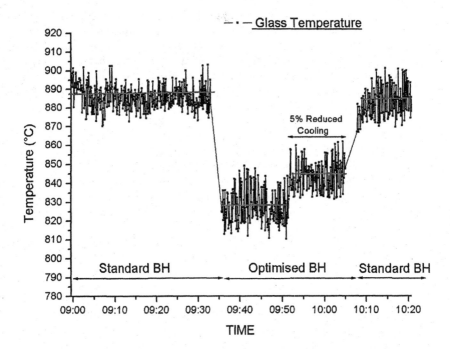

Fig. 3: IPMS online information clearly displays the change in glass temperature (cooling effect) between a standard blow head and an optimised design, together with the effect of implementing a reduction of 5% mould cooling.

Adopting the design of the "ideal" blow head will ensure maximum heat transfer between glass and mould together with a maximum internal convective cooling. This optimised cooling situation will provide increases in quality (less lean, reduce ovality) and provide great potential to implement a speed increase. When applying a machine speed increase, the IPMS is employed to ensure that temperatures remain at a level lower or equal to that of the original / standard situation.

Following the same blow-head optimisation concept, as discussed above, Steam & Glass Technology have recently developed a new design of blow head suited for bottles with long necks. This telescopic blow-head design incorporates a vertically moving blow tube, which provides far greater heat removal from within the container. A confirmed 10% increase in machine speed has been realised in our first long-running production trial.

SUMMARY

The application of intelligent temperature and pressure measurement tools provides important information for process stabilization and improvement.

Applying appropriate blow-head design changes and adopting the optimum forming pressure significantly maximises the heat-removal process gained from the final blowing medium. The maximised cooling provides an increase in quality levels together with the potential to increase machine speeds. Productivity increases have been realised from 2 – 8%. Using the same measurement equipment, optimisation to other parts of the process can be performed. For example:

- Verify pressures between sections / cavities
- Measure and monitor plunger temperatures
- Measure and monitor blank mould temperatures
- Measure machine manifold distribution pressures
- Validate glass temperature changes during process optimisation
- Monitoring the effect on glass and mould temperature during swabbing
- Compare cooling efficiencies between different neck ring materials / designs
- Verify bottle / mould temperatures whilst applying variations in mould cooling

Acknowledgement: Reprinted from the November 2003 edition of *Glass*, the journal for European glassmakers, with permission from DMG World Media

EFFECT OF MOLD TO GLASS HEAT TRANSFER ON GLASS CONTAINER FORMING

Matthew R. Hyre
Assistant Professor of Mechanical Engineering
Virginia Military Institute
Lexington, VA 24450

ABSTRACT

Recent advances in numerical simulation capabilities have made the modeling of glass container forming processes feasible. These forming models must include large free surface deformations, viscoelastic behavior, conjugate heat transfer, and complex contact phenomena between the glass and the forming molds. One of the most critical inputs to these models is the heat flux between the glass and mold. This boundary condition is crucial in determining the final thickness distribution of the container along with the required mold cooling.

INTRODUCTION

Industrial glass forming is a complex sequence of unit processes that includes melting and refining the raw material, cooling and conditioning the molten glass in forehearths, transitioning from a continuous glass stream to discrete gobs in the feeder, and the actual forming processes in the IS machine. The forming of glass containers by the 'press and blow' process can be roughly divided into five main steps: gob formation at the feeder, transfer and loading of the gob into the blank mold, pressing of the parison, inversion of the parison into the blow mold, and the reheat and final blowing of the container (see Figure 1).

After the glass gob is loaded into the blank mold, the plunger presses the gob to create the parison shape. At this point, the parison is upside down, and needs to be inverted before it is blown into its final shape. A swinging arm inverts the parison into the blow mold where the skin of the parison is allowed to reheat before being blown into the desired bottle shape (Figure 1). Once the container is blown and has cooled to the point where post-forming deformations will be small, the bottle is removed from the mold and swept onto a conveyor for inspection.

Figure 1. Press & Blow Forming Process.

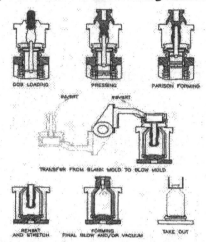

GOB LOADING   PRESSING   PARISON FORMING

INVERT   REVERT

TRANSFER FROM BLANK MOLD TO BLOW MOLD

REHEAT AND STRETCH   FORMING FINAL BLOW AND/OR VACUUM   TAKE OUT

FORMING MODELING AND HEAT TRANSFER

The sensitivity of glass to the processing history is such that it is critical to ensure comprehensive modeling to yield accurate design information. In particular, it is important to take an integrated approach that includes the effects of upstream processing conditions in order to control the final thickness distribution in the bottle.

Glass/mold heat transfer is well known in the literature where several studies have been carried out to evaluate the heat transfer within the glass and the contact layer to the mold. The models developed in conjunction with these studies generally employ one-dimensional numerical algorithms with more or less complex radiation models for the heat transfer in the glass and at the boundaries (references). While these models have provided a great deal of information concerning the glass/mold interface, they are all specific to the glass, mold and geometry selected for simulation. An additional problem with these models is that they do not account for the change in contact conductance due to pressure changes during pressing/blowing and thermal contraction of the glass at the mold interface. Moreover, one-dimensional models say nothing of the total heat transfer to the mold since this is heavily dependent on the flow behavior, glass/mold time of contact, and local temperatures within the mold cavity. However, if these models capture the basic physics of glass/mold heat transfer, they could be incorporated into three-dimensional forming models and applied to each glass node in contact with the mold. This would provide a comprehensive picture of the heat transfer between the parison or bottle and the molds at any given time for the forming conditions modeled while avoiding the limitations described above.

SEMI-INFINITE BODY HEAT TRANSFER MODELS

Since the penetration depth of the thermal boundary layer during the press is of the order $\delta \sim o(\sqrt{\alpha t})$, the ratio of parison thickness to thermal penetration distance is $O(10^{-2})$. This indicates that a semi-infinite approximation for the glass may be valid. Two models were developed to check how well the semi-infinite approximation holds for the heat transfer between the glass and blank mold. The first assumed perfect contact between the glass and mold, and the second approximated the thermal contact resistance as a convective heat transfer coefficient (the same approximation made by Storck et al. [1]). Both models resulted in explicit expressions for the heat flux and temperature distributions in the glass. Table I presents these results.

Table I. Semi-Infinite Body Model Analytical Expressions

	Perfect Contact	Contact Resistance
Transient Heat Flux (at surface)	$q_x'' = \dfrac{k(T_o - T_i)}{\sqrt{\pi \alpha t}}$	$q_x'' = (T_o - T_i)\left[h\exp\left(\dfrac{h^2\alpha t}{k^2}\right)\left[1 - erf\left(\dfrac{h\sqrt{\alpha t}}{k}\right)\right]\right]$
Temperature Distribution	$\dfrac{T(x,t)-T_o}{T_i-T_o} = erf\left(\dfrac{x}{2\sqrt{\alpha t}}\right)$	$\dfrac{T(x,t)-T_o}{T_i-T_o} = 1 - erf\left(\dfrac{x}{2\sqrt{\alpha t}}\right) - \exp\left(\dfrac{hx}{k} + \dfrac{h^2\alpha t}{k^2}\right)$ $* \left[1 - erf\left(\dfrac{x}{2\sqrt{\alpha t}} + \dfrac{h\sqrt{\alpha t}}{k}\right)\right]$
Contact Temperature	$\dfrac{T_c - T_m}{T_g - T_m} = \sqrt{\dfrac{(k\rho c_p)_g}{(k\rho c_p)_m}}$	$\dfrac{T_c - T_m}{T_g - T_m} = \dfrac{k_m}{k_g}\exp\left(\left(\dfrac{h^2\alpha t}{k^2}\right)_g + \left(\dfrac{h^2\alpha t}{k^2}\right)_m\right)$ $* \dfrac{1 - erf\left(\dfrac{h\sqrt{\alpha t}}{k}\right)_g}{1 - erf\left(\dfrac{h\sqrt{\alpha t}}{k}\right)_m}$

The advantage of having relatively simple expressions for the heat flux is that the result is generalized for all glass types and mold materials. Additionally, sensitivity and optimization studies can be completed very quickly by differentiating the flux expression with respect to the variable of interest.

Figure 2 shows a comparison of the perfect contact semi-infinite body model with the numerical data generated by FLUENT. The plots show almost no difference in either the instantaneous flux curve or the total heat transferred to the mold. Figure 3 shows a comparison of the numerical model of Storck et al. [1] and the semi-infinite body model that includes contact. Again, the agreement between curves is quite good. The deviation in plot 3 are mainly due to the time dependency of the contact resistance, which was included in the Storck data, but was not in the semi-infinite body model.

The close agreement between the semi-infinite body models and the more complex numerical algorithms points to several conclusions concerning glass to blank mold heat transfer. First, the temperature of the mold surface does not change nearly to the extent as that of the glass during press. This is evident from the expression for the contact temperature. Since the value of $(k\rho c_p)_g$ tends to be much smaller than $(k\rho c_p)_m$, the contact temperature remains much closer to the mold temperature than the initial glass temperature. Additionally, the heat flux to the mold is not very sensitive to mold properties. For a given glass composition and temperature, the only parameters that have a large influence on the instantaneous flux to the mold are the contact resistance and initial mold temperature.

Figure 2. Perfect Contact Model Comparison   Figure 3. Contact Resistance Model Comparison

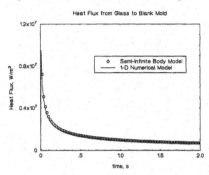 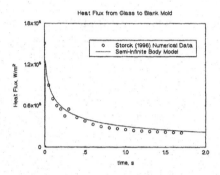

ESTIMATING THE CONTACT CONDUCTANCE

Experimental studies aimed at estimating the magnitude of the thermal contact conductance between the mold and glass have produced a wide range of results. For example, according to the data from various research workers, the step in temperature between the glass and the mold varies from 5 to 300 °C and the heat transfer coefficient in the contact layer changes from 2 kW/m²-K to 20 kW/m²-K. The wide variation in data is probably due to the different experimental conditions, the different properties of the glass being used, and the difference in mold surface properties.

In general, it is felt that the contact resistance between the glass and mold is the result of a thin gas layer consisting of air and the combustion products of the lubricant. This gas gap is normally separated into three components: the gap due to micrononuniformities of the mold surface, the increase in the gap due to thermal contraction of parison, and the decrease in the gap due to pressing/blowing pressure

At certain times during loading and pressing, the gap size is of the same order as the mean surface roughness of the mold (anywhere from 2 to 20 μm). Gaseous conduction across gaps of this size is a very different phenomenon than bulk conduction. This is because the mean free path of the molecules is of the same

Advances in Fusion and Processing of Glass III

order as the gap size. For example, the mean free molecular path of air is about 10 μm. In this case, molecular collisions are rare and heat transfer is the result of free-molecule conduction. Therefore, the contact conductance, h, needs to rewritten to take this effect into account:

$$h = \frac{k_{gap}}{(\delta_m + 2g)} \tag{1}$$

g is the temperature jump distance and depends on the accomodation coefficient and the thermophysical properties of the gas in the gap (see any text on kinetic theory of gases). δ_m is the mean gap size, and k_{gap} is the conducitivy of air between the mold and glass. The heat flux can then be determined through the standard convection heat transfer equation.

EXPERIMENTAL MEASUREMENTS

Experiments have been undertaken to identify the effects on various forming parameters on the contact conductance. The experimental requirements and objectives called for an apparatus and procedure that:

1. Allows the glass and mold temperatures, glass pressure and mold material to be independently varied.
2. Develops a consistent and rapid testing methodology to ensure accuracy and repeatability of results
3. Relatively simple geometry and setup to minimize effects which cannot be controlled

Figures 4 and 5 show the experimental setup. An off the shelf drill press was modified for the design to facilitate the lowering and clamping mechanisms needed for testing. Due to its high testing temperature, the glass was heated and tested in graphite crucibles. Small air holes were drilled in the bottom of the crucibles to allow the simulation of mold blow or pressing pressure on the glass. The crucibles were situated inside a cast iron vessel into which the air supply was piped. This vessel also served to provide structural support during clamping. A fast response erodable thermocouple was used to measure the glass surface temperatures during the contact between the glass and the simulated mold. A second thermocouple (also fast response) was imbedded in the mold 0.5 mm from the contact surface in order to measure the mold sub-surface temperature. A high temperature band heater was used to set the initial mold temperature. A high temperature controllable oven was used to set the initial glass temperature. Glass pressure, average mold temperature, mold surface temperature, and glass surface temperature were all recorded using high speed data acquisition.

Given the glass surface temperature and the mold subsurface temperature, the glass/mold heat flux can be determined. Equation 1 can then be used to calculate the contact conductance. Figures 6 and 7 show typical results. In Figure 7, the effect of glass pressure on the resultant contact conductance can be clearly seen.

As one would expect, as the pressure increases the resultant contact conductance increases. For all pressures, the characteristic exponential dropoff in contact conductance with time can also be seen. This is due to the thermal contraction of the glass away from the mold surface creating a larger gap between the mold and glass.

Figure 4. Experimental Apparatus   Figure 5. Thermocouple Placement

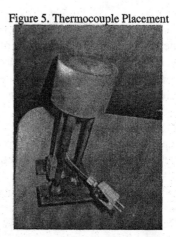

Figure 7 shows a comparison between the contact conductances of an amber and flint glass as a function of glass pressure for a contact time of 1 second. Both glasses show an increase in heat transfer with pressure. In general, the heat transfer coefficient tended to vary linearly with glass pressure for flint glass. This supports the hypothesis of Pchelyakov and Guloyan [2]. However, for amber glass, the function relationship between the contact conductance and glass pressure was $h \sim P^{0.7}$. Futher studies are currently underway to determine the effects of initial mold temperature, initial glass temperature, and mold material.

Figure 6. Contact Conductance vs. Time   Figure 7. Contact Conductance vs Pressure

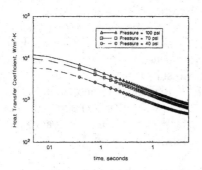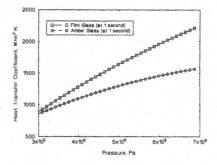

It is interesting to note that the amber glass results in a lower mold/glass heat transfer when compared to flint. This is most likely due to the radiative component of heat transfer which can make up a significant fraction of the overall

Advances in Fusion and Processing of Glass III

heat flux from the glass. The mean free path for radiation is much larger in the flint glass than the amber glass. This results in more radiative transfer from glass away from the mold surface in the flint glass when compared to the amber glass.

Figure 8 shows the estimated fractional contribution of heat flux due to radiation in a typical amber and flint glass. Figure 9 depicts the resultant glass surface viscosities and average glass temperatures. Note that while the surface temperature of the amber glass drops due to the lower radiative contribution from glass away from the surface, the overall temperature remains higher. The distance through which heat is transferred is effectively much smaller in the amber glass than in the flint glass.

Figure 8. Radiative Contribution to Heat Transfer   Figure 9. Surface Temperatures and Viscosities

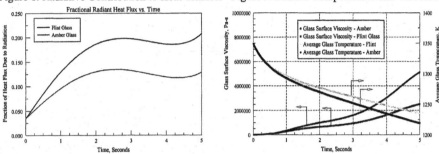

The higher surface viscosities in amber glass result in longer stress relaxation times. This is typical of what is widely known as a "quick-setting" glass. The advantage to using glasses that have relatively thin thermal boundary layers with large thermal gradients is that they produce a "stiff" parison which can be inverted without large deflections. Additionally, mold residence times can be reduced since the glass skin is created rapidly allowing for earlier movement of the parison or bottle. However, the primary negative effect is the larger stress relaxation times associated with the high viscosity region in the glass skin. High shear rates can create defects in the finish of the container, especially in narrow neck press and blow (NNPB) forming processes. This is because the glass is unable to relax the stresses created during forming. When the stresses in the glass exceed its yield stress, the glass structure will fracture. Typical defects resulting from "quick-setting" glasses include split finishes and pressure checks.

INCORPORATION OF HEAT TRANSFER DATA INTO FORMING MODELS

The semi-infinite body model combined with the experimental results allow the effects of mold properties and forming parameters on mold to glass heat transfer to be studied numerically. The spatial and temporal temperature distribution in the glass and mold are being determined by incorporating the resulting contact conductance heat transfer algorithms into three-dimensional forming models that simulate the fluid flow of the glass within the molds (see [3]).

A typical application of the heat transfer and forming models is the determination of effects of a change in heat transfer coefficient on the container forming process. The change in heat transfer coefficient may be due to swabbing the molds, a change in mold material, mold surface oxidation, or a reduction in pressing or blowing pressure. Table II shows the effects on IS machine timings and glass conditions after a drop in heat transfer coefficients due to a change in pressing pressure. In this case, the only change was the heat transfer coefficient between the glass and blank mold. The machine timings were left unchanged up to the inverted reheat. Here the timings were changed so that a bottle could be formed. If the inverted reheat timings were left unchanged, the amount of deformation would have been too great to fit the parison inside the blow mold without excessive defects. The blow side timings were also changed so that the container came out with the same average temperature.

Table II. Glass Conditions and Timings through IS Machine

	Normal IS Machine Operation			Reduced Blank Pressing Pressure		
	Time, s	Average Temp., K	Surface Temp., K	Time, s	Average Temp. K	Surface Temp., K
Press	0.785	1320	1168	0.785	1338	1235.5
Inverted Reheat	0.21	1315	1235	0.047	1337	1253
Reheat/Stretch	0.99	1296	1261	0.62	1320	1287
Final Blow	2.34	1109	990	2.94	1109	1001

As expected, the inverted reheat and reheat/stretch timings were reduced due to the softer parison skin. However, the final blow time was increased because of increased contact time requirements to remove sufficient heat for processing the bottle through ware handling. Overall, the forming time of the bottle did not change significantly (0.042 seconds). However, for this particular container, the production rate is limited by blow side contact time. The lower pressing pressure increases the required blow side residence time by 0.23 seconds. This effectively increases the time between bottles coming out of the blow mold by 0.23 seconds. A similar increase in required blank side residence time would have had little impact since it is not a critical path in the forming cycle. Therefore, the better strategy in this case would be to increase blank side residence time with the aim at keeping blow side residence time constant.

CONCLUSIONS

More validation of the forming models is needed to fine-tune the parameters with significant uncertainties (mold/glass slip, contact conductance, and glass effective conductivity). Additionally, the effects initial glass and mold temperatures, glass types and mold materials need to be included in the expression for the contact conductance. This is the focus of ongoing experimental work. It does appear, however, that a methodology now exists to test the effects of various disturbances on container forming.

REFERENCES

[1] Storck, K., Loyd, D., and Augustsson, B., "Heat Transfer Modelling of the Parison Forming in Glass Manufacturing," *Glass Technol.*, **39** 210-216 (1998).

[2] Pchelyakov, S.K., and Guloyan, Y.A., "Heat Transfer at the Glass Mold Interface," *Steko i Keramika*, **9** 14-15 (1985).

[3] Hyre, M.R., "Numerical Simulation of Glass Forming and Conditioning," *J. Am. Cer. Soc.*, **85** 1-10 (2002).

INVESTIGATIONS ON STICKING TEMPERATURE AND WEAR OF MOLD MATERIALS AND COATINGS

Daniel Rieser, Peter Manns, Gerd Spieß, Günter Kleer
Fraunhofer-Institut für Werkstoffmechanik IWM
Wöhlerstraße 11
79108 Freiburg i.Br.
Germany

ABSTRACT

For the characterization and assessment of the service behavior of mold materials and mold coatings for glass forming processes, like e.g. glass container manufacturing, a testing procedure was developed and a corresponding laboratory device was built. The testing procedure mimics industrial working conditions, where the mold material specimens are subjected to a cyclic loading by pressing small gobs from a melting crucible under defined non-isothermal conditions. The testing facility enables investigations with different glass types (soda-lime glass, lead-crystal glass) and variable pressing parameters.

In the present investigations, the sticking behavior as well as corrosion and wear of mold materials are investigated. The effect of sticking is quantified by recording the time interval necessary for separating the glass blank from the mold surface using a defined separation procedure. The sticking characteristics of the tested mold materials and coatings are described by two material specific values: the "lower" and the "upper" sticking temperatures, which are different for different glass types. The "lower sticking temperatures" were found to depend monotonically on the thermal effusivity of the mold materials, whereas the "upper sticking temperatures" should also depend on further quantities like surface roughness and possibly also on chemical and structural properties of the mold surfaces.

A simplified physical approximation leads to the calculation of the "contact viscosity" at sticking conditions. This model and the experimental results indicate that glass melts would stick to the molds, if a characteristic viscosity value was reached.

Corrosion and wear of the various mold materials are quantified and exhibit significant differences in the short time laboratory tests as the pressing experiments were conducted in the region of the respective "sticking temperatures".

INTRODUCTION

In glass article manufacturing processes like e.g. glass container production or optical lens molding the mold materials are subjected to high thermal, mechanical, chemical and tribological loadings which lead to wear and abrasion of the molds. Therefore, the glass manufacturing industry demands improved mold materials and coatings which permit increased service life time of the molds and the production of high quality molded glass articles. For the development of improved mold materials several physical properties of the materials, e.g. hardness, texture, thermal conductivity, and oxidation resistance can be characterized and assessed in the metallurgical laboratory. However, quality of the formed glass articles and service life of the molds are strongly determined by the contact behavior of the molds with the hot glass melts. Particularly, wear, hot corrosion and the so called "sticking temperature" are considered as important properties of the molds in the glass forming process [1]. For the systematic development of materials and coatings for molds with regard to higher "sticking temperature" and lower wear, a reliable testing method is required, by which the behavior of mold materials can be characterized in laboratory tests with small specimens.

Numerous investigations on sticking behavior of mold materials with glass melts have been conducted during the past decades [2-8] and often led to contradictory results and even to contradiction with industrial experience. However, a big disadvantage for materials research is that until now, neither a precise definition of "the sticking temperature" nor a commonly accepted or standardized testing procedure has been developed.

In a first step to overcome these difficulties, a new testing method and respective testing device were developed, by which the sticking behavior of mold materials and coatings can be characterized and which also comprises quantification of corrosion and wear behavior of mold materials on a laboratory scale.

EXPERIMENTAL METHOD

The applied testing method uses a non-isothermal process similar to industrial manufacturing processes. The mold material specimens under test are subjected to cyclic loading by pressing small glass gobs under precisely defined process conditions. Wear of the mold surfaces, quality and surface structure of the glass as well as the sticking behavior are determined as a function of the process parameters - especially as a function of the mold temperature.

The glass melt from a melting crucible is portioned to small gobs of constant mass (2.5 g) which fall through a transfer channel into the pressing unit. The pressing unit consists of a symmetrical setup of lower and upper die which hold the mold material specimens. The mold specimens have plane plate geometry and are clamped to heating units which are controlled to achieve preset steady state temperature in the mold specimens. The temperature of the mold specimens is measured with soldered-in thermocouples in the center of the pressing area, 2.5 mm below the mold surface. The sturdiness of sticking is quantified by recording the time necessary for separating a pressed glass blank from the mold surface using a special cooling procedure, which avoids damage to the mold surfaces. By adjusting a temperature difference of 20 K between lower and upper mold, sticking of the pressed glass blanks occurs only at the hotter upper mold. The testing device is shown schematically in Figure 1, further details are described in [9].

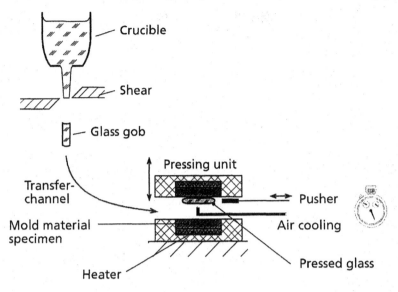

Figure 1: Schematic drawing of the testing device, which consists of a melting crucible, servo-electrical parallel shear, transfer channel, pressing unit with heated inserts, cooling and pushing mechanism for separating the pressed glass blanks from the mold, devices for controlling and recording the process parameters.

Pressing load and pressing time were kept constant throughout these investigations, i.e. 2.5 MPa, 5 s. The terms "mold temperature" and "sticking time" used in this paper denote the steady-state temperature of the upper mold before the glass gob touches the mold, and the time interval between lifting of the upper die and the separation of the glass from the mold surface, respectively.

RESULTS

WEAR AND ABRASION

Surface roughness of the mold material specimens was measured by stylometry before and after each pressing series for investigation of sticking behavior. Before the tests, the mold inserts were polished to a surface roughness of about $R_{rms} \approx 0.05$ μm. The pressing experiments were conducted in the region of the respective sticking temperatures of the mold materials. As shown in Figure 2, the increase of surface roughness, i.e. the wear of the various mold materials under test can be quantified as function of service time by this testing method in short time laboratory experiments. For further results on abrasion of mold materials and transfer of particles from the molds to the glass surfaces it is referred to [9].

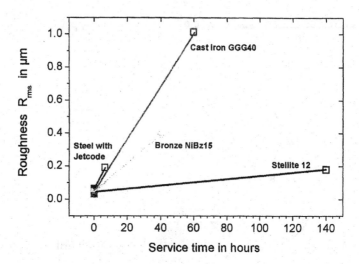

Figure 2: Increase of surface roughness of the mold materials under test plotted as a function of service time (linear interpolation).

STICKING TEMPERATURE

In the pressing experiments it was found that the extent of sticking of glass blanks to the surface of the pressing molds depends on several parameters: glass temperature, mold temperature, molding pressure, mold surface roughness, type of mold material, glass type, etc., with the strong indication that the mold temperature is a major parameter determining the sticking behavior. With keeping all other parameters constant, the intensity of sticking increases monotonically with mold temperature and increases steeply within a rather small temperature interval. A typical sticking curve of soda-lime glass (B270) with a commercial mold material, Stellite, is shown in Figure 3. All the mold materials and coatings tested in this investigation exhibit essentially a very similar behavior, with distinct differences in the temperature of onset of sticking as well as in the slope of the curves. In order to describe the specific variations in sticking behavior, the curves are characterized by two values: a "lower sticking temperature" and an "upper sticking temperature" (see Figure 3). The "lower sticking temperature" T_{Lower} was defined as the lowest mold temperature at which pressed glass gobs stick to the mold. Below this characteristic temperature no sticking of glass blanks to the mold was observed, above T_{Lower} all pressed glass blanks sticked to the mold. The "upper sticking temperature" T_{Upper} was defined as the lowest mold temperature at which glass blanks stick very firmly to the mold surface with sticking duration of more than 60 seconds [9].

Measurements on sticking behavior were conducted with various mold materials and coatings and with two different types of glass, i.e. soda-lime glass (B270) and lead-crystal glass containing 24% PbO. Results of the measurements of these mold materials are summarized in Figure 4. Note that the lower sticking temperatures increase monotonically with the thermal effusivity of the mold materials.

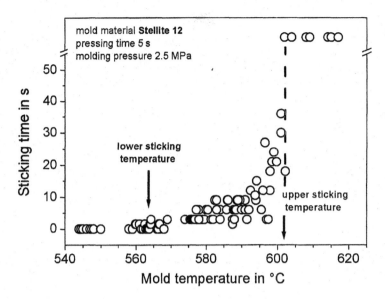

Figure 3: Experimental sticking curve of mold material Stellite 12 for pressing soda-lime glass melt with the pressing parameters indicated. The characteristic points of the sticking curve, upper and the lower sticking temperature are marked.

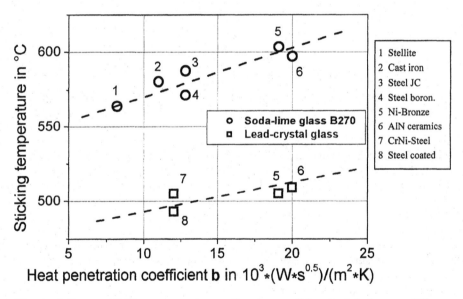

Figure 4: Lower sticking temperatures of various mold materials for two different glass types plotted as a function of the heat penetration coefficient b (thermal effusivity) of the mold materials (values for b at 500 °C are taken from literature).

DISCUSSION

The experimental results on lower sticking temperature of the mold materials tested in this investigation are in good qualitative agreement with empirical experiences concerning temperature regions where sticking is observed in industrial glass manufacturing, thus providing evidence for the suitability of this testing procedure.

The experimental result that sticking temperature increases with thermal effusivity (heat penetration coefficient) of the mold materials can be explained by an approximation assuming an "ideal contact" of glass melt and mold surface in the case of sticking. Starting with different temperatures of glass and mold, the resulting temperature distribution inside the glass and mold as function of time and distance from the interface can be derived by solving the differential Fourier Equation, which takes into account phononic heat transfer only [10]:

$$\frac{\partial^2 T(x,t)}{\partial^2 x} - \frac{1}{a} \cdot \frac{\partial T(x,t)}{\partial t} = 0 \tag{1}$$

With the boundary conditions at the interface, equality of temperature and heat flow, the analytical solution can be written as

$$T(x,t) = T_{mold} + \left(T_{contact} - T_{mold}\right) \cdot erfc\left(\frac{x}{2 \cdot \sqrt{a_{mold} \cdot t}}\right) \tag{2}$$

By dint of this approximation, the introduced parameter $T_{contact}$ the "contact temperature" in the interface, is independent of time and can be calculated:

$$T(0,t) = T_{contact} = \frac{b_{glass}}{b_{glass} + b_{mold}} T_{glass} + \frac{b_{mold}}{b_{glass} + b_{mold}} T_{mold} \tag{3}$$

with the thermal effusivity b of the respective materials, mold and glass

$$b = \sqrt{\lambda \cdot c_p \cdot \rho} \tag{4}$$

Using the experimentally determined values of the "lower sticking temperatures" from Figure 4 to calculate "contact temperatures" at the onset of sticking from Eqn. 3 yields almost equal values of contact temperatures for all mold materials tested, i.e. 630±8 °C with soda-lime glass and 567±8 °C with lead-crystal glass. Referring to the corresponding viscosity curves for these glasses, the respective "contact viscosity" η_C in the glass-mold interface can be calculated, resulting in $\eta_C = 10^{8.5}$ Pas for soda-lime glass and $\eta_C = 10^9$ Pas for lead-crystal glass as shown in Figure 5. The apparent difference in "contact viscosity" for the different glasses may be attributed to several factors which have to be analyzed further.

Putting up the hypothesis that the "contact viscosity" at onset of sticking is equal for all mold materials and glass types with other process parameters (molding pressure, mold surface roughness, etc.) kept fixed, transforming Eqn. 3 yields an equation for calculating the "lower sticking temperature" as a function of the thermal effusivity of mold materials:

$$T_{mold} = T_{contact} + (T_{contact} - T_{glass}) \cdot \frac{b_{glass}}{b_{mold}} \tag{5}$$

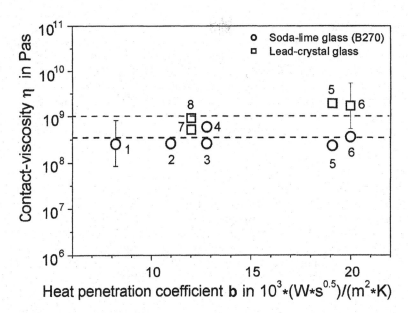

Figure 5: Calculated "contact-viscosity" from ideal contact approximation at the "lower sticking temperature" for two glass types and all mold materials tested in this investigation as function of the heat penetration coefficient of mold materials

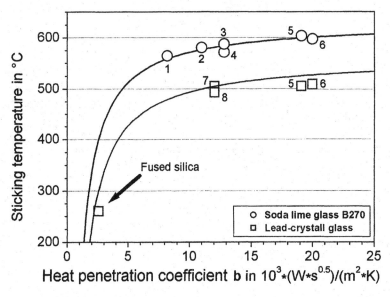

Figure 6: Experimental values of lower sticking temperatures of mold materials for two different glass types and calculated curves for "lower sticking temperature" as function of the heat penetration coefficient b of mold materials.

Using the mean value of the "contact viscosity" $\eta_C = 10^{8.8}$ Pas from the data shown in Figure 5, Eqn. 5 leads to non-linear curves for sticking temperatures as function of the heat penetration coefficient b of the mold materials for both glass types, as shown in Figure 6. According to this hypothesis of constant contact viscosity at onset of sticking, mold materials with small thermal effusivity should exhibit very low sticking temperatures compared to standard mold materials.

In a preliminary experiment a non-metallic model material with a low b value, fused silica, was tested and a "lower sticking temperature" of $T_{Lower} \leq 260$ °C was found, in good agreement with the prediction.

SUMMARY AND OUTLOOK

From measurements on contact behavior of two types of glass with various mold materials (metals, ceramics, and coatings) carried out with a new testing procedure, a definition of two characteristic temperatures was derived for the description of the sticking behavior of each mold material. The "lower sticking temperature" increases monotonically with thermal effusivity of the mold materials, which can be explained quantitatively by the simplified approximation of an "ideal contact". The experimental data support the hypothesis, that independent of the type of mold material (metal, ceramics, glass), glass melts will begin to stick to smooth mold surfaces if the "contact-viscosity" in the interface drops below the critical value of approximately $\eta_C \approx 10^{8.8}$ Pas.

Future investigations will be focused on the influence of process parameters like e.g. molding pressure, pressing time, surface roughness etc. and experiments with selected mold materials and coatings and additional types of glass melts in order to acquire precise data for refining the approximation of "contact viscosity" and for a better understanding of the sticking phenomena.

REFERENCES

[1] Tanhöfner, S.: Analysis of various plunger materials used in container glass industry. Glastech. Ber. Glass Sci. Technol. 72 (1999) No. 11, p. 335-340

[2] Kapnicky, J.A.; Fairbanks, H.V.; Koehler, W.A.: Study of adherence of molten glass to heated metals. J. Am. Ceram. Soc. 32 (1949) No. 10, p. 305-308

[3] Fairbanks, H.V.: Effect of surface conditions and chemical composition of metal and alloys on the adherence of glass to metal. In: Symposium sur le contact du verre chaud avec le métal; Scheveningen 26.-29.05.1964. Charleroi: Union Scientifique Continentale du Verre 1964, p. 575-595

[4] Smrcek, A.: Sticking temperature of glass to metals // Part 1: The influence of the solid phase // Part 2: The influence of the glass phase // Part 3: The influence of the contact conditions on sticking temperature. (Orig. Czech.) Silicáty 11 (1967) p. 267-277, 339-344, 345-351

[5] Kluge, W.-D.: Ein Beitrag zum Kontaktverhalten von Glas mit Formgebungswerkzeugen (A contribution to the contact behavior of glass with molding tools. Thesis. Sektion Verfahrenstechnik und Silicattechnik, Bergakademie Freiberg 1989

[6] Manns, P.; Döll, W.; Kleer, G.: Glass in contact with mould materials for container production. Glastech. Ber. Glass Sci. Technol. 68 (1995) No. 12, p. 389-399

[7] Falipou, M.; Donnet, C.; Maréchal, F.; Charenton, J.-C.: Sticking temperature investigations of glass/metal contacts — Determination of influencing parameters. Glastech. Ber. Glass Sci. Technol. 70 (1997) No. 5, p. 137-145

[8] Pahnke, K.: Wear mechanisms of forming tool materials in glass pressing process. Klei Glas Keram. 23 (2002) No. 3, p. 10-13

[9] Manns, P.; Rieser, D.; Döll, W.; Kleer, G.; Spieß, G.: Sticking and wear of mold materials. Part1: Experimental method and testing device. Glass Sci. Technol. (2003) , to be published

[10] Carslaw, H.S.; Jaeger, J.C.: Conduction of heat in solids. (2. Ed.) Oxford: Clarendon 1959

BASIC CONSIDERATIONS AND TECHNICAL ASPECTS CONCERNING GLASS CONDITIONING

H. Hessenkemper
TU Bergakademie Freiberg (Germany)

ABSTRACT

The conditioning of a glass melt is an important part of the glass production due to the decisive part of the temperature in the forming process. The classical view towards the aim and the possibilities of the glass conditioning is examined and several problems are worked out. New possibilities concerning different techniques in the conditioning and thermal influence of the gob are discussed as a tool for future increase in productivity and economic results.

I. INTRODUCTION

The forming process is strongly dependant on the heat extraction from the melt. This sensitive process is time and place dependant and has to be repeatable within a small variation of the physical parameters. The viscosity is one of the special properties of decisive importance to the forming of glass melts. The temperature dependence of the viscosity is given by the Vogel-Fulcher-Tammann equation:

$$\log \eta = A + B/(T-To) \quad (1)$$ where T = Temperature and A,B,To = constants.

The viscosity is proportional to the relaxation time τ of the stress relaxation in a first approximation and a simple time dependant stress is described by

$$S(T) = S_o \exp(-t/\tau) \quad (2)$$

The relaxation process is important for the time dependant stress during the forming process, the brittleness of the melt and possible surface damages which can result in a lower mechanical strength of the glass. With the nearly exponential coupling of the viscosity and the temperature Eq. (1), the approximation $\eta \sim \tau$ and regarding Eq. (2) it became obvious, that the temperature is a parameter strongly influencing the time dependent mechanical stresses during the forming process and all related attributes. In addition the time dependant viscosity is directly linked to the thickness of the glass layer by constant stresses.

It is obvious that the temperature situation of the glass melt is the most important parameter of the forming process. Different interconnected parameters do influence the heat balance, i.e. the design, heat capacity and cooling of the mould. The specific heat input by the melt, the heat transfer by heat conduction, by radiation and so on. Figure 1 shows different temperatures in a forming process just due to different spectroscopic properties of the melt. The importance of these changes is well known and is obvious regarding the discussion up to now.

In this complicated situation it is a strong demand to keep a stable temperature situation of the gob and therefore the glass conditioning is seen as one of the important fields in the glass industry to reach a good productivity, weight reduction and other aspects to earn money in complicated industrial surrounding.

Therefore it is useful to examine the situation and try to find additional solutions for optimising the situation.

II. THE GOB SITUATION

The aim of the glass conditioning should be a homogeneous temperature of the gob. This is the most common answer, but is this right? Just assuming that the gob has really this property this is not the situation which is seen in the beginning of the forming process. The heat losses by radiation and convection cooling are quite big during the gob forming, cutting and gob delivery to the blank mould. In these 1-3 seconds the heat losses for a dark melt are estimated and shown in Fig. 2. The integral temperature losses reach 10°C, but the losses are focussed in the surface area which is shown assuming the heat losses coming from 10% or 20% of the diameter. It is seen that even the shape of the gob has a certain impact to this instability.

Is the homogeneous temperature of the gob really the best situation for an optimised forming process? Probably not. The forming of the finish has to have other time dependence than the rest of the glass melt because the finish has to carry the bottle to the final blow process. This is known and it is tried to get a special heat situation by using different materials and forming tools for the finish.

The situation is that a certain gradient of the temperature along the gob could be ideal. Instead of this a different temperature gradient from to surface to the volume is reality.

Figure 3(a) and (b) show a different influence. By using the FLUENT program the temperature change under isothermal conditions is shown for different heat conduction due to changed spectroscopic properties of the melt. The starting point is a temperature inhomogenity as a step function of 10°C. Again strong time dependence and a difficult interaction can be seen.

A solution to come along with these difficulties has been presented[1] and is demonstrated Fig. 4. There are different possibilities existing for a better glass conditioning. First the temperature gradient between surface and volume can be minimised. Second a gradient can be produced for a faster cooling of the finish without mechanical problems. Third a special problem can be solved which is existing in the glass industry. Normally the chemical length defined in Fig. 5, is fitted to a special forming situation. In a furnace feeding several lines with products of different weight there is always not an ideal situation for all production lines concerning the length of the melt. Here the change of the heat situation using the tools being presented in Fig. 4 enables the operator to reach always an optimum and to improve the workability.

III. THE GLASS CONDITIONING

Regarding the situation of the feeder and glass conditioning it seems quite clear, that is a very complicated problem to reach a homogeneous temperature of the melt. Some own measurements and calculations demonstrated that even a new glass conditioning system looses a thermal power of about 200 kW. This energy is lost by different means through surface of the melt into the surrounding, on the other side there is an additional heating in a similar range. With about 50 tons of refractory material every regulating of changed temperature from the furnace or different heat losses into the surrounding is very complicated due to the long time reaction for any feedback. The classical thermocouple measurement and the calculation of the feeder efficiency is not the whole truth, because effects of the area between refractory and melt are not seen. Here a lot of difficulties are created. The situation is demonstrated in Fig. 6. In addition the situation became even more uncontrolled reaching the spout with a complicated system of geometry, heat losses and reheating.

A possible solution is again presented[1] and is apparent from Fig. 7. The idea is to combine different methods for a more stabilised temperature situation. Starting with passive thermal tools like a high emission coefficient of the refractory material and different kinds of mixing the melt, starting with specialised bubbling techniques and using an intelligent stirrer system. This system should be able to adapt to the result of the temperature distribution after the stirrer as a feedback and with several free physical parameters of the stirrer.

To reach the aim with such a combined system it is necessary to do a lot of fundamental work in the range of material science, measurement techniques and engineering.

IV. SUMMARY

It has been shown that the glass conditioning has a major impact to the success of the forming process, because the controlled extraction of heat is a decisive part of the forming. In spite of great efforts in the conditioning the aim and the results show different problems. Therefore new techniques are discussed to improve the glass conditioning for the higher demands of the future glass production

[1] H. Hessenkemper: Optimierung der Glaskonditionierung und Formgebung, Proceedings 75. Glastechnische Tagung Wernigerode, Page 221-226

Acknowledgement: This article was previously published in *The International Glass Journal*. It is published here with permission of *The International Glass Journal* and with the agreement of The American Ceramic Society.

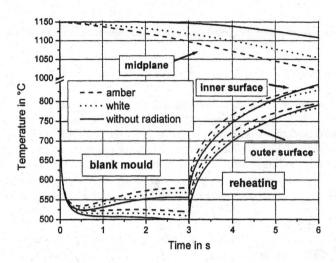

Fig. 1. Different Temperature profile due to spectroscopic changes

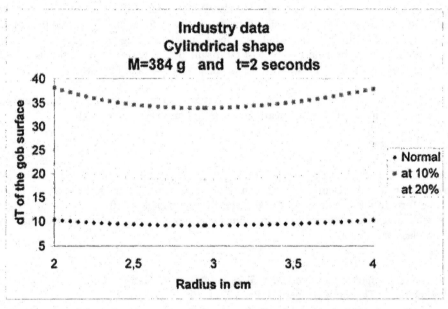

Fig. 2. Changes of the gob temperature due to heat losses, focused on different surface volume

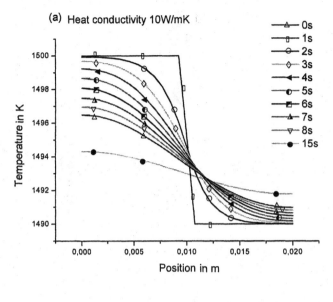

(a) Heat conductivity 10W/mK

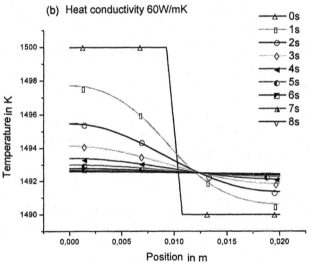

(b) Heat conductivity 60W/mK

Fig. 3 (a), (b). Time dependent temperature situation in a gob with a temperature disturbance for different heat conductivities

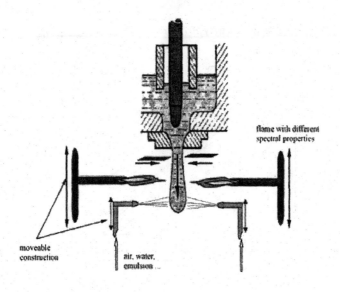

Fig. 4. Tools for changing the gob temperature situation

Correlation between viscosity and temperature of different glasses

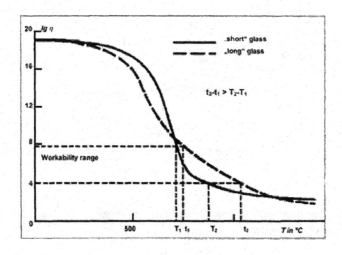

Fig. 5. Chemical length of a glass melt

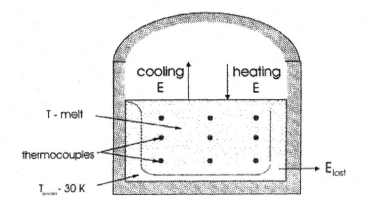

$$E_{lost} = f \text{ (convection, radiation, isolation)}$$
$$\eta_{border} = 25\ \% \text{ higher viscosity}$$

Fig. 6. Temperature situation in the glass conditioning

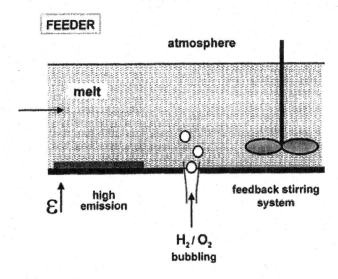

Fig. 7. Possible tools for an improved melt conditioning

Polyvalent Elements and Redox Behavior

REDOX-DEPENDENT GLASS PROPERTIES AND THEIR CONTROL UNDER INDUSTRIAL CONDITIONS

Hayo Müller-Simon
Research Association of the
German Glass Industry (HVG)
Siemensstraße 45
63071 Offenbach

INTRODUCTION

Polyvalent elements play an important role in coloring and refining of industrially produced glasses. Although even today industrial glass making is still done by intuition in large part, in many cases glass technology and the results of basic research are useful aids in glass production procedure. In this paper the basic principles of redox reactions in glass melts are compiled and applied to industrial glass production.

REDOX-DEPENDENT GLASS PROPERTIES

The most apparent redox-dependent property is color which characterizes the appearance of a glass. The absorption and emission behavior also plays an important role in heat conductivity of the melt in the glass melting tank as well as in the mould. Viscosity, visco-elastic behaviour and surface tension determine the flux behavior inside the glass melting tank and the glass´ workability. The refining behavior, i. e. the gas release with increasing temperature, of the glass melt critically affects the quality of the finished product. Obviously, redox reactions influence not only the product properties but also the glass production process in its many stages.

BASIC RELATIONS

Polyvalent elements can change their valence state by reacting with molecular oxygen. Iron may serve as a practical example because it is found in nearly all glasses, due to raw material contamination or additions as a coloring agent. Also, iron has the advantage that it behaves in a straightforward way in glass melts. The fundamental redox reaction of the pure iron oxides is

$$2\text{FeO} + \frac{1}{2}\text{O}_2 \underset{\leftarrow}{\overset{\rightarrow}{}} \text{Fe}_2\text{O}_3 \qquad (1)$$

If one assumes the complete dissociation of iron in glass melts according to

$$FeO \rightarrow Fe^{2+} + O^{2-} \tag{2}$$

and

$$Fe_2O_3 \rightarrow 2Fe^{3+} + 3O^{2-} \tag{3}$$

combination of reactions 1 to 3 gives

$$Fe^{2+} + \frac{1}{4}O_2 \underset{\leftarrow}{\overset{\rightarrow}{}} Fe^{3+} + \frac{1}{2}O^{2-} \tag{4}$$

The thermodynamic behavior of a reaction as given by equation 4 is described by van't Hoff's equation:

$$\Delta G = \Delta G^0 + RT \ln \left(\frac{a_{Fe^{2+}} \, a_{O_2}^{n/4}}{a_{Fe^{3+}} \, a_{O^{2-}}^{n/2}} \right) \tag{5}$$

In the context of this paper only equilibria will be discussed, i. e. $\Delta G = 0$. Furthermore, ideal dilution with respect to Equations (2) and (3) is assumed, i. e. the activity coefficients $\gamma_{Fe^{2+}} = \gamma_{Fe^{3+}} = 1$. Fundamental difficulties in the conclusive description of redox reactions in glass melts are the several bonding conditions of oxygen in oxidic glasses and glass melts:

- network formers are connected by bridging oxygen O^0. Both valences of the oxygen are bound by covalent bonds.
- some oxides dissociate completely when they are introduced into glass melts, for instance alkaline elements. These elements form free oxygen ions O^{2-}. The oxygen ion activity $a_{O^{2-}}$ is commonly related to basicity of the glass.
- bridging oxygen and free oxygen ions react according to

$$O^0 + O^{2-} \underset{\leftarrow}{\overset{\rightarrow}{}} 2O^- \tag{6}$$

providing non-bridging oxygens O^-, which combine covalent and ionic bonding behavior. Because of Equation (6), the basicity reflects the mixture of covalent and ionic bondings.
- molecular oxygen O_2 can be physically dissolved in glass melts. Experimental results show that the physically dissolved oxygen behaves nearly according to the Henry's Law [1]. Under typical glass

melting conditions the pressure is sufficiently low so that the equality $a_{O_2}=p_{O_2}$ is valid.

The fact has often been stressed that the relation between basicity and redox ratio in Equation (4) contradicts the experimental findings where the concentrations of the polyvalent elements in their oxidized state increase with increasing basicity. The reason is that the basicity is not controlled by Equation (4) but by the alkaline and earth alkaline concentrations and Equation (6). The contribution of reaction (4) to the oxygen ion activity is only small as the concentrations of the polyvalent elements are about 100 times smaller than the concentrations of network modifiers. Additionally, the oxygen ion activity is buffered by reaction (6), thus, with respect to reaction (5), $a_{O^{2-}}=1$ can be assumed. In summary, Equation (5) becomes

$$\Delta G^0 = -RT \ln \frac{[Fe^{2+}]}{[Fe^{3+}]} \cdot p_{O_2}^{\frac{1}{4}} \qquad (7)$$

or in a more convenient way and extended to any polyvalent element A,

$$K(T) = \frac{[A^{x+}]}{[A^{(x+n)+}]} \cdot p_{O_2}^{\frac{n}{4}} = e^{-\frac{\Delta G^0}{RT}} = e^{-\frac{\Delta H^0 - T \cdot \Delta S^0}{RT}} \qquad (8)$$

The validity of this relation has been experimentally confirmed for many polyvalent elements in glass melts by different measuring techniques [2, 3].

The elimination of the basicity from the thermodynamic considerations means that each ΔG^0 is related only to a particular glass composition. This is a reasonable limitation because most of the industrially melted glasses such as fiber, float and container glasses have a very similar composition. In these cases the dependence of the redox ratio on oxygen partial pressure and temperature can be simply described by one set of ΔH^0 and ΔS^0 for every polyvalent element. In case of special glass production such as TV or optical glass the basicity varies considerably. Some practical scales have been developed which describe the influence of the basicity on the redox ratios of polyvalent elements [4, 5, 6].

DEFINITION OF REDOX SERIES
According to Equation (8), the properties imparted by a particular polyvalent element are determined by the oxygen partial pressure at melting temperature. However, during cooling, pairs of polyvalent elements can exchange electrons, thus, the redox states may change. In practice both the oxygen partial pressure and the interaction during cooling define the state in the final glass product. The respective influence can be associated with two redox series: an isothermal and a non-isothermal redox series.
Isothermal redox series

Redox series are well-known for redox reactions in aqueous solutions. These reactions can occur either between a metal electrode and a salt of the metal which is dissolved in the solution, or as the reaction of two different redox states of a dissolved polyvalent element at the surface of an inert electrode. For any electrode reaction the electrode potential is given by the Nernst equation

$$E = E^0 + \frac{RT}{nF}\ln\frac{a_{ox}}{a_{red}} \qquad (9)$$

According to Equation (9) the measured potential is equal to the standard potential E^0, if the activities $a_{ox}=a_{red}$. This can be achieved if a setup is chosen where all activities are unity, for instance a metal electrode in a solution which is saturated with the salt of the metal. If this is not possible, amperometric methods can be used to measure E^0. However, in practice it is only possible to measure a voltage between two electrodes, thus, E^0 in Equation (9) can not be measured absolutely. In aqueous solutions E^0 is usually related to the standard hydrogen electrode. The resulting redox series shows which metal will dissolve if two metals are short-circuited and which salt will corrode which metal.

The same principles are valid in glass melts, only the standard conditions for temperature and pressure must be chosen appropriately. A useful reference is the oxygen flushed zirconia electrode, which relates the electrode reaction to the cathodic partial reaction of Equation (4). Using an inert electrode in combination with a zirconia reference, Equation (9) characterizes the principle of oxygen partial pressure measurement in glass melts.

In glass melts metal electrodes are immediately covered with an oxide layer. Simultaneously, in the range of the oxygen partial pressures observed in oxidic glass melts most polyvalent elements exhibit more than one oxidation state. Thus, a direct measurement of E^0 is impossible. However, numerous standard potentials are provided by voltammetric studies [3]. Table I shows an electrochemical redox series of polyvalent elements in soda-lime-silica glass melts related to the oxygen flushed zirconia electrode. The series is referred to a fixed standard temperature, thus, it may be called isothermal redox series. Because of the relation

$$-nFE^0 = \Delta G^0 \qquad (10)$$

the basic quantity of the series defined by the standard potentials is the standard free enthalpy.

Table I. Isothermal redox series at 1100 °C. Data taken from [3].

Ce^{3+}/Ce^{4+}	Sb^{3+}/Sb^{5+}	Cr^{3+}/Cr^{6+}	Sn^{2+}/Sn^{4+}	As^{3+}/As^{5+}	Fe^{2+}/Fe^{3+}
+80	+50	+30	-110	-170	-560

Non-isothermal redox series

A different redox series can be defined, if the optical spectra of glasses containing a polyvalent element alone and together with a second one are compared [7, 8]. Even if the glasses have been melted at the same temperature and the same oxygen partial pressure, in most cases an electron exchange reaction between polyvalent elements A and B of the type

$$mA^{x+} + nB^{(y+m)+} \underset{\leftarrow}{\overset{\rightarrow}{}} mA^{(x+n)+} + nB^{y+} \qquad (11)$$

can be observed. Comparing all pairs of polyvalent elements a redox series can be established using the interaction matrix shown in Table II. In the interaction matrix every cell is associated with two redox couples. The series in the columns and lines are the same. Using optical data the matrix is filled in such a way that the column is determined by the redox couple which is oxidized and the line by the redox couple which is reduced. The advantage of this interaction matrix is that information of pairs of polyvalent elements are also considered which are not adjacent in the series. If only the silicate glasses are considered, the matrix obviously shows the true redox series. Some elements exchange their places if the composition changes. In table II, glasses with a large variation of sodium [8] are marked by 1) and sodium borate glasses with very low sodium content by 2).

This type of redox series compares the properties of glasses at room temperature while the redox state is fixed at melting temperature, thus, this type of redox series may be called non-isothermal. It can be shown that the non-isothermal redox series can be related to the standard enthalpy, ΔH^0, of the redox couples [9]

$$\frac{\Delta H^0_{red}}{n_{red}} < \frac{\Delta H^0_{ox}}{n_{ox}} \qquad (12)$$

The non-isothermal redox series in Table II is completed by sulfur [10] and selenium [11] in Table III for soda-lime-silica glasses. This redox series has been developed by considering thermodynamic data as well as chemical and optical behavior. If the results of optical data were available, they have been preferred, otherwise the series has been established using $\Delta H^0/n$ values, which are also shown in Table III.

Table II. Interaction matrix of interaction between pairs of polyvalent elements.
● soda-lime-silica glass, O binary silicate, ■ borosilicate, □ binary borate, ◎ phosphate, ◆ alumosilicate.

oxidized→ ↓ reduced	Cr^{3+}/Cr^{6+}	Mn^{2+}/Mn^{3+}	Ce^{3+}/Ce^{4+}	As^{3+}/As^{5+}	Sb^{3+}/Sb^{5+}	Fe^{2+}/Fe^{3+}	Sn^{2+}/Sn^{4+}
Cr^{3+}/Cr^{6+}		●O	■1)	O		●□	
Mn^{2+}/Mn^{3+}	□2)		■●	●	●	■●□◎	
Ce^{3+}/Ce^{4+}	■1)	□2)				■●◆	●
As^{3+}/As^{5+}						●	
Sb^{3+}/Sb^{5+}						●	
Fe^{2+}/Fe^{3+}							O
Sn^{2+}/Sn^{4+}							

Table III. Non-isothermal redox series and $\Delta H^0/n$ values in soda-lime-silica glasses.

Se^{4+}/Se^{6+}	Cr^{3+}/Cr^{6+}	Mn^{2+}/Mn^{3+}	Se^{2-}/Se^{0}	Ce^{3+}/Ce^{4+}	As^{3+}/As^{5+}
23	25-28	37-71	54-63	40-78	58

Sb^{3+}/Sb^{5+}	Se^{0}/Se^{4+}	S^{2-}/S^{4+}	Fe^{2+}/Fe^{3+}	Sn^{2+}/Sn^{4+}	S^{4+}/S^{6+}
72-80	81	87-90	81-121	107	118-125

Relation between isothermal and non-isothermal redox series

Obviously, the two redox series in Tables I and III provide slightly different orders. This is not surprising as the information which is contained in the isothermal redox series is determined at melting temperature while in the non-isothermal redox series the state of the system is defined at the melting temperature and the detected property is measured at room temperature. This difference is expressed by the different basic quantities: the isothermal redox series is governed by ΔG^0 and the non-isothermal redox series by $\Delta H^0/n$.

Nevertheless, it has been widely discussed in the past whether optical redox series may be related to an isothermal redox series or not, if for instance the activity coefficients change due to certain interactions. From the extent of the redox shift of iron in combination with manganese or arsenic, the change of the standard potential E^0 can be estimated to be between some tens up to several hundreds of mV. However, shifts of the standard potential in this order of magnitude can be excluded according to experimental results [12, 13]. On the other hand the occurrence of electron exchange reactions during temperature changes is obvious also from optical and oxygen partial pressure measurements [12, 14]. These results are very important with respect to industrial glass melting: properties such as viscosity and absorption do not only vary due to the thermal influence on the basic network but also when the concentrations of polyvalent elements vary. Accordingly, the properties associated with particular oxidation states change with temperature because of electron exchange reactions.

ADJUSTMENT OF REDOX-DEPENDING PROPERTIES IN INDUSTRIAL GLASS MELTING

Redox number concepts

It is an important concern in industrial glass melting to predict the final glass properties as early as possible. For that purpose a characterization of the batch with respect to redox-dependent properties would be desirable.

In the case of oxygen refining, the amount of released oxygen of oxidic batch components is the quantity in question. The concentration of a polyvalent element in its oxidized state can be understood as chemically bonded oxygen which may be released by complete reduction. In this regard, Equation (8) combines chemically and physically dissolved oxygen. Conversely, if a higher oxide is dissolved in the glass melt its excess oxygen increases the oxygen partial pressure up to 1 bar. Since the physical solubility of molecular oxygen is very low, the reduced state is equal to the released oxygen

$$[O_2] = \frac{n}{4}[A^{x+}] = \frac{[A]}{\dfrac{p_{O_2}^{n/4}}{K(T)} + 1} = \frac{[A]}{\dfrac{p_{O_2}^{n/4}}{e^{-\frac{\Delta G^0(T)}{RT}}} + 1} \tag{13}$$

For instance in typical soda-lime-silica glass melts, about 5 % of all iron is in the Fe^{2+} state. According to Equations (10) and (13), E^0 would be a measure of the released oxygen related to the number of transferred electrons $[O_2]/n$, i. e., in principle, the isothermal redox series can be used to compare the oxygen release of oxides of different polyvalent elements. A more positive potential in the isothermal redox series in Table I means increased oxygen release. The validity of the redox series implies that the oxygen is totally released in the melt and no oxygen loss occurs during batch melting. However, this is not always true, as for instance, manganese releases the oxygen at an early stage of the batch reaction when the oxygen can still leave the batch, while cerium releases most of its oxygen at high temperatures when a primary melt is already formed.

In order to predict the oxidation state of a sulfur refined glass, redox number concepts have been developed [15, 16, 17]. These concepts denote oxidizing agents with positive factors and reducing agents with negative factors. Summarizing the contributions of all batch components with respect to a certain amount of sand yields the redox number of the batch. However, in practice a quantitative prediction of properties such as Fe^{2+}/Fe^{3+} ratio is not possible because the dependence between the Fe^{2+}/Fe^{3+} ratio and the redox number is not a linear function. The reason is the behavior of coke. At constant sulfate addition the dependence of the Fe^{2+}/Fe^{3+} ratio on the carbon addition separates into three different regions, as shown by Figure 1. The three regions can be related to the three equilibria given in Figure 1 [18, 19].

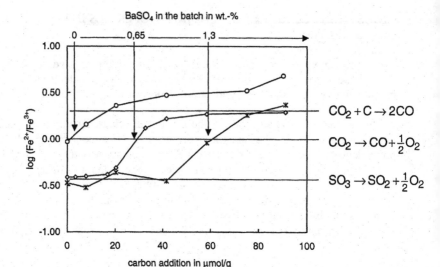

Figure 1. Redox ratio of iron depending on sulfate and carbon addition to the batch according to [19].

The basis for the concept of redox numbers is the reaction [17]

$$Na_2SO_4 + C \underset{\leftarrow}{\overset{\rightarrow}{}} Na_2O + SO_2 + CO_2 \qquad (14)$$

Assuming stoichiometric conditions one mole sodium sulfate compensates for one mole carbon. However, the observed effect of carbon is less than Equation (14) predicts. The reason is that the oxidation of carbon stops in a first stage with the formation of CO. Before the primary melt is formed, this CO can leave the melt.

In summary, in oxygen as well as in sulfur refining the establishment of the redox state is governed by balances at lower temperatures and equilibria at higher temperatures. This makes a characterization of the final redox state based on the batch properties impossible. Possibly, advanced mathematical batch models will eventually be able to solve the discussed problems.

Redox control

Since isothermal redox series and redox number concepts cannot exactly predict the redox state of the final glass product, monitoring tools are required. Typically, two measuring quantities are available: the redox ratio of iron and the oxygen partial pressure.

The redox ratio of iron is mostly used in order to determine the oxygen partial pressure according to Equation (8). However, the Fe^{2+}/Fe^{3+} ratio can be changed by electron exchange reactions with other polyvalent elements according to the non-isothermal redox series in Table III. Thus, the calculated oxygen partial pressure only agrees with the real oxygen partial pressure if the total iron concentration is much larger than the concentrations of other polyvalent elements. A particular influence is observed in sulfur refined glasses compared to glasses without sulfur. With a known oxygen partial pressure the iron redox ratio can be calculated by means of Equation (8). Figure 2 compares the redox ratio of iron calculated from the oxygen partial pressure based on the reaction constants in sulfur free glasses. In industrial sulfur refined glasses more Fe^{2+} is measured than would correspond to the measured oxygen partial pressure. According to Table III this is obviously due to an electron exchange reaction between iron and sulfur. If the redox ratio of sulfur refined glasses is used to calculate the oxygen partial pressure in flint glasses, the results can be up to two orders of magnitude too low.

Today the oxidation state of a glass can be determined on-line in the glass melt by means of oxygen sensors. The advantage of such a sensor compared to the iron redox state is the continuous stream of data, which allows faster recognition of trends. However, the oxidation state of a glass is determined by the chemically dissolved oxygen rather than by the oxygen partial pressure. Since both are related by Equation (8), the chemically bonded oxygen can be calculated if the concentrations of all polyvalent elements are known, which can be measured by means of voltammetric sensors [20].

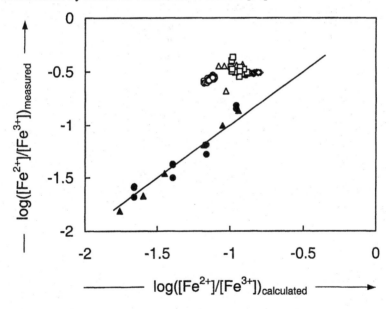

Figure 2. Comparison of iron redox ratio as calculated from oxygen partial pressure and measured, respectively.

Decolorization

Raw materials, especially sand, are often contaminated with iron which yields a green tint. The intense blue coloration of Fe^{2+} is more disturbing than the yellow tint of Fe^{3+}, thus, the redox state should be as oxidizing as possible. On the other hand, under industrial conditions the redox state of a glass melt can rarely be adjusted only to glass properties. For instance, sulfur refining is known to improve with reducing conditions, thus, the oxygen partial pressure is not available as a decolorizing measure. Moreover, Figure 2 shows that there is an additional shift of the Fe^{2+}/Fe^{3+} ratio at a given oxygen partial pressure due to the electron exchange reaction between iron and sulfur.

Under these conditions a target Fe^{2+}/Fe^{3+} ratio can only be established after refining, i.e. by electron exchange reactions during cooling. The non-isothermal redox series shows which redox couple can be used to decolorize the blue tint of Fe^{2+}. This can be achieved by all elements to the right of iron in Table III. Addition of any of these elements provides chemical decolorization. Additionally, the reddish tint of Mn^{3+} or Se^0 provides a physical decolorization. The non-isothermal redox series reveals an advantage of selenium decolorization compared to manganese: the electron exchange increases the Se^0 coloration but decreases the concentration of the colorant Mn^{3+}.

The non-isothermal redox series gives no information about the extent of the decolorization. However, according to Equation (8) elements can provide more chemically bonded oxygen the more to the right they are positioned in the isothermal redox series, provided the decolorization reaction goes to completeness during cooling.

Modelling of redox reactions in glass melts

Advanced use of electrochemical sensors in industrial glass melting requires a more elaborate model which relates measured quantities and required properties. For example, oxygen balance calculations can explain some characteristics of sulfur refining [10, 20].

The conventional attempt of a description of sulfur refining by the simple reaction

$$SO_3 \rightarrow SO_2 + \tfrac{1}{2} O_2 \qquad (15)$$

would result in a refining temperature of 1400 °C and higher. Under industrial conditions this temperature is only achieved for small parts of the melt. Under slightly reducing conditions the bubble generation temperature is considerably lower. Under these conditions the bubbles contain only SO_2 [21]. Since SO_2 is not chemically dissolved in the glass, oxygen must be provided by another glass component in order to produce SO_3 during cooling. This can be accomplished by all redox couples which stand to the right of the S^{4+}/S^{6+}

redox couple in the non-isothermal redox series, where the reaction with iron plays a particular role [10]

$$2\,Fe^{2+} + S^{6+} \underset{\leftarrow}{\overset{\rightarrow}{}} 2\,Fe^{3+} + S^{4+} \tag{16}$$

It can be assumed that the formation of SO_2 from S^{4+} occurs without inhibition. Calculations based on Equation (16) explain the shift of the Fe^{2+}/Fe^{3+} ratio in Figure 2. Considering the redox shift due to the interaction between iron and sulfur in the calculated redox ratio leads to the agreement of the calculated and measured Fe^{2+}/Fe^{3+} ratios (Figure 3).

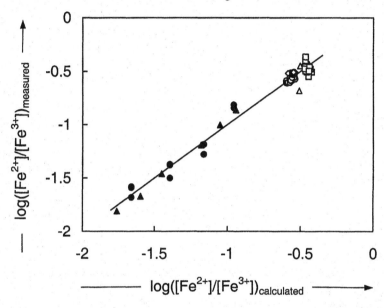

Figure 3.   Comparison of iron redox ratio as calculated from oxygen partial pressure considering the interaction between iron and sulfur during cooling and measured, respectively.

Conversely, iron increases the formation of SO_2 with increasing temperature. In this context it has often been observed that changing the sand to one containing low iron results in problems with sulfur refining. Obviously, in these cases sulfur refining is no more aided by reaction (16) but reacts according to (15). A possible countermeasure is an increase in the furnace temperature until the direct sulfate dissociation according to Equation (15) takes place. Another possible measure is the introduction of a non-coloring redox couple to the right of the S^{4+}/S^{6+} couple.

Also in the case of amber glass the influence of iron is well-known. From the non-isothermal redox series the refining reaction

$$6\,Fe^{3+} + S^{2-} \underset{\leftarrow}{\overset{\rightarrow}{}} 6\,Fe^{2+} + S^{4+} \qquad (17)$$

can be derived for amber glass. According to Equation (17) decreasing the iron content leads to refining problems, which is often observed in industrial amber glass melting.

SUMMARY

Glass melting is a very complex process especially under industrial conditions and with respect to redox-dependent glass properties. The redox state of a glass melt is primarily determined by the reducing and oxidizing agents in the batch. However, balances and equilibria compete during batch reaction, thus, an accurate prediction of the glass properties by means of a batch characterization is impossible. In practice control has to be based on measurements of the iron redox ratio or oxygen partial pressure in the melt. For a satisfactory control an appropriate model is required relating the measured quantities to the desired glass properties. Consideration of influences of viscosity, foaming or fluxes in the glass melting tank requires a greater effort such as is provided by computer based modelling. The oxygen partial pressure is not only adjusted with respect to glass properties but also the refining behavior. A subsequent adjustment of glass properties such as decolorization or improvement of refining is possible using additives according to the non-isothermal redox series. A nearly accurate description is only possible by mathematical models [22, 23]. However, mathematical models require large computer capacities which are not available in every production site. When production problems arise, use of the isothermal and non-isothermal redox series can then provide guidance.

REFERENCES

[1] Nair, K. M.; White, W. B.; Roy, R.: Solubility of oxygen in glasses. J. Am. Ceram. Soc. 48 (1965) 52.

[2] Schreiber, H. D.; Hockman, A. L.: Redox chemistry in candidate glasses for nuclear waste immobilization. J. Am. Ceram. Soc. 70 (1987) 591-594.

[3] Rüssel, C.; Freude, E.: Voltammetric studies of the redox behaviour of various multivalent ions in soda-lime-silica glass melts. Phys. Chem. Glass (1989) 62-68.

[4] Sun, K.-H.: A scale of acidity and basicity in glass. Glass Ind. 29 (1948) 73-74, 98.

[5] Duffy, J. A.; Ingram, M. D.: An interpretation of glass chemistry in terms of the optical basicity concept. J. Non-Cryst. Sol. 21 (1976) 373-410.

[6] Krämer, F.: Contribution to basicity of technical glass melts in relation to redox equilibria and gas solubilities. Glastech. Ber. 64 (1991) 71-80.

[7] Kühl, C.; Rudow, H.; Weyl, W.: Oxydations- und Reduktionsgleich-gewichte in Farbgläsern. Sprechsaal 71 (1938) 91-93, 104-106, 117-118.

[8] Lee, J.-H.; Brückner, R.: The electrochemical series of the 3d transition metal ions in alkali borate glasses. Glastech. Ber. 59 (1986) 233-251.

[9] Müller-Simon, H.: Electron exchange reactions between polyvalent elements in soda-lime-silica and sodium borate glasses. Glastech. Ber. Glass Sci. Technol. 67 (1996) no. 12, p. 387-395.

[10] Müller-Simon, H.: Oxygen balance in sulfur-containing glass melts. Glastech. Ber. Glass Sci. Technol. 71 (1998) 6, 157-165.

[11] Müller-Simon, H.; Bauer, J.; Baumann, P.: Redox behavior of selenium in industrial soda-lime-silica glasses. Glastech. Ber. Glass Sci. Technol. 74 (2001) 283-291.

[12] Lenhart, A.; Schaeffer, H. A.: The determination of oxidation states and redox behavior of glass melts using electrochemical sensors. Collected papers, XIV. Int. Congress on Glass, New Delhi (1986) 147-154.

[13] Rüssel, C.; Freude, E.: Voltammetric studies in a soda-lime-silica glass melt containing two different polyvalent ions. Glastech. Ber. 63 (1990) 149-153

[14] Schirmer, H. Müller, M.; Rüssel, C.: High-temperature spectroscopic study of redox reactions in iron- and arsenic-doped melts. Glass Sci. Technol. 76 (2003) 49-55

[15] Manring, W. H.; Davis, R. E.: Controlling redox conditions in glass melting. Glass Ind. (1978) 5, 13-16, 23-24, 30.

[16] Simpson, W., Myers, D.D.: The redox number concept and its use by the glass technologist. Glass Technol. 19 (1978) 4, 82-85.

[17] Manring, W. H.; Hopkins, R. W.: Use of sulfates in glass. Glass Ind. 39 (1958) 139-170.

[18] Nölle, G.; Al Hamdan, M.: Kohlenstoff in Glasrohstoffgemengen. Silikattechnik (1990) 192-193.

[19] Bassine, J. F.; Mestdagh, M. M.; Rouxhet, P. G.: Redox buffering by sulphate and carbonate during the melting of reduced soda-lime-silica glasses. Glass Technol. 28 (1987) 50-56.

[20] Müller-Simon, H.: Electrochemical online-sensors in industrial glass melting tanks. The GlassResearcher 10 (2000) 10-11,18,22

[21] Chopinet, M.-H.; Massol, J. J.; Barton, J. L.:La relation entre la teneur en sulfate et l'état d'oxydation de verres fondus en présence de réducteurs. Rev. Staz. Sper. Vetro (1982) 5, 200-201.

[22] Beerkens, R.G.C.: Modeling of the melting process in industrial glass furnaces. In: Krause, D.; Loch, H.: Mathematical Simulation in Glass Technology. Schott Series on Glass and Ceramics, Science, Technology and Applications, Springer Verlag 2002, 17-73.

[23] Chmelar, J.; Bodi, R.; Muysenberg, E.: Supervisory advanced control of glass melters and forehearth by the GS expert system. Glastech. Ber. Glass Sci. Technol. 73 (2000) 276-284.

USING ADDITIVES FOR COLOR CONTROL IN COPPER-CONTAINING GLASSES

Henry D. Schreiber, Mary E. Stokes, and Amy M. Swink
Department of Chemistry
Virginia Military Institute
Lexington, VA 24450

ABSTRACT

When added to glass-forming melts, copper establishes the Cu^{2+}-Cu^+ redox equilibrium under oxidizing conditions, and the Cu^+-Cu^0 equilibrium at reducing conditions. Cu^{2+} typically introduces a blue coloration to glass, while Cu^0 imparts a ruby red color (Cu^+ is colorless). A systematic study investigated the ability of additives (iron; and cerium, chromium, europium, manganese, nickel, lead, uranium, and vanadium, individually or in concert with iron) to undergo redox reactions with copper during the processing of an alkali borosilicate glass. Such interactions were determined in glass melts with Cu^{2+}, Cu^+, or Cu^0 as copper's predominant redox state and were monitored by spectral changes. Depending on the synthesis conditions, a particular additive could be a reducing agent, oxidizing agent, or redox neutral with respect to the prevailing copper redox states. The results illustrate that an electromotive force series of redox couples can be used to predict the effect of an additive on the copper redox state(s) in the glass.

INTRODUCTION

Copper is an important colorant in commercial glass-making.[1] Its oxidized redox state, Cu^{2+}, usually gives glass a characteristic blue color; whereas the reduced state, Cu^0, provides a ruby red coloration attributed to colloidal copper. Accordingly, the redox chemistry of the glass melt plays a critical role in defining the prevailing copper redox state, and thus the color of the resulting glass. Under relatively oxidizing conditions, copper establishes the redox equilibrium:

$$4Cu^{2+} + 2O^{2-} \leftrightarrow 4Cu^+ + O_2 \qquad (1)$$

and under relatively reducing conditions, the redox equilibrium:

$$4Cu^+ + 2O^{2-} \leftrightarrow 4Cu^0 + O_2 \qquad (2)$$

Colorless Cu^+ is involved in both equilibria. Therefore, copper colors an oxidized glass blue and a reduced glass red, with the intensity dependent on the amount of Cu^{2+} and Cu^0, respectively. Each of the redox equilibria, and thus the resulting color of the glass, are controlled by the processing temperature and oxygen content, as well as by the melt composition (melt basicity and additives).[2]

Although copper ruby glass can be prepared by production of the Cu^0 colloid under reducing conditions, the conventional way is to first synthesize the melt at intermediate redox conditions to stabilize the Cu^+ ion, with the melt also containing a reducing agent.[1] Subsequent heat treatment of the glass to about 500°C reduces Cu^+ to Cu^0 *in situ*, generating the red color. This study focuses on the primary coloration of the glass, not that due to subsequent heat treatment.

OBJECTIVE

Under processing conditions, the redox state of copper in the glass-forming melt can be adjusted by redox-active additives, those species acting as oxidizing or reducing agents. The goal of this study is to determine the applicability of an electromotive force (EMF) series of redox couples to predict changes in copper's redox equilibria in such melts, and predict changes in the colors of the resulting glasses.

EXPERIMENTAL PROCEDURES

An EMF series of redox couples has previously been developed for an alkali borosilicate glass composition (SRL-131) at 1150°C.[2] Even though SRL-131 has been used as a model composition for glasses in nuclear waste immobilization, its redox chemistry is also applicable to commercial glasses.[3] SRL-131's base composition (in wt%) is 57.9 SiO_2, 1.0 TiO_2, 0.5 ZrO_2, 14.7 B_2O_3, 0.5 La_2O_3, 2.0 MgO, 17.7 Na_2O, and 5.7 Li_2O. Redox equilibria of copper in SRL-131 at 1150°C were previously calibrated to the melt's atmosphere (oxidizing or reducing conditions controlled by gas flow and expressed as oxygen fugacity, fO_2) during synthesis, as shown by Table I.[4]

Table I. Dependence of copper redox states in SRL-131 on atmosphere

redox condition	gas atmosphere of sample	$-\log fO_2$	% of all copper as Cu^{2+}	Cu^+	Cu^0
oxidizing	air	0.7	40	60	
intermediate	$CO_2/CO = 7$	9.6		100	
reducing	C/CO	14.5		30	70

The base composition included 0.20 or 0.25 wt% Cu, added as CuO. Visible absorption spectra of individual copper redox states in SRL-131 glass at 1150°C are provided in Figure 1, illustrating the characteristic signature of each state at each of the three synthesis conditions indicated in Table I.

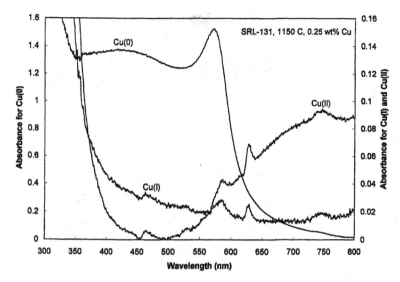

Figure 1. Visible spectra of copper-containing glasses prepared under oxidizing conditions (Cu^{2+}, -log fO_2 = 0.7), at intermediate conditions (Cu^+, -log fO_2 = 9.6), and under reducing conditions (Cu^0, -log fO_2 = 14.5). Spectra are normalized to 1 mm thickness

Other components (M = Ce, Cr, Eu, Mn, Ni, Pb, U, or V) were added to the copper-doped base composition at a concentration of 1 wt% M. Such additives were included individually as well as in concert with iron (0.5 wt% Fe). The additives were predicted to be oxidizing agents, reducing agents, or redox neutral with respect to Cu^{2+}-Cu^+ or to Cu^+-Cu^0 according to an EMF series, as shown in Figure 2.[2] In this diagram, those redox couples further to the right are easier to reduce, and are accordingly oxidizing agents to any couple to its left. If the additive does not have an operational redox couple at a particular atmosphere, it cannot operate as an oxidizing or reducing agent but is redox neutral, not interacting with the redox equilibrium of copper. Individual samples at air were synthesized in Pt crucibles. In order to minimize melt-container interactions, samples prepared at CO_2/CO atmospheres were made in alumina crucibles, and those at CO in graphite crucibles.

Samples were equilibrated for 24 hours at 1150°C, previously shown to be sufficient for equilibration.[4] Spectrophotometric analyses of polished slabs of the glasses were used to determine the effect of the additives on the Cu^{2+} absorption at 740 nm, or the absorptions of colloidal Cu^0 centered at 440 nm and 560 nm. Figure 3 interprets the broad and narrow peaks for this copper ruby signature as Cu^0 atoms and $(Cu^0)_n$ particulates, respectively.[5] Changes in the 560 nm peak are related to the particle size and size distribution. In some cases, changes in copper's spectrum were obscured by absorptions introduced by the additive.

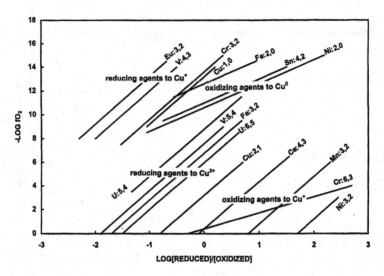

Figure 2. EMF series of redox couples in SRL-131 at 1150°C.

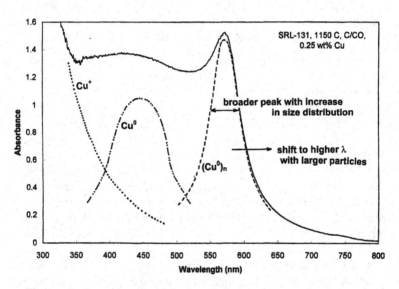

Figure 3. Interpretation of the visible spectrum of copper ruby glass (analogous to other colloidal colorations of glass).

RESULTS AND DISCUSSION

Oxidizing Conditions (Cu^{2+}-Cu^+ Equilibrium)

According to Figure 2, additives that should be reducing agents with respect to the Cu^{2+}-Cu^+ equilibrium are Fe as Fe^{2+}, V as V^{4+}, and U as U^{5+}. In agreement with this prediction, the Cu^{2+} absorbance decreased by 25 to 50% in all cases upon reaction with the additive in the melt. Cu^{2+}, however, was not fully reduced. Even though the additives are effective reducing agents, there is a limited amount of that additive available as its reduced species under the synthesis conditions. Cu^{2+} was even more effectively reduced when Fe acted in concert with V or U.

Redox neutral species were identified as Pb, Ni, Sn, and Eu, all existing as only one redox state at air. Experiments confirm that there was indeed no change in the 740 nm absorbance of Cu^{2+}.

Figure 2 illustrates that potential oxidants of Cu^+ in the melt at air are Mn as Mn^{3+}, Cr as Cr^{6+}, and Ce as Ce^{4+}. Addition of cerium to the copper-containing melt under relatively oxidizing conditions clearly accentuated the color of Cu^{2+}, in agreement with previous work.[6] Interestingly, Mn did not oxidize the copper as much as expected; in fact, Mn^{3+} coexisted with Cu^{2+} in the glass samples. Figure 4 shows that the visible spectra of manganese (Mn^{3+}) and copper (Cu^{2+}) were additive, resulting in a gray glass. Additional stability of Mn^{3+} may have been due to complexes forming between the Mn and Cu redox ions.[7]

Intermediate Conditions (Cu^+)

Most additives to the Cu^+-containing melt did not oxidize ($Cu^+ \rightarrow Cu^{2+}$) or reduce ($Cu^+ \rightarrow Cu^0$) the Cu^+. Evidently, the predicted reducing agents (V as V^{3+}) and oxidizing agents (Ni as Ni^{2+}, U as U^{5+}, Fe as Fe^{3+}) were either in insufficient quantities or too weak to react with the Cu^+. Ce as Ce^{3+}, Mn as Mn^{2+}, and Cr as Cr^{2+}-Cr^{3+} were predicted to be redox inactive and, indeed, did not react with Cu^+.

The EMF series suggested that Eu as Eu^{2+} would be a reducing agent under these conditions. In fact, Eu additions did produce some colloidal Cu^0 (and a red color) in the resulting glass, according to Equation 3.

$$Eu^{2+} + Cu^+ \rightarrow Eu^{3+} + Cu^0 \qquad (3)$$

In agreement with these experiments, Figure 2 identifies Eu to be a stronger reducing agent than V. Sn, a predicted marginal oxidizer (but very dependent on fO_2, as shown in Figure 2), was also shown to be an effective reducing agent, producing swirls of copper ruby coloration due to Cu^0, as shown by Equation 4.

$$Sn^{2+} + 2Cu^+ \rightarrow Sn^{4+} + 2Cu^0 \qquad (4)$$

For both Eu and Sn, additions of Fe to the Cu^+-containing system did not affect their reducing ability with respect to Cu^+. Also in both cases, the copper ruby coloration was not homogeneous. The ability of Sn to alloy with Cu is probably the underlying reason for the EMF series to provide an inaccurate prediction in this case.

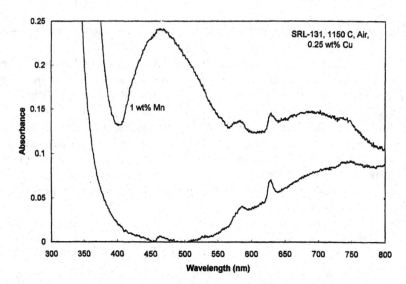

Figure 4. Visible spectra of SRL-131 with 0.25 wt% Cu and with 0.25 wt% Cu + 1 wt% Mn, both synthesized at 1150°C in air.

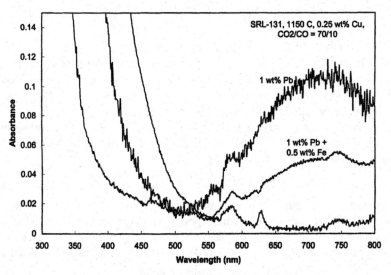

Figure 5. Visible spectra of SRL-131 with 0.25 wt% Cu and with 0.25 wt% Cu + 1 wt% Pb (with and without Fe), all synthesized at 1150°C in an atmosphere of CO_2/CO of 7.

Also interestingly, Pb was a very effective oxidizing agent, producing a blue glass characteristic of Cu^{2+}, as shown in Figure 5. One possible explanation is that lead is indeed redox active with multiple redox states (for example, Pb^{4+}-Pb^{2+}).[8]

Reducing Conditions (Cu^{+}-Cu^{0} Equilibrium)

All predicted oxidizing agents (Ni as Ni^{2+}, Sn as Sn^{4+}, Fe as Fe^{2+}, and U as U^{5+}) indeed did oxidize Cu^{0}, decreasing the intensity of the ruby red color, as shown by Figure 6 and Equation 5 for Fe.

$$Fe^{2+} + 2Cu^{0} \rightarrow Fe^{0} + 2Cu^{+} \qquad (5)$$

Even though Eu^{2+} and V^{3+} should be reducing agents and enhance the copper ruby signal, both diminished the copper ruby color when Eu or V were added at 1 wt%. However, there does seem to be a dramatic concentration dependence of the effect of Eu and Sn on the ruby red coloration. Also, both Sn and Eu additions, individually and in concert with Fe, changed the particulate distribution of Cu^{0} and $(Cu^{0})_n$ in the ruby red glass, as illustrated by changes in absorption and location of the Cu^{0} and $(Cu^{0})_n$ peaks.

Interestingly, all predicted redox neutral species (Pb, Ce, and Mn) operated as very effective oxidizers, eliminating all Cu^{0} from the glass.

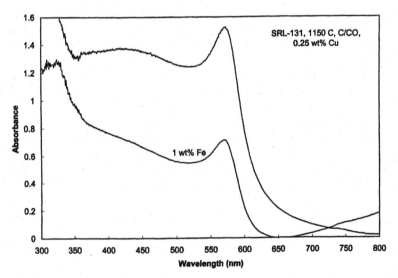

Figure 6. Visible spectra of SRL-131 with 0.25 wt% Cu and with 0.25 wt% Cu + 1 wt% Fe, both synthesized at 1150°C in C/CO.

CONCLUSIONS

The electromotive force series has been shown to be a reliable predictive tool for indicating the potential oxidizing and reducing reactions that occur between copper and other redox additives in a glass-forming melt. This is particularly true for copper (Cu^{2+}-Cu^+) under relatively oxidizing conditions. However, there are some interesting anomalies, especially in the effect of redox additives on the Cu^+-Cu^0 equilibrium. Both tin and europium can act as an oxidant or a reductant to Cu^+-Cu^0, depending on the prevailing redox conditions. The resulting red color, due to changes in the particulate size and size distribution, is also a function of the concentration of the tin or europium additive. Lead is an effective oxidizing agent to Cu^0 under reducing conditions, indicating that lead may be redox active in the melt.

REFERENCES

[1]M.B. Volf, *Chemical Approach to Glass*; pp. 477-484. Elsevier, Amsterdam, 1984.

[2]H.D. Schreiber and A. L. Hockman, "Redox Chemistry in Candidate Glasses for Nuclear Waste Immobilization," *Journal of American Ceramic Society*, **70** [8] 591-594 (1987).

[3]H.D. Schreiber and M.E. Stokes, "From Fundamentals to Applications; Redox Stages in Commercial Glassmaking," *The Glass Researcher*, **11** [1] 13-15, 50 (2001).

[4]A.B. Morgan and H.D. Schreiber, "Recovery of Recyclable Metals from Waste Glass Melts," *Ceramic Transactions*, **45**, 145-153 (1994).

[5]A.A. Ahmed and E.W. Abd Allah, "Origin of Absorption Bands Observed in the Spectra of Silver Ion-Exchanged Soda-Lime-Silca Glass," *Journal of American Ceramic Society*, **78** [10] 2777-2784 (1995).

[6]Z.D. Xiang and M. Cable, "Redox Interactions between Cu and Ce, Sn, As, Sb in a Soda-Lime-Silica Glass," *Physics and Chemistry of Glasses*, **38** [4] 167-172 (1997).

[7]A.K. Bandyopadhyay, "A Study of Interaction between Copper and Manganese in a Soda-Borate Glass by ESR," *Journal of Materials Science*, **15**, 1605-1608 (1980).

[8]A.M. Hofmeister and George R. Rossman, "Exsolution of Metallic Copper from Lake County Laboradorite," *Geology*, **13**, 644-647 (1985).

DECOLORIZATION OF AMBER GLASS

Douglas B. Rapp, Melissa A. Dorsey, Melissann M. Ashton-Patton, and James E. Shelby
New York State College of Ceramics
Alfred University
Alfred, NY 14802

ABSTRACT
    The amber color of many container glasses presents a problem in recycling of these glasses. Although there are other problems associated with recycling these glasses, decolorization of the amber glass would aid in promoting the use of mixed cullet in recycled glass production.
    Amber glass has been decolorized, i.e. converted to the green color typical of soda-lime-silica glasses containing similar amounts of iron oxide, by the addition of small amounts of ZnO or NaF to amber cullet. The effect of ZnO or NaF concentration, melting time, and melting temperature on the optical spectra of these glasses has been determined. Additions of a few wt% of ZnO completely eliminate the amber coloration of these glasses. NaF is less effective, resulting in glasses with regions of complete decolorization, mixed with regions which retain the amber color. Changes in other properties of interest, e.g. density, refractive index, thermal expansion coefficient, glass transformation temperature, etc. have also been measured. Additions of oxidation agents such as $NaNO_3$ were less successful in decolorizing amber glass.

INTRODUCTION
    An increase in the recycling of container glass would be of great benefit to the environment. Unfortunately, the recycling of glass containers is complicated by the common use of three colors of glass containers: amber, or brown, green, and clear [1]. Mixing of cullet of these glasses produces melts with streaks of varying color unless great care is taken to homogenize the melt. A change in the ratio of the various glasses would also result in variations in color of the product even if it is homogeneous. Decolorization of the amber glass would eliminate, or at least reduce, the color control problems which result from mixing of amber and green glasses. Since the clear glass simply acts to dilute the color and does not introduce a color by itself, the problems associated with clear glass are much less than those of the other two glasses.
    A separate issue in use of mixed container cullet lies in the reactions which occur during melting of batches containing both amber and green glasses. These reactions result in the rapid release of large quantities of gas, which leads to foaming of the melt [2]. While it will also be necessary to solve this problem before mixed cullet can be routinely used in commercial production, this issue is not addressed in the

current study. It is possible, however, that the methods discussed here will also affect the foaming problem.

Both the color and foaming problems require precise control of the redox equilibria of the melt. These problems can be addressed by adjusting the batch chemistry. The cause of the amber color of these glasses, for example, has been widely debated [1,3-8]. The most commonly accepted view assumes that the chromophore consists of a ferric ion in tetrahedral coordination with three O^{2-} ions and one S^{2-} ion. The resulting negative charge on the tetrahedron is compensated by a neighboring alkali or alkaline earth cation. The amber color results from a charge transfer process which causes an absorption band at about 400 nm. An additional, much more intense band is located at about 295 nm. The assignment of these absorption bands to a specific photo-induced reaction remains ambiguous, although the presence of both ferric and sulfide ions is believed to be necessary to produce the amber color. Evidence for the existence of this chromophore is found primarily based on the results of spectroscopic studies of glasses with various Fe^{2+}/Fe^{3+} and S^{2-}/SO_4^{2-} ratios and total concentrations. Supporting evidence can be found in results of electron spin resonance studies of these glasses. It has also been found that the intensity of the band near 400 nm is proportional to the product of the concentrations of Fe^{3+} and S^{2-}, which supports the contention that formation of the chromophore requires the presence of both ions.

Production of a consistent amber color thus requires careful control of both the ferric and sulfide concentrations in the melt. If the melt is produced under highly reducing conditions, the iron will be mostly in the ferrous form, which prevents formation of the chromophore described above. On the other hand, if the melt is too oxidizing, the sulfide ions will be oxidized to form sulfate ions and the chromophore will also be eliminated. Control of the concentrations of both the ferric and sulfide ions is achieved by addition of reducing agents, such as carbon, to the batch and by maintaining a specific partial pressure of oxygen above the melt. Typical amber glasses are made under a partial pressure of oxygen 10^{-8} to 10^{-10} atm. Most of the iron in the melts is in the ferrous form under these conditions, but enough ferric iron is present to produce an adequate concentration of amber chromophores.

These arguments lead to the obvious hypothesis that amber glasses can be decolorized by either oxidizing the melt to eliminate the sulfide ions or by reducing the melt to eliminate the ferric ions. A third possibility also exists. Replacement of the ferric ion by another species which will preferentially react with the sulfide ions, thus removing them from the possible formation of the iron/sulfur chromophore would also eliminate the amber color of the glass. In 1927, Jackson mentioned that additions of ZnO to amber glass would eliminate the amber color. A few patents have since been based on this idea [9-11]. Bamford [4] proposed that Zn^{2+} ions preferentially bond to the sulfide ions, displacing the Fe^{3+} ions from the chromophore. It is also, however, possible that the Zn^{2+} ions alter the redox of the melt, eliminating the S^{2-} ions and thus eliminating the chromophore.

The present paper presents a detailed study of the effect of ZnO additions on the color of amber glass. Since oxidation of the melt can also potentially lead to decolorization of these melts, a number of oxidizing agents, including NaF and $NaNO_3$, were also examined. In addition to a study of the optical effects, several properties of these glasses, e.g. density, refractive index, glass transformation temperature, and thermal expansion coefficient were also measured.

EXPERIMENTAL PROCEDURES

Samples used in this study were produced by remelting finely crushed amber container glass with the desired dopant in air in an electrically heated furnace in Pt alloy crucibles. Batches were melted in uncovered crucibles at 1500°C for 30 minutes. A test of the effect of remelting the base container glass was carried out by making melts for various times to determine if simply remelting in air would significantly alter the color of the glass. At the end of the melting period, the crucible was removed from the furnace and cooled to room temperature. The glass was removed and annealed by heating to the glass transformation temperature, holding for 30 minutes, and then cooling to room temperature at 3 K/min. Annealed samples were cut into ≈ 1 mm thick plates for the optical measurements and to the appropriate dimensions for the thermal expansion measurements. The plates were ground and polished prior to the optical measurements.

Uv-vis spectra were measured using a Perkin-Elmer Lambda 40 spectrometer. Infrared spectra were measured on the same samples using a Thermo Nicolet Avatar FTIR spectrometer. Refractive indices were also determined on these samples using an Abbe refractometer with resolution of ±0.0001.

Densities were determined using the Archimedes' Method with kerosene as the immersion fluid. Values are reproducible to ±0.002 gm/cm^3. Thermal expansion curves were measured using a single push-rod vitreous silica dilatometer with a heating rate of 4 K/min. Average thermal expansion coefficients, which are reproducible to ±0.3 ppm/K, were determined over the range from 100 to 400°C. Glass transformation temperatures (T_g) were determined using the onset of increased expansion method, while dilatometric softening temperatures (T_d) were determined from the point of maximum expansion on the curve [1]. The glass transformation temperature was also determined using a differential scanning calorimeter (DSC) with a heating rate of 20 K/min.

RESULTS

Initial tests were made to determine the effect of simply remelting the glass in air on the color. A series of melts were made with melting time ranging from 15 minutes to one hour. While small changes in visual appearance and in the uv-vis spectra indicate that the intensity of the absorption due to the amber chromophore was progressively reduced by remelting, the effect was minimal

Addition of ZnO to the base glass results in rapid elimination of the amber color. The sample containing only a 1 wt% addition of ZnO is obviously less brown to the eye. The sample containing 2 wt% ZnO appears to be a green glass, with only hints of remaining amber color. Further additions of ZnO result in minor, but progressive, changes in the appearance of the glass, which now closely resembles green container glass. A few specks of amber color are still present, but this is to be expected due to the difficulty in producing a thoroughly homogeneous melt in such small quantities, especially considering the lack of convective mixing in these small crucibles.

Uv-vis spectra for the series of glasses doped with ZnO are shown in Figure 1. These spectra were obtained from typical areas of the samples, i.e. were not selected to deliberately exclude any regions of remaining amber color. The height of the band at 415 nm is shown as a function of the ZnO concentration in Figure 2. The rapid decrease in the height of this band corresponds with the visual observations of these glasses.

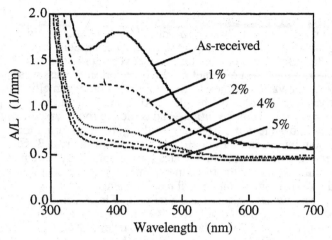

Figure 1: Uv-vis spectra of ZnO-doped amber glasses.

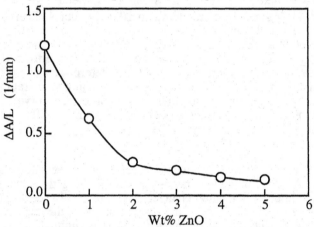

Figure 2: Effect of ZnO concentration on the intensity of the 415 nm absorption band.

All of these glasses contain a significant amount of chemically bound water, or hydroxyl, as can be seen from their infrared spectra. Since the concentration of hydroxyl does not vary significantly or systematically with ZnO content, those spectra are not shown.

Additions of NaF also decolor the amber glass. While the best samples exhibit behavior similar to that of ZnO (Figure 3), the overall results are not as impressive. Spectra shown in Figure 3 are representative of sections of glasses selected to have the largest reduction in amber color and hence do not represent the typical region of the melt. The glasses produced by addition of NaF do exhibit large regions of virtually complete decolorization, where the glass has a blue-green color. These glasses, however, also still contain large regions which are still obviously amber. If samples taken from those regions are selected for uv-vis spectra measurements, the

Advances in Fusion and Processing of Glass III

spectra show little improvement over that of the base glass. The cause of the difference in behavior between ZnO and NaF is not understood at this time. Since the batches were produced using identical methods, the initial degree of mixing prior to remelting should be the same for each additive. Some other process must be contributing to the variegated appearance of the NaF-doped glasses.

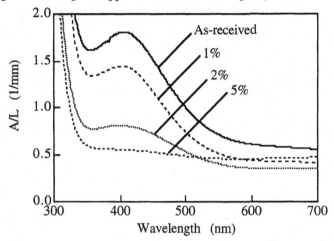

Figure 3: Uv-vis spectra of selected NaF-doped amber glasses.

The effect of additions of ZnO and NaF on the T_g of these glasses is shown in Figure 4. ZnO slightly increases the T_g, while NaF strongly decreases T_g. This finding is not surprising when one considers the roles of these materials in the glass network. Since ZnO closely resembles CaO in nature, addition of ZnO would be expected to behave in a manner similar to addition of CaO, i.e. to cause a small increase in T_g and viscosity. NaF would be expected to lower T_g and viscosity due to the increase in soda concentration and to the fluxing action of fluorine. Both the cation and anion in NaF, independently, are expected to act as fluxes in oxide glasses, so it should not be surprising that the combination reduces T_g.

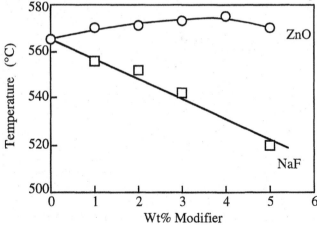

Figure 4: Effect of additions of ZnO and NaF on T_g of decolored glasses.

The effect of additions of these dopants on the density and refractive index of these glasses are shown in Figures 5 and 6, respectively. ZnO would be expected to increase the density by a significant factor due to the higher atomic weight of Zn relative to the other constituents of this glass. The small increase in density caused by addition of NaF agrees with the well-known small increase in density of sodium silicate glasses with increasing soda content. The role of fluorine in this behavior is not known. In a similar manner, one would expect an increase in refractive index as ZnO is added to the glasses, since ZnO is known to increase the refractive index of many silicate glasses. The decrease in refractive index which results from the NaF additions is probably caused by the replacement of oxygen by fluorine, whose ions are considerably less polarizable than those of oxygen. A reduction in refractive index has been reported for other systems where NaF is used as a glass component.

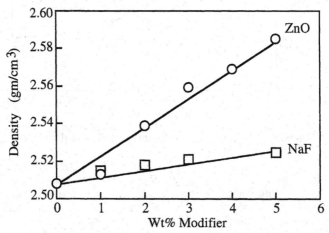

Figure 5: Effect of additions of ZnO and NaF on the density of decolored glasses.

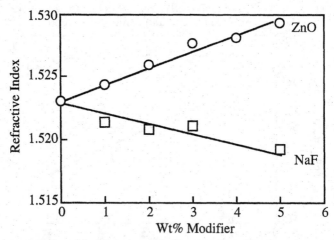

Figure 6: Effect of additions of ZnO and NaF on the refractive index of decolored glasses.

The effect of addition of ZnO and NaF on the thermal expansion coefficient of these glasses is shown in Figure 7. The very small decrease in the thermal expansion coefficient which appears to occur for ZnO additions is considered to be smaller than the detection limit of our measurements. The somewhat larger increase in thermal expansion coefficient with increasing NaF concentration is believed to be real and is consistent with the arguments presented earlier to explain the behavior of T_g, i.e. the addition of Na^+ ions increases the number of non-bridging oxygens in the network, which increases the thermal expansion coefficient of the glass. In a similar manner, F^- ions replace bridging oxygens, which will also increase the thermal expansion coefficient [1].

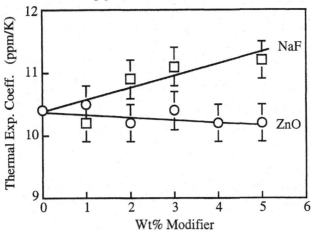

Figure 7: Effect of additions of ZnO and NaF on the thermal expansion coefficient of decolored glasses.

One additional melt was made using a 2 wt% addition of $NaNO_3$ to the amber cullet. While this addition did cause a small reduction in the amber color, the effect was considerably less than that found for ZnO. No further work using $NaNO_3$ or other strong oxidizing agents has been carried out at this time.

DISCUSSION

Amber glass can be decolorized by reducing the melt to eliminate ferric ions, by oxidizing the melt to eliminate sulfide ions, or by replacing the FeS chromophore with another species which will not cause the amber color. Changes in the uv-vis spectra do not indicate that the ferrous iron concentration increases significantly with additions of ZnO. Since addition of Zn^{2+} ions increases the basicity of the melt, it may be possible that these additions oxidize a sufficient number of sulfide ions to eliminate the amber color. Oxidation of the sulfur might then result in the formation of SO_2, which would evolve as a gas, creating bubbles in the melt. No such bubble evolution was found to be associated with addition of ZnO to these melts. It follows that the decolorization mechanism alluded to by Bamford [4] may be correct. If Zn preferentially bonds to sulfide ions, FeS bonds will be replaced by ZnS bonds. If this complex either does not exhibit a charge transfer band, or if that band absorbs at much higher energies than the FeS complex, the amber color will be eliminated. The small amount of ZnO needed to eliminate the amber color supports this contention, as does the fact that elimination of the amber chromophore quickly approaches saturation with increasing ZnO concentration, as shown in Figure 2.

The effect of NaF is more difficult to explain. The highly localized nature of the decolorization effect makes it difficult to decide if NaF really is effective as a decolorization agent. Addition of NaF will not create a new chromophore in preference to the FeS complex, so the explanation used for ZnO is not useful. It is possible that NaF oxidizes the sulfur, but creation of bubbles was also not observed in these glasses. The failure of $NaNO_3$ to eliminate the amber color also argues against simple oxidation of sulfur by these additives as a cause of decolorization. Perhaps the lower viscosity and higher non-bridging oxygen contents of melts containing NaF reduce the tendency of ferric ions to preferentially associate with sulfide ions. It has been shown in other systems that transition metal ions will preferentially associate with halide ions over oxygen ions [5]. A similar effect may occur in these melts, where association of ferric ions with fluorine ions reduces the tendency to form the FeS complex. If this effect is not very strong, it is possible that the decolorization effect will vary spatially within the melt.

ACKNOWLEDGMENTS

Portions of this work were funded by the EPA-supported Center for Environmental and Energy Research at Alfred University.

REFERENCES

[1] J. E. Shelby, "Optical Properties"; pp. 195-216 in *Introduction to Glass Science and Technology*, by J. E. Shelby, Royal Society of Chemistry, Cambridge, 1997.

[2] J. E. Shelby, "Glass Melting"; pp. 25-47 in *Introduction to Glass Science and Technology*, by J. E. Shelby, Royal Society of Chemistry, Cambridge, 1997.

[3] R. W. Hopkins and W. H. Manring, "Factors Influencing Control of Color in Amber Glasses," *Glass Ind.*, 34 251-4 (1953).

[4] C. R. Bamford, "The Sulphur-amber Effect", pp. 106-9 in *Colour Generation and Control in Glass*, by C. R. Bamford, Elsevier, New York, 1977.

[5] A. Paul, "Coloured Glasses", pp. 204-76 in *Chemistry of Glasses*, by A. Paul, Chapman and Hall, London, 1982.

[6] F. L. Harding and R. J. Ryder, "Amber Colour in Commercial Silicate Glasses," *J. Can. Ceram. Soc.*, 39 59-63 (1970).

[7] R. W. Douglas and M. S. Zaman, "The Chromophore in Iron-Sulfur Amber Glasses," *Phys. Chem. Glasses*, 10[4] 125-32 (1969).

[8] H. Jackson, "Some Colouring Agents in Glasses and Glazes," *Nature*, 120[3016] 264-6 (1927).

[9] W. H. Rising, "Heat Absorbing Borosilicate Glass and Method of Making the Same," U.S. Patent 1737685 (1929).

[10] W. H. Rising, "Method of Producing Heat Absorbing Glasses and Batches Therefor," U.S. Patent 1737686 (1929).

[11] H. P. Hood, "Ultra-Violet Transmitting Substance," U.S. Patent 1830902 (1931).

REDOX COUPLES IN GLASS – A SERIES OF NEW DATA

S. Michon, A. Littner, J. Di Martino and B. Gaillard-Allemand
Laboratoire de Chimie du Solide Minéral
UMR7555, UHP-Nancy I, 54 506
Vandoeuvre-Les-Nancy, France

ABSTRACT

The knowledge of basic data such as redox couples values in glass melts is a useful tool for the prediction of metals' corrosion reactions. A series of data concerning iron, nickel, chromium, cobalt, zinc, titanium, molybdenum, and tellurium have been determined in basic borosilicate or soda-lime glasses of various compositions. Classical electrochemical techniques adapted to glass studies (such as polarization curves, cyclic and square wave voltammetries) have been carried out in molten glass previously enriched in the desired metal ions. Enrichment of glass into metallic species was obtained by 1) partial dissolution of a metal piece in the glass or by 2) dissolution of a small amount of a selected metal oxide in the glass. The main advances of this work consist in the determination of the following half-wave potentials or standard potentials of redox couples: $E^{\frac{1}{2}}(Ti^{II}/Ti^{0}) < -1.65V$, $E^{0}(Ti^{III}/Ti^{II}) = -1.0V$, $E^{0}(Ti^{IV}/Ti^{III}) = -0.7V$, $E^{0}(Ni^{III}/Ni^{II}) = -0.36V$, $E^{\frac{1}{2}}(Co^{II}/Co^{0}) = -0.91V$, $E^{0}(Co^{III}/Co^{II}) = -0.68V$, $E^{\frac{1}{2}}(Mo^{III}/Mo^{0}) = -0.96V$, $E^{0}(Mo^{VI}/Mo^{III}) = -0.83V$, $E^{\frac{1}{2}}(Te^{IV}/Te^{0}) = -0.33V$. All the potentials are given versus an yttria-stabilized-zirconia reference electrode.

INTRODUCTION

In the glass industry, the lifetime of many metallic parts in contact with molten glass is mainly controlled by high temperature corrosion. The damages caused to alloys are governed by oxidation-reduction reactions. The determination of the corrosion mechanisms is related to the knowledge of the couple redox potential involved in the reaction. Preliminary studies [1, 2, 3] have demonstrated that electrochemical methods can be used for both the characterization of alloy corrosion in molten glasses and the determination of redox potentials of oxidant species present in the glass [4-7]. A typical experimental apparatus was developed to work in the 950-1300°C temperature range [2, 3, 8, 9]. It allows studies under controlled atmosphere, using static or rotating working electrodes. The aim of this paper is to determine some potential of redox couples in molten glass, and to give some applications concerning the corrosion mechanisms.

EXPERIMENTAL

C-glass - The results presented here were performed in an industrial borosilicate C-glass whose composition is 64.5% SiO_2 – 0.15% Fe_2O_3 – 3.40% Al_2O_3 – 7.2%CaO – 16.0% Na_2O – 1.20% K_2O – 4.5%B_2O_3 – 3%MgO, expressed in weight percent.

Apparatus – The experimental apparatus was already described by Di Martino et al. [9]. A furnace in which a Pt-10%Rh crucible is placed in air atmosphere comprises the experimental equipment. The electrodes' tips are immersed in 1.5 kg of molten glass.

The working temperature is in the 1050°C - 1300°C range. The working electrode for the glass study is a Pt wire (1 mm diameter), whereas the working electrode for metal and alloy study is a 5.5-mm-diameter rod polished with 1200 grit SiC paper. The working electrodes are cleaned in alcohol before use. The counter electrode is a Pt plate (dimensions: 25 X 2 X 10 mm). A reference electrode, called a zirconia stabilized electrode was used. It was constructed using a stabilized zirconia rod ($\phi = 5$ mm) cemented with zirconia inside a mullite tube. The electrode was flushed with air as the reference gas introduced with a syringe needle. All potentials given in this paper are referred to the potential of the zirconia stabilized electrode flushed by air. These high-temperature-adapted electrodes are extensively detailed in Ref.[1, 10, 11]

Electrochemical measurements – Parc M273 and M263A potentiostats were used to perform the electrochemical measurements. Linear polarization, cyclic and square-wave voltammetry methods were used. All potentials reported in this paper refer to the yttria-stabilized-zirconia reference electrode.

RESULTS AND DISCUSSION

Glass electrochemical study

Preliminary studies showed that the solvent's electro -activity domain extends from −1.2 to +0.1 V. It is limited in the negative potentials by the reduction of the silicate network, in accordance with $SiO_4^{4-} + 4e^- \rightarrow Si + 4 O^{2-}$ [1, 2, 9]. The oxidation anodic limit of the solvent corresponds to oxygen gas formation, due to the oxidation of the O^{2-} ions ($O^{2-} \rightarrow \frac{1}{2} O_{2(g)} + 2e^-$), then of the silicate network ($SiO_4^{4-} \rightarrow SiO_3^{2-} + \frac{1}{2} O_{2(g)} + 2e^-$).

Electrochemical study of metal redox systems

If the glass does not contain impurities, the electro -active redox systems are Si^{IV}/Si^0 and O_2/O^{2-}. In order to predict the corrosion reactions of pure metals and alloys in the glass, formal potentials of the main constituents of alloys (Cr, Fe, Co and Ni) were determined at T=1050°C.

Iron systems – The C-glass used contains 0.15% Fe_2O_3 as an impurity. The Fe^{II}/Fe^0 and Fe^{III}/Fe^{II} redox systems potentials were determined by cyclic and square-wave voltammetries (Figure 1). The obtained values are $E_{1/2}$ (Fe^{II}/Fe^0) = -1.05 V and E_0 (Fe^{III}/Fe^{II}) = -0.41V.

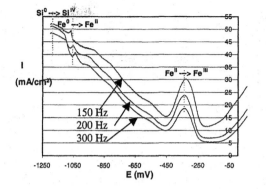

Figure 1. Square-wave voltammetry in C glass, anodic curve, platinum electrode, initial potential: -1.2V, ΔE=25mV, T=1050°C.

Cobalt systems – The results concerning cobalt systems in the literature are disparate. Co^{III}/Co^{II}, Co^{II}/Co^0, and Co^{II}/Co^I systems with non-self-consistent potential values are reported by different authors [4, 5, 6]. In order to determine the Co redox systems occurring in the C-glass, electrochemistry was carried out in a 0.30-wt%-Co_3O_4-doped C-glass. The glass has been homogenized by a 24-hour heat treatment at T=1150°C before further measurements.

Cyclic voltammetry suffers from a poor sensitivity; therefore, results are not given here. Current-potential curves obtained by square wave voltammetry are shown in Figure 2.

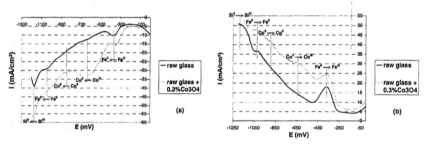

Figure 2. Square-wave voltammetry curves recorded in a 0.3-%-Co_3O_4-doped C-glass at 1050°C, comparison with a non-doped glass, platinum electrode.
(a) cathodic curve: initial potential = +0.1V, ΔE=25mV, f = 100 Hz.
(b) anodic curve: initial potential = -1.2V. ΔE=25mV, f = 100 Hz.

The Fe^{II}/Fe^0 and Fe^{III}/Fe^{II} peaks can be noted on these curves respectively at – 1.05 V and – 0.41 V. The curves obtained with the Co_3O_4-doped C-glass exhibit two additional peaks, which can be allotted to two different Co systems at E_1 = - 0.91V and E_1 = - 0.65V. The immersion of a Co-metal electrode in the C-glass yields a – 0.85V corrosion potential. The blue coloration of the glass surrounding this electrode indicates the presence of Co^{II} cations dissolved in the glass. In consequence, the E_1 potential is assigned to the Co^{II}/Co^0 system. Thus, the E_2 potential is allotted to the Co^{III}/Co^{II} system. The presence of Co^{III} cations in the C-glass at potentials higher than –0.65V is consistent with results reported in the literature, which indicate that Co^{III} can be the main oxidation state of Co under these conditions [12]. This recent study attributes the $-0.65V$ peak to a mono-electronic reaction. It is in good agreement with our results. Nevertheless, the authors [12] attributed this peak to the Co^{II}/Co^{I} system. The results obtained in the present study are summarized in Table II and compared with data from literature.

Titanium system

The Ti^{IV}/Ti^{III} formal potential is positioned at $-0.860V$ by Von der Gonna et al. [4]. A specific electrochemical study was carried out in the G-S glass in order to confirm this value and to determine the Ti^{III}/Ti^{II} formal potential. A platinum electrode is introduced in a titanium-enriched glass. The classical electrochemical techniques are used to characterize dissolved titanium ions [5, 6, 7]. Cyclic voltammetry suffers from a poor sensitivity and therefore results are not given here. Anodic and cathodic current – potential curves obtained by square wave voltammetry are reported in the Figure 3.

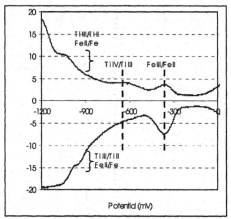

Figure 3. Curves of anodic and cathodic current *versus* potential recorded on Ti-enriched G-S glass by square-wave voltammetry.

The Fe^{II}/Fe^{0} and Fe^{III}/Fe^{II} characteristic peaks can be noted on these curves at -1.05 V and -0.41 V respectively. The anodic curve clearly exhibits an additional peak close to $-0.7V$ that can be assigned to the Ti^{IV}/Ti^{III} system. Considering results reported in the literature [4] and the high current value, the peak located at -1.0 V is assigned to both Fe^{II}/Fe^{0} and Ti^{III}/Ti^{II} systems.

These results were confirmed by polarization of a titanium electrode at different potential values (-0.3, -0.7, -1.0, -1.25, Ecorr= -1.4 and -1.6 V).

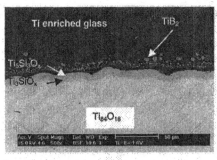 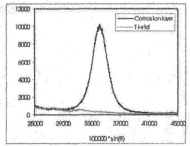

A/ -1.6 V   B/

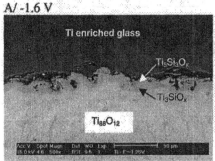 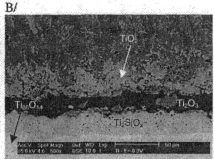

C/ -1.25 V   D/ -0.3 V

Figure 4. Sections of the 24-hour-polarized rods presented as a function of the imposed potential to the working electrode (T=1050°C) ; A/ -1.6V, , C/ -1.25V, D/ -0.3V B/Wavelength dispersion spectra of Ti and TiB_2 recorded with a PC3 spectrometer (SX100 electron microprobe) at the characteristic position of boron.

Table I : SEM observations of titanium rods polarized 24 h at different potential values in soda-lime glass at 1050°C

Potentials (V)	-1.60	Ecorr = -1.40	-1.25	-1.00	-0.70	-0.30
Corrosion products	Ti-enriched glass (no titanium oxide)	Ti-enriched glass (no titanium oxide)	Ti-enriched glass (no titanium oxide)	Ti_2O_3	Ti_2O_3	TiO_2
Reduction products	TiB_2 and Titanium silicide	Titanium silicide	Titanium silicide			
Reaction	$B^{III} \rightarrow B°$ $Si^{IV} \rightarrow Si$	$Si^{IV} \rightarrow Si$	$Ti \rightarrow Ti^{II}$	$Ti^{II} \rightarrow Ti^{III}$	$Ti^{III} \rightarrow Ti^{IV}$	$Ti \rightarrow Ti^{IV}$

After polarization, the rods were observed by SEM (Figure 4). The main observations of titanium after 24 hours of polarization are reported in the Table I. Ti^{IV} appears for potentials higher then –0.7 V. A corrosion potential equal to –

1.4V and the presence of titanium ions in the glass for a potential of −1.6V indicate that the Ti^{II}/Ti potential couple is located at a value lower than −1.6V. This experiment clearly shows that the B^{III} is oxidant at −1.6V.

Other electrochemical studies in molten glass have been devoted to the determination of redox couple potential. The main results are reported in Table II and compared with data from the literature.

Table II: Formal potentials or half-wave potentials of metallic couples in glass. Comparison with literature data. Values expressed in Volt. * Details of determination of these values are published in [13, 14]

Couples	Lizarazu [1]	Rüssel [5]	Claussen [6]	Medlin [7]	This work
Glass type	Soda-lime	Ternary soda-lime-silica	AR® Borosilicate	SRL-131	Borosilicate
T (°C)	1100	1100	1300	1150	**1050**
O^0/O^{-II}	>+0.15				**+0.1**
Cr^{VI}/Cr^{III}	+0.03	+0.03		−0.15	
Te^{IV}/Te^0					**−0.33***
Ni^{III}/Ni^{II}				−0.59	**−0.36***
Fe^{III}/Fe^{II}	−0.53	−0.59	−0.39	−0.10	**−0.41**
Co^{III}/Co^{II}					**−0.68**
Cr^{III}/Cr^{II}	−0.79	−0.86		−0.76	
Ti^{IV}/Ti^{III}					**−0.7**
Ni^{II}/Ni^0			−0.30	−0.83	**−0.70 ***
Mo^{III}/Mo^0					**−0.83***
Co^{II}/Co^0		−0.60	−0.42		**−0.91**
Mo^{VI}/Mo^{III}					**−0.96***
Ti^{III}/Ti^{II}					**−1.0**
Fe^{II}/Fe^0	−0.81			−0.9	**−1.05**
Si^{IV}/Si^0	<−1.02				**−1.18**
Ti^{II}/Ti^0					**<−1.6**
B^{III}/B^0					**−1.6/−1.4**

Concluding remarks: Application of redox potential; determination of corrosion mechanism

The corrosion mechanism of metals and alloys in molten glass is mainly conditioned by the nature of the oxidant present. For example Figure 5 shows the polarization curve of a soda-lime glass containing many oxidant species (such as Fe^{III}, Zn^{II}, Sb^V, ...). The knowledge of standard potential or half-wave potential values allows the assignment of the voltamperogram. The reduction wave of Fe^{III} is located at −500 mV and this of Zn^{II} at −750 mV.

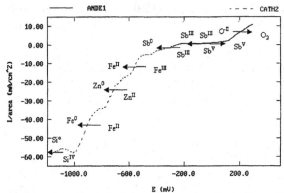

Figure 5. Polarization curve on a platinum electrode – Soda-lime glass – T=1200°C , v= 0.166mV/s

This polarization curve clearly explains the corrosion morphology of chromium immersed in glass containing iron (Figure 6). The corrosion reactions lead to the dissolution of chromium in Cr^{II} or Cr^{III} and the reduction of iron species in metallic iron. In the case of the corrosion of metals like tantalum in a glass containing zinc, the major oxidant is Zn^{II}, in accordance with the potential scale [14].

Figure 6. Micrograph of chromium immersed 2 hours in a soda-lime glass containing 1%wt of Fe_2O_3 at T = 1300°C

REFERENCES

[1]D.Lizarazu, P. Steinmetz, J.L. Bernard, "Corrosion of nickel-chromium alloys by molten glass at 1100°C: An electrochemical study," *Materials Sciences Forum*, 251 & 254, 709-720 (1997)

[2]J. Di Martino, C. Rapin, P. Berthod, R. Podor, P. Steinmetz, "Corrosion of metals and alloys in molten glasses: Part 1: Glass electrochemical properties and pure metals (Fe, Co, Ni, Cr) behaviours," *Corrosion Science*, in press

[3]J. Di Martino, C. Rapin, P. Berthod, R. Podor, P. Steinmetz, "Corrosion of metals and alloys in molten glasses: Part 2: Nickel and Cobalt high chromium superalloys behaviour and protection," *Corrosion Science*, in press

[4]G. Von Der Gonna and C. Rüssel, "Redox equilibria and polyvalent elements in binary $Na_2O.xSiO_2$ melts," *Glastech. Ber. Glass Sci. Technol.*, **73** [4] 105-110 (2000)

[5]C. Rüssel, "The electrochemical behavior of some polyvalent elements in a soda-lime-silica glass melt," *J. of Non Crystalline Solids*, **119** 303-309 (1990)

[6]O. Claussen and C. Rüssel, "Voltammetric investigation of the Redox behaviour of Fe, Ni, Co and Sn, doped glass melts of AR and BK7 type," *Glastech. Ber. Glass Sci. Technol*, **73** 33-38 (2000)

[7]M.W. Medlin, K.D. Sienerth, H.B. Schreiber, " Electrochemical determination of reduction potentials in glass forming melts," *J. of Non Crystalline Solids*, **240** 193-201 (1998)

[8]J. Di Martino, C. Rapin, P. Berthod, R. Podor, P. Steinmetz, "Electrochemical study of metals and alloys corrosion by molten glasses," 15th International Corrosion Congress, *Frontiers in Corrosion Science and Technology*, Grenade (Espagne), 22-27 Septembre 2002.

[9]J. Di Martino, C. Rapin , P. Berthod, R. Podor, P. Steinmetz, "Use of electrochemical techniques for the characterisation of alloys corrosion in molten glasses," *Glass Odyssey*, 6th ESG Conference, June 2 – 6, Montpellier, 2002

[10] F. Baucke, "High-Temperature Sensor for Oxidic Glass-Forming Melts," in *Sensors: A comprehensive Survey*, Volume 3, part II, p 1155-1198 Edited by W. Göpel, VCH Pub. (1992).

[11] F. Baucke, "Electrochemical cells for the online measurements of oxygen fugacities in glass forming melts," *Glastech. Ber.*, **61** [4] 87-90 (1988)

[12]J. De Strycker, P. Westbroek, E. Temmerman, "Electrochemical behaviour and detection of Co(II) in molten glass by cyclic and square wave voltammetry," *Electrochemistry Communications*, 4 41- 46 (2002).

[13] B. Gaillard-Allemand , A. Littner , R. Podor , C. Rapin, P. Steinmetz, M. Vilasi, "Electrochemical study in molten glasses of the multivalent systems of nickel, molybdenum and tellurium," *Glass Odyssey*, 6th ESG Conference, June 2 – 6, Montpellier, 2002

[14]R. Podor, C. Rapin, N. David, S. Mathieu, "Kinetics and mechanisms of tantalum corrosion in glass melts," *Glass Science and Technology*, submitted

ELECTROCHEMICAL STUDY IN MOLTEN GLASSES OF THE MULTIVALENT SYSTEMS OF NICKEL

Bruno Gaillard-Allemand, Arnaud Littner, Renaud Podor, Christophe Rapin, Anne Vernière and Michel Vilasi.
Laboratoire de Chimie du Solide Minéral, UMR7555
Université Henri Poincaré - Nancy I
54 506 Vandoeuvre-Les-Nancy, France

ABSTRACT

The chemical features of Ni multivalent systems have been characterized in molten glasses at T = 1423 K. Electrochemical methods, polarization curves, cyclic and square-wave voltammetries have been mainly used. Two redox systems have beenshown to be present. The first one is clearly attributed to Ni^{+II}/Ni^0 (-0.78V / YSZE), while the second remains uncertain with regards to electrochemical results. In order to overcome this indetermination, other analytical techniques, such as EELS, EPMA, and magnetic susceptibility measurements were carried out. Then, Ni^{+III} has been identified in oxidized molten glasses ($E^0_{(Ni+III/Ni+II)}$ = -0.36V / YSZE).

INTRODUCTION

Glass-making industries need metallic and intermetallic alloys with improved mechanical and chemical characteristics at high temperatures. As regards to chemical properties, a good resistance to the corrosion by molten glass is required and, in consequence, platinum and platinum alloys are widely used even if they are very expensive. In order to find new suitable materials, less expensive than the noble metal alloys, the study of the corrosion resistance of several materials was performed at high temperature (1473K) in a simplified glass melt (noted SG). The electrochemical techniques were used, because, as mentioned by many authors (1), they enable the measurements of the redox properties of metallic compounds as well as those of dissolved multivalent elements.

In the present work, the results of the chemical characterization of the Ni multivalent system are detailed. In particular, the results of the Energy Electron Loss Spectroscopy (EELS), magnetic susceptibility measurements (MSM) and electron-probe micro-analyses (EPMA) are presented, which are helpful for interpreting the electrochemical data.

EXPERIMENTAL DETAILS

The experimental procedures, as well as the high-temperature equipment used for the determination of voltammograms and corrosion kinetics in molten glass, have been already described in detail in (2, 3). The working electrodes were either a platinum wire ($\phi = 1$ mm) or a rod ($\phi = 5.5$ mm) of pure nickel. The counter electrodes consisted of a platinum plate ($S = 6$ cm^2). An yttria-stabilized-zirconia electrode (YSZE) flushed with air was used as reference electrode. Thus, all the potentials mentioned in this paper are given in comparison to the reference potential fixed by the ZrO_2 electrode. The electrochemical investigations were performed in a platinum crucible using a SG with the composition 16.55 wt% Na_2O, 58.6 wt% SiO_2, 24.75 wt% B_2O_3, equilibrated in argon U (max. O_2 content = 50ppm) during 15 hours at 1423K. This inert atmosphere was chosen in order to prevent the atmospheric corrosion of the metallic materials.

The EELS was performed using the electron beam produced by a transmission electron microscope, Philips CM20, equipped with a spectrometer, GATAN 666. The magnetic susceptibility measurements were carried out on a MANICS magneto-susceptometer (between 4.2 and 300 K) in an applied field up to 12 kOe. The data collected in the paramagnetic state have been treated by a least-square procedure. The measurements were performed on quenched polarized glasses containing NiO powder.

A part of each powder material was embedded in an epoxy resin, polished, and micro-analyzed by electron probe (SX 50 Cameca), using the PAP correction program (4).

RESULTS AND DISCUSSION

The redox behavior of nickel species was investigated by cyclic voltammetry using a platinum working electrode. The results revealed clearly the existence of two distinct redox systems at -0.780V and -0.360V, respectively (Figure 1).

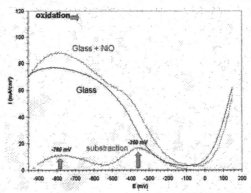

Figure 1. Square wave voltammogram recorded at 1423K in SG melt, doped with 1wt% NiO ; $\Delta E = 100$mV, f = 50 Hz, $v_b = 0.1$ V.s^{-1}

Thanks to the linear polarization of a nickel electrode, the first value was accurately attributed to the Ni^{+II}/Ni^0 system, because the free potential of metal is equal to -0.650V (Figure 2).

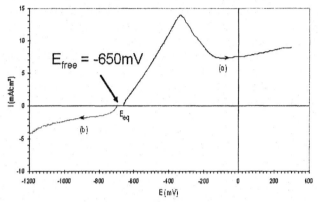

Figure 2. Dynamic polarization of a Ni electrode recorded at 1423K in SG doped with 5 wt% NiO ; vb = 0.16 mV.s^{-1}

The use of complementary analytical techniques was helpful for the identification of the second system. In a first step, simplified glasses, in which NiO powder was introduced, were polarized respectively at -0.4V and -0.1V during 1 hour, then quenched in water. In a second step, EELS, MSM, and EPMA were carried out on each kind of glass sample.

EELS - The results obtained by EELS are reported in Figure 3: i) both spectra of NiO standard and polarized glass at -0.4V are similar, ii) the spectra of the two types of glasses exhibit a shift of 2 eV, equal to that indicated in the literature related to NiO and Ni_2O_3 spectra (5), iii) the LIII/LII peak ratio is lowered in the case of the -0.1V polarized glass, as is the case for the characteristic ratios of d^8 and d^7 ions (6).

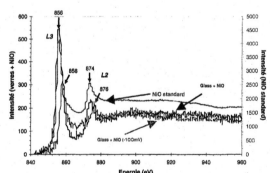

Figure 3. EEL spectra of pure NiO (NiO standard), glass with NiO polarized at -0.4V (glass+NiO) and -0.1V (glass+NiO (-100mV))

These first complementary results tend to demonstrate that Ni^{+III} could be stabilized in glass at a potential equal to -0.1V.

MSM - Concerning magnetic measurements, the results obtained allowed us to estimate the oxidation state and the content of nickel species dissolved in each glass. Their comparison with the composition values determined by EPMA led us to retain the appropriate solutions among the different assumptions taken into account for the calculation. In fact, the variation of the magnetic susceptibility (χ) versus temperature rules as an extended Curie law, from which the Curie constant (C) can be determined (Figure 4). Then, the effective magnetic moment can be calculated from equation (1):

$$\mu_{magn.}^{measured} \approx \sqrt{8C} \quad (\mu_B) \tag{1}$$

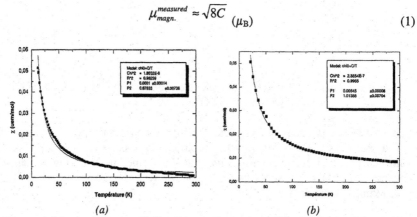

(a)   (b)

Figure 4. Magnetic susceptibility (χ) *vs* temperature (K) of SG containing NiO and polarized at -0.1V (a) and -0.4V (b)

Moreover, because nickel is the unique magnetogene species present in the SG tested, the effective magnetic moment can be deduced from the number of unpaired d electrons (n) of Ni ions, according to the following equation (2):

$$\mu_{Ni^{n+}} = \sqrt{n(n+2)} \quad (\mu_B/\text{mole of Ni ions}) \tag{2}$$

When considering all the cases of electronic configurations imposed by the crystal field and by the related tetra- or octahedral coordination of Ni ions, it is possible to reckon the associated magnetic moments of one mole of Ni ion ($\mu_{Ni^{n+}}$). Then, the Ni^{+II} and Ni^{+III} contents can be deduced by taking into account the following ratio $\left(\dfrac{\mu_{magn.}^{measured}}{\mu_{Ni^{n+}}} \right)$ and the Ni composition of the glass determined by EPMA. The results of this estimation are reported in Table I.

Table I. Results of MSM performed on glasses containing NiO and polarized at -0.1V and -0.4V, respectively.

	Polarization at – 0.1V	Polarization at – 0.4V
Determination of Curie Constant (molar quantity)	$C = 0.676(7)$	$C = 1.014(7)$
$\mu_{magn.}^{measured} \approx \sqrt{8C}$ (μ_B)	2.325(12)	2.848(10)
$\mu_{Ni^{n+}} = \sqrt{n(n+2)}$ (μ_B/mole Ni ions)	Assumption : Ni^{+III} is the unique species present in SG. *Tétra- or Octahedral High Spin coordination (n=3):* $\mu_{Ni+III} = \sqrt{15} = 3.873$ *Octahedral Low Spin coordination (n=1) :* $\mu_{Ni+III} = \sqrt{3} = 1.732$	Assumption : Ni^{+II} is the unique species present in SG. *Tétra- or Octahedral coordination (n=2):* $\mu_{Ni+II} = \sqrt{8} = 2.828$
Nickel content calculation when only one Ni ion species is considered : $\left(\dfrac{\mu_{magn.}^{measured}}{\mu_{Ni^{n+}}}\right)^2 = x\ mole$	$\left(x_{Ni^{+II}}\right)_{octa}^{étra} = 0.67570(14)$ $\left(x_{Ni^{+III}}\right)_{octa\ HS}^{étra} = 0.36040(14)$ $\left(x_{Ni^{+III}}\right)_{octa\ LS} = 1.80190(14)$	$\left(x_{Ni^{+II}}\right)_{octa}^{étra} = 1.01390(14)$ $\left(x_{Ni^{+III}}\right)_{octa\ LS} = 2.70370(14)$ $\left(x_{Ni^{+III}}\right)_{octa\ HS}^{étra} = 0.54070(14)$
EPMA results (mole of Ni ions)	$x = 0.430(100)$	$x = 1.045(100)$
Nickel content calculation when a mixing of Ni^{+II} and Ni^{+III} is considered	x1 = 0.14920(14) x2 = 0.28080(14) x1 = 0.03649(14) x2 = 0.39310(14)	All Ni ions are Ni^{+II}
Expression used	$\mu_{measured}^2 = x_1 \cdot \mu_{Ni^{+II}}^2 + x_2 \cdot \mu_{Ni^{+III}}^2$	

The comparison of the composition values obtained by MSM and EPMA indicates that i) Ni^{+II} is the unique ionic species existing in the glass polarized at -0.4V and conversely, ii) in the case of SG polarized at -0.1V, Ni^{+III} or a mixture of Ni^{+II} and Ni^{+III} are stabilized.

Finally, all these experimental features lead to the following conclusions: i) the value of the characteristic potential $E_{1/2}$ of the Ni^{+II}/Ni^0 couple is equal to – 0.780V and is lower than those mentioned by K. Takahashi et al. (7) and Rüssel et al. (8), ii) Ni^{+III} can be stabilized in glass melts, contradicting the conclusions of many authors who calculate that the characteristic potential $E_{1/2}$ of the Ni^{+III}/Ni^{+II} couple is higher than the potential of glass melts' oxidation (9). Moreover, one can note that the interpretations of the experimental results ido not refer to the possible existence of tetra- or octa-coordinated Ni^{+II} in the glass (10), but they do not contradict this assumption.

ACKNOWLEDGEMENTS

The authors gratefully acknowledge Dr. F. Diot and S. Barda (Service Commun d'Analyses par Sondes Electroniques – Université H. Poincaré) for the electron microprobe micro-analyses and Dr. A. Le Louarn for her advice.

REFERENCES

[1] H.D. Schreiber, N.R. Wilk Jr and C.W. Schreiber, "A comprehensive electromotive force series of redox couples in soda-lime-silicate glass," *Journal of Non-Crystalline Solids*, **253**, 68-75 (1999).

[2] D. Lizarazu, P. Steinmetz, J.L. Bernard, "Corrosion of nickel-chromium alloys by molten glass at 1100°C: an electrochemical study," *Material Science Forum*, **251 & 254**, 709-720 (1997).

[3] E. Freude and C. Rüssel, "Voltammetric methods for determining polyvalent ions in glass melts," *Glastech. Ber.*, **60** [6], 202-204 (1987).

[4] J.L. Pouchou, F. Pichoir, *La Recherche aérospatiale*, **3**, 167 (1984).

[5] K.S. Kim, R.E. Davis, "Electron spectroscopy of the nickel-oxygen system," *Journal of electron Spectroscopy and related Phenomena*, **1**, 251-257 (1973)

[6] W.G. Waddington, P. Rez, J.P. Grant, C.J. Humphreys, "White lines in the L2,3 electron-energy-loss and x-ray absorption spectra od 3d transition metals," *The American Physical Society*, **34** [3], 1467-1473 (1986).

[7] K. Takahashi and Y. Miura, "Electrochemical studies on diffusion and redox behavior of various metal ions in some molten glasses," *Journal of Non-Crystalline Solids*, **38 & 39**, 527-532 (1980).

[8] M.W. Medlin, K.D. Sienerth, H.D. Schreiber, "Electrochemical determination of reduction potentials in glass-forming melts," *Journal of Non-Crystalline Solids*, **240**, 193-201 (1998).

[9] F.GK. Baucke and J.A. Duffy, "Redox reactions between cations of different polyvalent elements in glass melts: an optical basicity study," *Physics and chemistry of Glasses*, **34** [4],158-163 (1993).

[10] M. Shibata, M. Ookawa and T. Yokokawa, "Studies of NiO dissolved in alkali silicate melts based on redox potential and visible absorption spectra," *Journal of Non-Crystalline Solids*, **190**, 226-232 (1995).

Effects of Composition and Forming on Structure and Properties

EFFECT OF WATER IN THE MELTING ATMOSPHERE ON THE TRANSFORMATION TEMPERATURE OF COMMERCIAL GLASSES

E. M. Birtch and J. E. Shelby
New York State College of Ceramics
Alfred University
Alfred, NY 14802

ABSTRACT

Results of an earlier study of the effect of bound water concentration, in the form of hydroxyl, on the properties of three commercial soda-lime-silica glasses (float, container, microscope slide) revealed that the glass transformation and dilatometric softening temperatures and viscosity isokoms in the transformation range decrease with increasing water concentration. The maximum water concentrations obtained in this study were limited to less than 600 wtppm by the apparatus available at that time.

Recent improvements in the apparatus used to produce glasses with varying water concentrations now allow production of samples containing approximately twice as much hydroxyl as was possible at the time of the earlier study, i.e. the range of study has doubled as compared to the earlier work. Additional glasses of commercial interest have been added to the study (wool, E-glass, and TV panel and funnel glasses). A linear relation exists between the glass transformation and dilatometric softening temperatures and the water content of the glass for float, container, and TV panel glasses, which is contrary to the earlier results, that suggested the effect of water concentration decreases with increasing water content. Results for the other 3 glasses indicate that melting under an atmosphere containing high concentrations of water vapor removes components from the glass, resulting in unusual behavior in the results of these measurements.

INTRODUCTION

A number of studies have considered the effect of bound water, in the form of hydroxyl, concentration on the properties of silicate glasses [1-7]. While many of the properties of glasses have been found to be altered slightly by variations in hydroxyl content, the largest changes are found for properties involving relaxation or flow of the network [1]. These properties include the glass transformation temperature, as measured by either differential scanning calorimetry or dilatometry, the dilatometric softening temperature, and the viscosity. Studies of binary alkali silicate [6-8] and ternary soda-lime-silicate glasses [9], for example, indicate that the glass transformation temperature (Tg) varies by about 0.04 K/wtppm of water in the glass. Since the water content can easily be varied by over 1000 wtppm for these glasses, values of Tg can vary by as much as 40 K for otherwise identical glasses. Simply melting these glasses at different times of the year, with different humidity in the laboratory, can vary Tg by 10 K or more.

A previous study in this laboratory considered the effect of water content on the properties of float, container, and microscope slide glasses [2]. In each case, Tg, Td, and the isokom temperatures decrease with increasing water content. The results, which cover the range from about 100 to 500 wtppm of water, suggest that the effect of water concentration may gradually decrease as the total water concentration increases, i.e. the slope of the Tg versus water concentration curve decreases with increasing water concentration.

Redesign of the apparatus used in the earlier work to produce glasses with varying water content has doubled the range of water contents available for the current study. The present paper presents additional data for the float and container glasses, covering this extended range of water concentrations, along with new data for several other commercial compositions (wool and E-glasses and TV panel and funnel glasses). Results indicate that the loss of volatile species can counter the effects of water content in some cases, which prevents a simple correlation between water content and Tg for those glasses.

EXPERIMENTAL PROCEDURES
Samples used in this study are commercial glasses. Different water contents were obtained by either remelting the glass in an electrically heat furnace in ambient laboratory air or by remelting under controlled water vapor partial pressures. Since the commercial glasses are produced in gas-fired tanks, they contain more water than identical glasses melted in electrically-heated furnaces. As a result, remelting in our laboratory in ambient air reduces the water content of the glass. Remelting under partial pressures of water vapor which are greater than those of the commercial tank produces glasses with increased water contents. Use of different partial pressures of water vapor produces glasses with different water contents. Details of the experimental method used to produce the glasses can be found in numerous other publications [2, 8,9].

Glass transformation temperatures were determined using a differential scanning calorimeter (DSC) with a heating rate of 20 K/m. Glass transformation temperatures (T_g) were also determined from thermal expansion curves using the onset of increased expansion method, while dilatometric softening temperatures were determined from the point of maximum expansion on the curve [2,8,9]. Water contents were determined using infrared spectroscopy and the Beer-Lambert Law, with extinction coefficients of 41.3, 42.4, 18.2, 46.8, and 49.1 for the float, container, TV panel, wool, and E-glass compositions, respectively [10-12]. The water content of the TV funnel glasses was determined using the 2 band method discussed elsewhere [1,13], with values of 70 and 150 L/mol-cm for the extinction coefficients.

RESULTS
Results of the present study for the float glasses are compared with those of the earlier work by Jewell, et al. [2] in Figure 1. The agreement between results of the two studies is excellent. A value obtained for the as-received float glass (solid point) falls on the best-fit line through the combined data, indicating that the laboratory results are in excellent agreement with those expected from industrial

practice. The best-fit to these data is a straight line with a slope of ≈0.025 K/wtppm, indicating that the effect of water concentration on Tg is independent of the water concentration, i.e. no saturation effect occurs. The slope of this line is less than the value of 0.042 K/wtppm of water recently reported for a ternary soda-lime-silica glass with a composition similar to float glass.

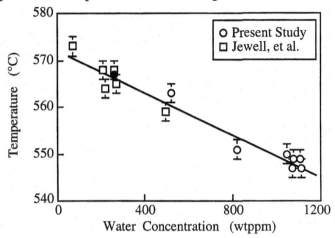

Figure 1: Effect of water concentration on the glass transformation temperature of float glasses.

Similar results for container glasses are shown in Figure 2. In this case, however, results for the present study do not agree with those of Jewell, et al. The values of Tg found in the present study are about 10 K greater than those of the earlier work [2]. This difference is probably due to the use of different sources of "container" glass, which may have slightly different compositions. The solid points on this figure are from as-received container glasses. The two solid points at lower water contents are from two different containers (one amber, one clear) melted in a

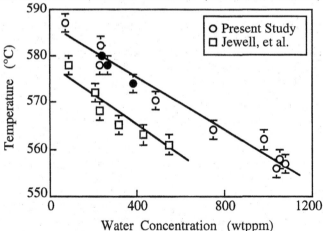

Figure 2: Effect of water concentration on the glass transformation temperature of container glasses.

air/fuel tank. The solid point at higher water content is from a clear container from the same manufacturer melted in a oxy/fuel tank. The agreement between the results for our laboratory melts and the commercially produced water variations confirms the reliability of the laboratory studies in predicting the effect of variations of water content produced in commercial practice. The slope of the best-fit line through these data is ≈0.028 K/wtppm water.

The effect of water concentration on the glass transformation temperature of TV panel glass is similar to that for the soda-lime-silica glasses, as is shown in Figure 3 (the solid point represents the as-received glass). Replacement of a portion of the soda by potash and use of strontium and barium instead of magnesium and calcium has no significant effect on the relation between Tg and water concentration. The slope of the best-fit line through the data is ≈0.02 K/wtppm.

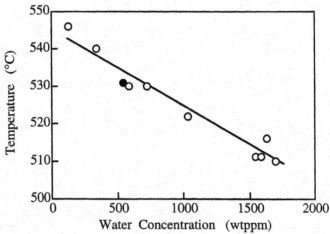

Figure 3: Effect of water concentration on the glass transformation temperature of TV panel glasses.

The relation between the glass transformation temperature and the water content of the glass is more complex for the TV funnel glass, as is shown in Figure 4. The glasses produced at lower water contents display similar behavior to that of the float, container, and TV panel glasses, i.e. Tg decreases with increasing water concentration (the solid point represents the as-received glass). Results for the glasses having the highest water contents, however, are quite different from those of these other glasses. The Tg of the glasses containing between 800 and 900 wtppm of water increase with increasing water content. These glasses were melted under 1 atm of water vapor. This behavior becomes more understandable if we consider the melting temperatures, in °C, used to produce these glasses, as shown by the values on the figure next to the points for the glasses with high water contents. It immediately becomes obvious that Tg increases with increasing melting temperature, even though the water content of the glasses is increasing as well. These results imply that the bulk composition of the glass is changing with changes in melting temperature when a high water vapor partial pressure is used. It seems reasonable to speculate that some component of the glass is removed from the melt

Advances in Fusion and Processing of Glass II

by an interaction with the water vapor in the melting atmosphere. It seems reasonable to suggest that removal of Pb may be responsible for these results. Lead is know to be quite volatile and to have a large effect on Tg. An increase in melting temperature could increase the rate of removal of lead, which would offset the effect of the water on Tg, resulting in the increase in Tg with increasing melting temperature.

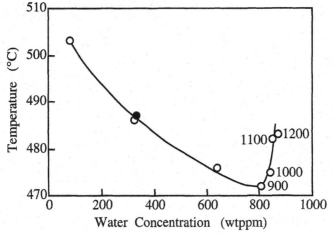

Figure 4: Effect of water concentration on the glass transformation temperature of TV funnel glasses.

The effect of water vapor on volatilization of a component and the consequent change in Tg is especially evident for the wool and E-glasses. The Tg measured for a series of E-glasses produced by melting under different partial pressures of water vapor and at different temperatures is shown in Figure 5. There is very little effect of water content on Tg for the glasses produced under lower water vapor partial

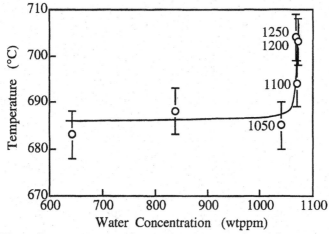

Figure 5: Effect of water concentration on the glass transformation temperature of E-glasses.

pressures, as indicated by the essentially flat line region at lower water contents. The effect of melting temperature for those samples produced under one atmosphere of water vapor, i.e. the data on the right side of the figure, is quite large even though these glasses have very similar water contents. Tg increases by 19 K while the water content is essentially constant for these glasses.

Once again, these results suggest that the composition of the melt changes during the water treatment. In this case, however, we have a method to detect the change in the concentration of a component which can explain these results. The infrared spectrum of E-glass has an absorption band at ≈ 2680 cm^{-1} which is due to an overtone of a fundamental B-O vibration [12]. The height of this band is directly related to the concentration of boron in the glass. A decrease in the height of this band thus indicates a loss of boron from the melt. If the Tg data from Figure 5 are replotted as a function of the height of the B-O band instead of the water concentration, we obtain the results shown in Figure 6. These results strongly support the contention that the loss of boron from the melt is accelerated by the presence of a high partial pressure of water vapor and by an increase in melting temperature. The overall effect of melting under an atmosphere containing water vapor on Tg is thus a combination of the effects of addition of water with increasing water vapor partial pressure in the melting atmosphere and a decrease in boron concentration with increasing water vapor partial pressure and melting temperature.

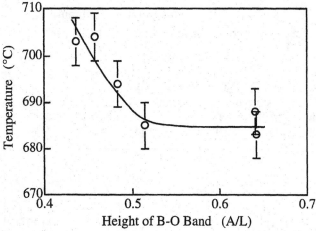

Figure 6: Effect of boron concentration on the glass transformation temperature of E-glasses.

A similar, but less dramatic effect of water vapor in the melting atmosphere is found for wool glass, as is shown in Figure 7. Tg decreases with increasing water concentration for melting temperatures of up to 1000°C even for one atmosphere of water vapor. If the melting temperature is raised to 1100°C or above, the water concentration in the saturated glass actually decreases with increasing melting temperature for melts made under a partial pressure of water vapor of one atmosphere. The Tg of these glasses, however, increases with increasing melting temperature, even though the water content decreases. Examination of the infrared spectra of these glasses indicates that the glasses melted under one atmosphere of water vapor and at higher melting temperatures have lost a significant portion of the original boron, as indicated by changes in the spectrum in the region around 2670 cm^{-1}, with the boron loss increasing with increasing melting temperature. Unfortunately, the high degree of overlap between the hydroxyl absorption band at 2880 cm^{-1} and the B-O band at 2670 cm^{-1} prevents a detailed quantitative analysis similar to that for E-glass. It follows that the changes in Tg observed for these glasses are due to the combined effects of hydroxyl formation and boron loss.

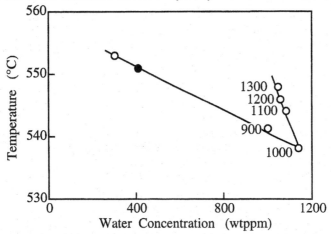

Figure 7: Effect of water concentration on the glass transformation temperature of wool glasses.

DISCUSSION

The effect of water content on the glass transformation temperature of the float, container, and TV panel glasses is consistent with that found for other silicate glasses. It is generally assumed [1,2] that the formation of terminal hydroxyl units is very effective in disrupting the network structure, allowing for more rapid relaxation and a decrease in Tg. A similar effect is found for the other glasses when melted under lower partial pressures of water vapor or at lower temperatures, as shown in Figures 4, 5, and 7. The changes in Tg for those glasses can also be explained by the formation of terminal hydroxyl units.

The effect of melting under high partial pressures of water vapor and at higher temperatures found for the TV funnel, E-glass, and wool glass is due to enhanced volatilization of specific components from the melts. Infrared spectra for the E-glasses clearly show the enhanced loss of boron under the more extreme melting conditions. While somewhat less dramatic, it is also clear that wool glasses also lose boron when melted under high partial pressures of water vapor and at higher temperatures.

The species removed from the TV funnel glass during melting under the more extreme conditions cannot be determined from this work. It is suggested, however, since lead is known to be a volatile species in silicate melts, that the loss of lead is a likely cause of the behavior observed in this study.

CONCLUSIONS

Melting commercial glasses under varying partial pressures of water vapor produces glasses with different hydroxyl contents. Changes in hydroxyl content cause changes in Tg of these glasses. High concentrations of water vapor in the atmosphere increases the rate of volatilization of boron from commercial fiber glasses and possibly of lead from TV funnel glasses.

ACKNOWLEDGMENTS

Portions of this work were funded by the NSF University/Industry Center for Glass Research at Alfred University (CGR) and portions by the US Department of Energy, Office of Industrial Technologies, grant #DE-FG07-96EE41262.

REFERENCES

[1] J. E. Shelby, "Water in Glasses and Melts"; pp. 217-34 in *Handbook of Gas Diffusion in Solids and Melts*, by J. E. Shelby, ASM International, Materials Park, OH, 1996.

[2] J. M. Jewell, M. S. Spess, and J. E. Shelby, "Effect of Water Concentration on the Properties of Commercial Soda-Lime-Silica Glasses," *J. Am. Ceram. Soc.*, **73**[1] 132-5(1990).

[3] J. E. Fenstermacher, R. C. Lesser, and R. J. Reiden, "A Study of the Water Content of Container Glass," *Glass Industry*, **46**[9] 518-21 (1965).

[4] P. W. McMillan and A. Chlebik, "The Effect of Water Content on the Mechanical and Other Properties of Soda-Lime-Silica Glasses," *J. Non-Cryst. Solids*, **38/39** 509-14 (1980).

[5] J. E. Shelby, "Viscosity of Glassforming Melts"; pp. 107-32 in *Introduction to Glass Science and Technology*, by J. E. Shelby, Royal Society of Chemistry, Cambridge, 1997.

[6] H. Scholze, H. Franz, and L. Merker, "The Nature of Water in Glass: VI," *Glastechn. Ber.*, **32**[10] 421-9 (1959).

[7] J. E. Shelby and G. L. McVay, "Influence of Water on the Viscosity and Thermal Expansion of Sodium Silicate Glasses," *J. Non-Cryst. Solids*, **20** 439-49 (1976).

[8] M. G. Mesko, P. A. Schader, and J. E. Shelby, "Water Solubility and Diffusion in Sodium Silicate Melts," *Phys. Chem. Glasses*, **43**[6] 283-90 (2002).

[9] D. B. Rapp and J. E. Shelby, "Water Diffusion and Solubility in Soda-Lime-Silica Melts," accepted by *Phys. Chem. Glasses*

[10] M. G. Mesko and J. E. Shelby, "Water Solubility and Diffusion in Melts of Commercial Silicate Glasses," *Glastech. Ber. Glass Sci. Technol.*, **73**[C2] 13-22 (2000).

[11] M. G. Mesko and J. E. Shelby, "Solubility and Diffusion of Water in Melts of a TV panel Glass," *Phys. Chem. Glasses*, **42**[1] 17-22 (2001).

[12] M. G. Mesko and J. E. Shelby, "Solubility and Diffusion of Water in Melts of E and Wool Glasses," *Phys. Chem. Glasses*, **42**[6] 389-96 (2001).

[13] H. Scholze, "The Influence of Water in Glasses: I," *Glastechn. Ber.*, **32**[4] 81-8 (1959).

DEPENDENCE BETWEEN THE COLOR OF TITANIUM CRYSTAL
GLASSES AND THE OPTICAL BASICITY

Marco Seufert and Armin Lenhart
University of Applied Sciences, Nuremberg
Wassertorstrasse 8-10
90489 Nuremberg, Germany

INTRODUCTION
Lead crystal glasses which fulfill the standard of the "Crystal glass identification law" [1] have many excellent properties. Melting and thermal processing are comparatively simple. They show outstanding optical properties like a high refractive index of about 1.545, a low Abbé number less than 47 and an optical transmittance high in the visible range. Furthermore grinding lead crystal glasses on machines is easy because of their low hardness. A successive acid polishing is possible.

The outstanding optical properties of the lead crystal glass distinguish this glass clearly from the other crystal glasses on the market. The high refractive index leads to a high reflection of light on the surface of the glasses. Because of this, lead crystal glasses shine very nicely. The Abbé number expresses the dependence between the refractive index and the wavelength of the light [2]. Glasses with a low Abbé number refract the light very strongly depending on the wavelength. This effect leads to the so-called "fire", which means the sparkle of the lead crystal glass.

Up to now many efforts were undertaken to substitute the toxic lead oxide content [3] by other non-toxic oxides which have the same effect on the optical properties of the glass. As yet, those efforts were not successful. There are some different oxides which effect a high refractive index and a low Abbé number in glasses like bismuth oxide, tantalum oxide, lanthanum oxide, zirconium oxide, titanium oxide and others [4]. After some economic and technological considerations it was decided to take titanium oxide to substitute the lead oxide content. Titanium oxide is comparatively inexpensive and available in the necessary quality and amount. It is known that the addition of titanium oxide to silicate glasses leads to a yellow to green discoloration [5]. This discoloration depends on the iron and titanium concentration and other basic influences. The cause for this coloration is the shifting of the ultra-violet-absorption-band of the glass towards the visible range and the increase of the absorption coefficient of the iron with increasing titanium oxide concentration. The iron absorptions are caused by d-d transitions of the electrons which are effected by neighboring oxygens and the following ligands. These transitions have a discrete transition-probability. Due to the fact, that the location of the absorption bands of the iron does not change with changing titanium oxide concentration it can be noticed that the transition energy does not change either. Because of this the titanium ion should influence the transition probability of the iron d-d transitions and due to this the absorption coefficient of the iron ion.

In this work it was noticed that the discoloration depends on many factors like the redox equilibrium, the melting conditions, the basic glass composition, the choice of raw materials, the optical basicity of the glass and many other basic conditions.

At the University of Applied Sciences in Nuremberg it was possible to melt titanium crystal glasses which fulfill the major properties of the lead crystal glasses including the optical properties like an Abbe number less than 47, a refractive index higher than 1.55 and a high transmittance. We now want to concentrate on the dependence between the glass color and the optical basicity of the glasses.

The optical basicity, L, developed by Duffy [6] is a figure of merit for the polarizability parameter of oxygen in glass. Duffy observed the location and intensity of absorption bands depending on the glass composition. It is known that the location and intensity of absorption bands depend on the basic glass composition and therefore on the optical basicity. He noticed that every oxide contributes a share to the sum basicity of the glass. In the periodic table of the elements there are several linear dependencies of the optical basicity, like within the alkaline and the earth alkaline series. The values of the optical basicity for various oxides are published in different works of Duffy. By adding the values for the optical basicity from the glass composition, a "sum basicity" of the glass itself can be formulated.

For some oxides like rubidium and cesium oxide there are no published values. These values can be estimated by plotting the optical basicity of lithium, sodium and potassium oxide over characteristics of these ions like their ionic diameter, electron negativity or other values. Normally it is possible to observe linear correlation and therefore it is possible to assess the unknown values.

RESULTS
Different glass samples were melted under identical melting conditions. The glasses were melted in an inductive-heated platinum-rhodium pot. The used raw

materials had an iron content of less than 5ppm. If necessary the mixture was doped with iron or titanium oxide. The glasses were melted for four hours with a temperature of 1450 °C. After that the samples were poured out and cooled down in a stove in a defined way to room temperature. Thereafter samples for the spectrometer were prepared. A λ19-UV/VIS/NIR-spectrometer was used for the measurements of the transmission curves. Using the measured transmission curves the CieLab-color values of the glasses were calculated [7]. All shown color values are corrected, standardized to 10 mm and calculated for the 2° standard observer by using D_{65} standard light in the range of 380 to 780 nm.

The following Table I shows the basic glass composition for the first melting series (alkaline series). For R the elements Li, Na, K, Rb and Cs were introduced. During the alkaline melting series Zn was used for M. In the earth alkaline series which is described in the Figure I the elements Mg, Ca, Sr, Ba and Zn were introduced for M, Na was used for R.

Table I: Basic composition of the melted glass samples (series I + II).

Oxide	SiO_2	TiO_2	R_2O	MO	Sb_2O_3	Fe_2O_3
concentr. in mol-%	63.89	8.00	18.00	10.00	0.10	0.01

It is known that rubidium and cesium oxides are very expensive. They can't be used for a mass production of crystal glasses. The crystal glass industry normally uses sodium and potassium either alone or in different mixtures for their glass production. To adjust the viscosity of the new glass and to detect the effect of different mixtures on the color values, a third melting series was prepared. In this mixed alkaline series different mixtures of Na and K were examined in the basic titanium glass system. Table II shows the basic composition of the mixed alkaline series.

Table II: Basic composition of the melted glass samples (series III).

Oxide	SiO_2	TiO_2	R_2O	ZnO	Sb_2O_3	Fe_2O_3
concentr. in mol-%	64.89	8.00	20.00	7.00	0.10	0.01

The mixed alkaline series starts with R equal K. After that Na was introduced for K in steps of 5 mol-%. Five glasses were prepared, one with twenty mol-% K_2O, one with 20 mol-% Na_2O and three glasses with K_2O and Na_2O in the ratios K to Na equal 3 to 1, 1 to 1 and 1 to 3.

Table III contains the calculated optical sum basicities for the alkaline-, the earth alkaline- and the mixed alkaline series. The optical basicity increases in the alkaline series from the Li- to the Cs-, and in the earth alkaline series from the Mg- to the Ba- glass. In the mixed alkaline melting series the optical basicity increases from the Na- to the K-glass.

Table III: Calculated optical basicities of the glass series I + II + III

R	Li	Na	K	Rb	Cs
optical basicity Λ	0.458	0.489	0.539	0.563	0.582
M	Mg	Ca	Sr	Ba	Zn
optical basicity Λ	0.493	0.506	0.512	0.517	0.490
R	K	K/Na=3/1	K/Na=1/1	K/Na=1/1	Na
optical basicity Λ	0.562	0.548	0.535	0.518	0.504

It was noted earlier that the optical basicity characterizes the oxygen polarizability of a glass. It is possible to melt glasses with identical optical basicities but different compositions. For example, the glasses 1 and 2 shown in Table IV have the same optical basicity Λ but a different glass composition.

Table IV: Two glass samples with different basic compositions.

sample	SiO_2	MgO	CaO	Na_2O	K_2O	Λ
1	74	-	10	16	-	≈ 0.47
2	71	10	-	-	19	≈ 0.47

During the development work, about 400 to 500 glass samples with different basic glass compositions were melted. Therefore glasses with similar optical basicities and completely different basic compositions were produced. From these 500 glass samples some with identical titanium and iron oxide content (9 weight-% titanium oxide and 100 ppm iron oxide) were taken. This means that there are glasses with different basic compositions and similar basicities. The transmission curves of these glasses were measured and the Lab values were calculated and the results were plotted over the optical basicities.

Figure 1 shows the transmission curves of the melting samples (series I) plotted over the wavelength in the range of 400 to 800 nm. The uv-edge of the curves is shifting to a longer wavelength from the Cs to the Na-glass. Furthermore an absorption in the range of about 430 nm can be seen. This absorption increases from the Cs to the Na-glass. Both effects lead to the absorption of blue light which results in a yellow discoloration. From Figure 1 it can be seen that the Cs-glass has a lower discoloration than the Na-glass. The following Figure 2 shows the Lab-description of the melted glass samples (series I + II). We can see that we have a mixture discoloration existing of a yellow (+b) and a green (-a) color share. Furthermore it can be seen that the discoloration increases from the Na- to the Cs- and from the Mg- to the Ba-glasses. The increase of the discoloration seems to be linear.

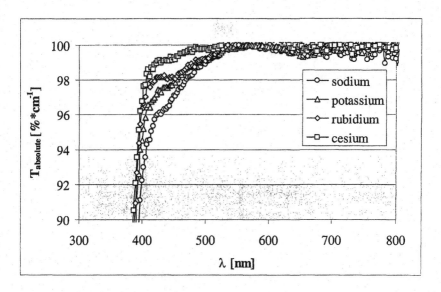

Figure 1: Transmission curves of the alkaline glass samples (series I).

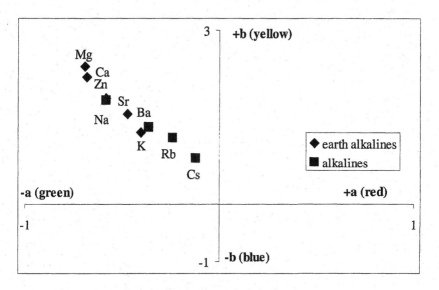

Figure 2: Color values of the melting samples (series I + II)

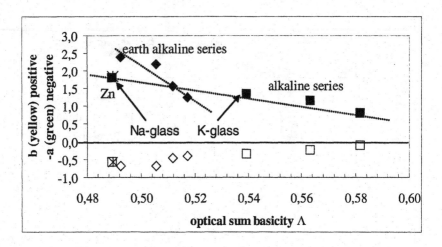

Figure 3: Color values of the alkaline- and earth alkaline melting series plotted over the optical sum basicity.

Figure 3 shows that the color values within a single melting series can be correlated linearly to the optical basicity. This can be done for the alkaline and earth alkaline series. It is interesting that Zn leads to a comparatively lesser discoloration although it shows a low optical basicity number.

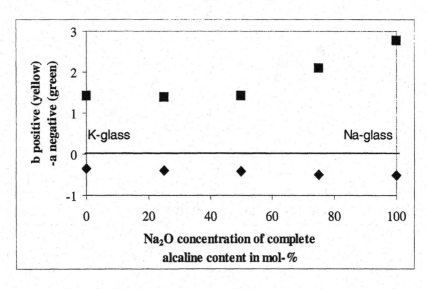

Figure 4: Color values b (positive) and a (negative) plotted over the concentration of Na_2O of complete alkaline content in mol-%.

Because of the results from the alkaline- and earth alkaline melting series it could be expected that there is a linear dependence between the optical basicity in the mixed alkaline melting series beginning from K to Na-glass. From Figure 4 it

Advances in Fusion and Processing of Glass III

can be seen that the dependence between the color values and the different alkaline mixtures is not as linear as expected. From the K_2O-glass to the glass with the K/Na ratio equal 1/1 the discoloration is similar. After increasing the Na content to a K/Na ratio smaller than 1/1 a volatile growing of the discoloration, especially the yellow color, share can be seen. The effect is reminiscent of the well known "mixed oxide effect". For the design of the new glass it is important to consider this effect. Furthermore, this effect shows that there is no linear correlation between the optical basicity and the glass color values in this mixed alkaline series.

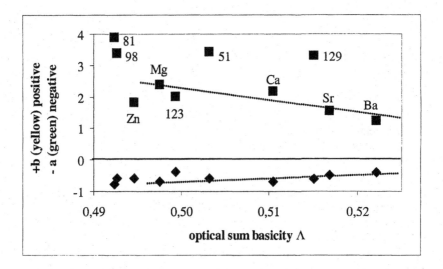

Figure 5: Color values a and b of the earth alkaline series and some different glass samples plotted over the calculated optical sum basicities.

In Figure 5, the optical basicities and color values of the earth alkaline series and some other glass samples (51, 81, 98, 123, 129) are shown. It can be noticed that it is possible to melt a lot of glasses with similar optical basicities, different basic compositions and completely different color values. For example, the yellow color values b of the Zn- and the Sr-glass are identical but the optical sum basicity is completely different.

On the basis of the theory of optical basicity it could be expected that glasses with the same optical sum basicity show the same location and intensity of the coloring absorptions and therefore the same color values. The conclusion here is, that there is no dependence between the color values and the "optical sum basicity".

SUMMARY

There is a measured context between the optical basicity and the color values of titanium oxide glasses within the alkaline- and earth-alkaline series. If the mentioned oxides are replaced within the series by each other the calculated optical sum basicity and the measured color values can be correlated.

If there is a mixture for example within the alkaline oxide series like shown in Figure 4, there is no linear dependence between the glass color and the optical sum basicity. The observed effect is reminiscent of the so-called "mixed oxide effect".

In Figure 4 it is shown that it is possible to melt different glasses with similar or same optical basicities, but completely different color values. Linear dependencies between the glass color and the optical basicity could be observed only in the alkaline- and earth-alkaline melting series. Therefore, the use of the optical basicity theorem was not useful for our glass development. It is not possible to forecast the color of titanium crystal glasses by using the theory of optical basicity.

REFERENCES
[1] B. Francken: Glas+Kristall, Warenkunde für den Fachhandel, Fachverband des Deutschen Eisenwaren- und Hausrathandels e. V., Duesseldorf, 1992

[2] G. Schroeder: Technische Optik, 8. überarbeitetet Auflage, Vogel Buchverlag, Wuerzburg 1998

[3] Sicherheitsdatenblatt: Bleioxid, PbO, CAS-Nr. 1317-36-8

[4] H. Scholze: Glas, Natur, Struktur und Eigenschaften, p. 208-209, dritte neubearbeitete Auflage, Springer Verlag, Berlin, 1988

[5] M. D. Beals, J. H. Strimple: Effects of Titanium Dioxide in Glass, in *The Glass Industry*, Part One-September 1963, p. 495-501, Part Two-October 1963, p. 569-573, Part Three-November 1963, p.625-629, Part Four-December 1963, p. 679-683

[6] A. Duffy: Bonding, Energie Levels & Bands in Inorganic solids, Longman Scientific & Technical, Longman House, Burnt Mill, Harlow Essex, England, 1990

[7] A. Berger-Schunn: Praktische Farbmessung, Musterschmidt Verlag, 1994

STUDY OF THE KINETICS OF PHASE SEPARATION IN 3.25 Na$_2$O - 3.25 Li$_2$O/Na$_2$O - 33.5 B$_2$O$_3$ - 60 SiO$_2$ GLASSES BY SKELETONIZATION

Alexander Fluegel, Ian M. Spinelli, and William LaCourse
New York State College of Ceramics
Alfred University
2 Pine Street
Alfred, NY 14802

ABSTRACT

Glasses with the composition (mol%) 6.5 Na$_2$O - 33.5 B$_2$O$_3$ - 60 SiO$_2$, and 3.25 Na$_2$O - 3.25 Li$_2$O - 33.5 B$_2$O$_3$ - 60 SiO$_2$ were thermally treated at temperatures in the range of 600°C to 800°C for 1 h to 128 h. This resulted in phase separation, i.e., in the formation of a silica- and an alkali borate-rich phase with an interconnected microstructure. Both, the volume content of the borate-rich phase and the mean structure thickness increased with time as well as with temperature. The mean structure thickness was determined through SEM image analysis by skeletonization, which is described in detail. In addition, the mean intercept length was examined. The mean structure thickness increased with time according to a power law ($\sim t^{1/n}$). In contrast to previous studies, n was in the range of 1 to 1.3 within the temperature range and time scale studied. The kinetic rate constants were calculated with n approximated to 1. The mean structure thicknesses were much larger (up to 12 μm) and the viscosities much lower than in most previous studies. The kinetic law was explained as being controlled by viscous flow. Lithium decreased the phase-separation rate.

INTRODUCTION

Silica-rich porous glasses, prepared by leaching phase-separated borosilicate glass, have many applications in chemical technologies, e.g., for separation techniques, enzyme immobilization, and catalyst support. The thicknesses of the phase-separated structure, and hence the pore diameter of the silica framework after leaching, are a function of the thermal history and the glass composition [1]. Image analysis is an efficient and powerful method for studying phase-separation kinetics.

EXPERIMENTAL TECHNIQUES

The investigated sodium borosilicate glass (in mol%) 6.5 Na_2O - 33.5 B_2O_3 - 60 SiO_2 (NaBSi) was prepared and characterized by Scanning Electron Microscopy (SEM) as described in [2]. The lithium sodium borosilicate glass 3.25 Na_2O - 3.25 Li_2O - 33.5 B_2O_3 - 60 SiO_2 (LiNaBSi) was prepared from analytical-grade chemicals (SiO_2, H_3BO_3, Na_2CO_3, and Li_2CO_3 as a 1.5 kg batch) and melted at 1450°C for 2 h in a platinum crucible under stirring. The glass was poured into a copper mold (~100 x 150 x 20 mm), and cooled in 15 min to room temperature without annealing. All samples were cut from the inner part of the produced glass block (at least 5 mm thickness from all edges was cut off), so that the samples had a relatively homogeneous cooling rate. Following the heat-treatment times and temperatures described below, the cooling rate was ~150 K/h for all glasses presented in this paper. The NaLiBSi glass was characterized using a Philips SEM 515, after leaching the fractured surface for 5 minutes in water. The conversion of the 256-grayscale SEM images into black/ white (B/W) images according to the two phases present was done manually, where the secondary phase-separation effect during cooling [1] was neglected. Image analysis was performed through a Basic program described below and published in [3].

IMAGE ANALYSIS

Phase-separation kinetics in borosilicate glasses may be analyzed, e.g., using N_2 adsorption and mercury porosimetry of the leached glass, or Small Angle X-ray Scattering (SAXS). Image analysis has several advantages over such methods:
- In principle the accuracy is not limited by very small or large structures. It depends on the quality and resolution of the microscope only.
- The relative volumes of the phases can be detected, even if it is not possible to leach one phase out selectively.
- The secondary phase-separation interference can be effectively eliminated.
- The relative volumes are calculated independently.
- Small amounts of the sample are sufficient for analysis.
- The sample preparation is fast and simple.
- In principle, even more than two phases can be analyzed independently.

The mean intercept method

The mean intercept method was developed for analyzing the grain-size distribution and grain orientation in ceramics and metals. In principle, lines are drawn through the image randomly (or with defined orientation), and the distances between the phase intercepts that the lines pass are averaged. In this work it was done in Microsoft Excel, after importing the image as binary "portable bitmap" (pbm format). The mean intercept method is useful for analyzing the dimensions of relatively simple morphological features (e.g., spheres, oriented crystals, lamellae) [4]. If the structures analyzed are complex in shape and relative volume, as in phase-separated glasses, further image analysis techniques need to be considered.

Skeletonization based on simple erosion

The principle of the erosion technique is taking off the outer pixel layer of one phase step by step, except when doing so would cause a separation of one region into two, until a skeleton (medial axis) remains [5, 6]. However, erosion is not very accurate, as seen below in Figure 1 (for orientation, the pixels of the phase of interest contain ten times the distance to the opposite phase in rounded numbers). Horizontally the erosion proceeds faster than at the 45° angle, i.e., a circle would become a square after several erosion steps. More exact erosion techniques have been developed (summarized in [7]), but still a stepwise erosion process accumulates errors. Therefore, it was not used in this work.

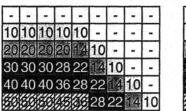

-	- opposite phase
10	- 1st erosion step
14	- 2nd erosion step
30	- 3rd erosion step
36	- 4th erosion step
50	- 5th erosion step

Figure 1: The erosion technique

Skeletonization based on the Euclidean distance

A skeletonization technique based on the shortest (Euclidean) distance to the opposite phase was developed in the way explained below. To the authors' knowledge, no research has been reported in the literature up to now regarding studies on phase-separation kinetics in glass through skeletonization based on the Euclidean distance.

1. Euclidean Distance Map (EDM) generation

The shortest distance for each pixel of the considered phase to the opposite phase was calculated, and assigned as the pixel's value or height. Figure 2 on the following page shows an example* [7]. For convenience, in all figures in this paper the pixel center-to-center distances are displayed in rounded numbers and multiplied by ten. The image analysis program used 7 decimal points, and the EDM was based on pixel center-to-edge distances.

* The skeleton pixels defined later in the paper are marked, where endpoints have a grey background, and nodes are black. The arrows illustrate the fact that skeleton pixels have the same distance to at least two pixels of the opposite phase, considering a tolerance of one pixel thickness.

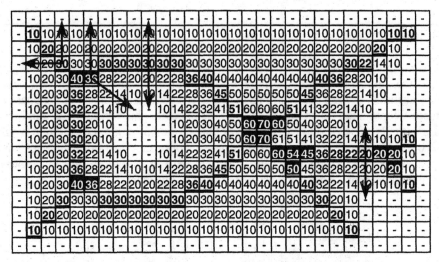

Figure 2: Euclidean Distance Map

2. Ridge-point network generation

Next, the direct 3x3 and 3x4 neighborhoods of each pixel were analyzed. If the neighboring pixels on opposite sides were closer to the opposite phase than the pixel in consideration (i.e., their values were lower), it was marked as a ridge point. The four cases shown in Figure 3 and the 90° rotated counterparts were analyzed:

· - Ridge-point pixel
⊠ - Pixel not analyzed
▨ - Pixel with lower value than ridge point

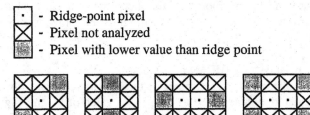

Figure 3: Patterns for ridge-point generation

3. Thinning

Because the as-determined ridge-point network was two pixels in thickness in many cases, a thinning algorithm had to be applied: If three skeleton pixels are found in the configuration displayed in Figure 4 (or the 90°, 180°, 270° rotated counterparts), the ridge point at the center is unmarked. This procedure was performed stepwise several times, until no further thinning was possible:

Figure 4: Pattern for thinning algorithm

4. Directional Uphill Generation

The ridge-point network generated through the procedure described above at point 2 is still discontinuous in those few locations which are not covered by the patterns in Figure 3. Therefore a procedure called "Directional Uphill Generation" [7] had to be applied: If at least one of the three pixels marked by "?" as seen in Figure 5 (or the 90°, 180°, 270° rotated counterparts) was of higher value than the marked ridge points, the pixel with the highest value among the three "?" pixels was marked as a ridge point as well. The procedure was repeated stepwise, until no further uphill ridge point could be generated:

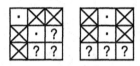

Figure 5: Patterns for Directional Uphill Generation

5. Endpoint corrections

Finally, endpoint corrections of the ridge-point network were performed. Endpoints are those that have one ridge-point neighbor only (marked with grey background in Figure 2). All endpoints with values considerably lower than the average were unmarked; in general, those had values lower than two times the pixel thickness. This step is necessary to correct pixel scattering at the phase boundaries.

All ridge points have the same distance (with a tolerance of one pixel thickness) to at least two pixels of the opposite phase, and angles between those pixels of the opposite phase and the ridge point can be established (see example arrows in Figure 2). If the largest of those angles for endpoints is lower than 90°, those endpoints are not a significant part of the ridge network, and they were unmarked. Now a definition of the term skeleton pixel used in this work can be given: Skeleton pixels form one continuous median line network (line thickness one pixel) per morphological feature analyzed. They have the same distance to at least two pixels of the opposite phase, considering a tolerance of one pixel thickness. If a skeleton pixel has one direct skeleton pixel neighbor only within its 3x3 neighborhood (i.e. endpoint), the largest angle between two closest pixels of the opposite phase and the skeleton pixel must be 90° or higher. Skeleton pixel endpoints can not be established directly adjacent to the opposite phase due to pixel scattering.

As soon as the skeleton was established, the two-dimensional thickness of any structure could be calculated, as well as further topological and metric characteristics.

As seen in the example given in Figure 6, the images analyzed in this study consisted of two interconnected structures, a silica- and a borate-rich phase, with different relative volumes and structure thicknesses [2]. At the top left (A1) is the original B/W image, white being the silica-rich phase. To its right (A2) is the analyzed image, with the mean structure thickness of 378 nm and a relative volume of 76.7%. Below is the same image in inverted colors (B1) and analyzed (B2). The mean structure thickness is 161 nm, and the relative volume is 23.3%:

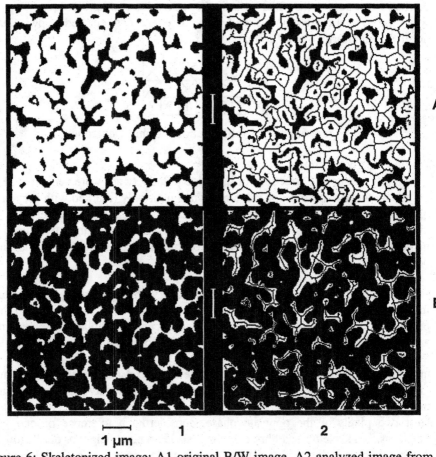

Figure 6: Skeletonized image; A1 original B/W image, A2 analyzed image from A1 with skeleton superimposed, B1 inverted image from A1, B2 analyzed image from B1

In the end, the average of both structure thicknesses was calculated, weighted by the relative volumes:

$$S = V_S \cdot S_S + V_B \cdot S_B \qquad (1)$$

with V_S = relative volume of the silica-rich phase, V_B = relative volume of the alkali borate-rich phase, S_S = structure thickness of the silica-rich phase, S_B = structure thickness of the alkali borate-rich phase, and $V_S + V_B = 1$. With this, the structure thickness S in the example given above is: $0.767 \cdot 378$ nm + $0.233 \cdot 161$ nm = 327 nm. No further room correction was considered, e.g. the factor $4/\pi$ as in [2, 3] assuming a cylinder/sphere-like structure towards the third dimension; or in [5, 8] assuming muralia ("intertwined sheets").

PHASE-SEPARATION KINETICS

The relative volume of the alkali borate-rich phase approached a limiting value at constant temperature. The mean structure thickness increased with time according to a power law ($\sim t^{1/n}$). By contrast to previous studies (summarized in [3] with n = 2-3), n was in the range of 1 to 1.3 within the temperature range (600-800°C) and time scale (1-128 h) studied [2]. The rate constants k, for n approximated to 1 according to $S = k \cdot t + S_o$, with S_o being the initial structure thickness at t = 0, and t being the time in hours, were:

Table I. Phase-separation rate constants

Sodium Borosilicate		Lithium Sodium Borosilicate	
T in °C	k in nm/h	T in °C	k in nm/h
660	10.3	650	4.3
680	26.5	660	8.4
700	68.7	670	12.3
720	142	700	59.6
S_o = 336 nm		S_o = 250 nm	

The average of the linear regression coefficients R^2 was 0.99, with a standard deviation of 0.01. No trend of R^2 was observed in temperature and time. If the mean intercept length (as described above) was analyzed, instead of the mean structure thickness, the linearity of the kinetic law did not change. However, the mean intercept length was larger than the mean structure thickness by a factor of ~1.2-1.5, depending on the relative volumes. In [2, 9] the linear kinetic law is explained as being controlled by viscous flow, since the mean structure thicknesses investigated were much larger (up to 12 µm) and the viscosities much lower than in most previous studies. The activation energy after Arrhenius was 340 kJ/mol for NaBSi, and 386 kJ/mol for LiNaBSi.

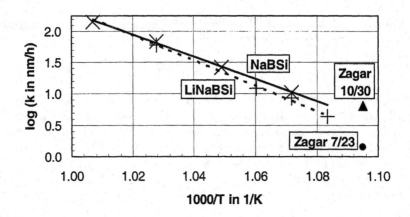

Figure 7: log (k in nm/h) = f (1000 / T in 1/K)

Figure 7 shows a comparison with borosilicate glasses of different compositions, investigated through mercury porosimetry by Zagar [9] (mol%): 10 Na_2O, 30 B_2O_3, 60 SiO_2 (10/30), and 7 Na_2O, 23 B_2O_3, 70 SiO_2 (7/23). The lithium in the LiNaBSi glass decreases the phase-separation rate, presumably because it increases the phase-separation tendency [1] (i.e. the highly viscous silica rich phase separates with higher relative volume), compared to the NaBSi glass. Increasing the silica concentration decreases the phase-separation rate considerably [9], as seen in Figure 7.

REFERENCES
[1]W. Vogel, "Glass Chemistry", Springer-Verlag, Berlin, New York, 1994
[2]A. Fluegel, and Ch. Ruessel, "Kinetics of phase separation in a 6.5 Na_2O - 33.5 B_2O_3 - 60 SiO_2 glass"; Glass Sci. Technol. 73 (2000), no. 3, pp 73-78
[3]A. Fluegel, Dissertation, 1999, Jena, Germany (in German)
[4]J. C. Russ and R. T. Dehoff, "Practical Stereology"; 2nd Edition, 2000, Kluwer Academic / Plenum Publishers, New York
[5]J. C. Russ, "The Image Processing Handbook", 2nd Edition, 1995, CRC Press
[6]T. Pavlidis, "A thinning algorithm for discrete binary images", Comp. Graph. Im. Proc. 13 (1980), 142-157
[7]F. Y. Shih, and C. C. Pu, "A Skeletonization Algorithm by Maxima Tracking on Euclidean Distance Transform", Pattern Recognition, 1995, Vol. 28, No. 3, pp 331-341
[8]H. J. G. Gundersen, E. B. Jensen, and R. Osterby, "Distribution of membrane thickness determined by lineal analysis", J. Microscopy 113 (1978), 27
[9]L. Žagar, "Über die Porenstruktur von entmischten ausgelaugten Natrium-Borosilicatgläsern", Glastechnische Berichte 48 (1975), 248-255 (in German)

PROPERTIES OF SODA-LIME-SILICA GLASSES

James E. Shelby, Melissa G. Mesko, Holly Shulman, and Daniel L. Edson
New York State College of Ceramics
Alfred University
Alfred, NY 14802

ABSTRACT
A number of properties have been measured for 50 soda-lime-silica glasses based on the compositions of typical commercial container and float glasses. The glasses studied are part of a broad program to create a data base for the effect of compositional variations around those of several common commercial glasses on the properties of those types of glasses and melts.

Properties studied for glasses include density and refractive index at room temperature, d.c. electrical conductivity at 200 and 300°C, thermal expansion coefficient below the glass transformation region, and water content of the glasses. The behavior of these materials in the glass transformation region has been characterized by measurements of their transformation range viscosity, glass transformation temperatures using both differential scanning calorimetry and dilatometry, and their dilatometric softening temperatures. Relations among these properties, e.g. the relation between density and refractive index, are presented here. Values for these properties are tabulated for the container-variation glasses. Comparisons are made between the properties of the more complex commercial compositions and those of simple ternary soda-lime-silica glasses.

INTRODUCTION
A large number of glasses based on the compositions of typical container and float soda-lime-silica glasses have been made. These glasses were prepared as a portion of an effort to develop a melt property database which can be used to predict the effect of small compositional changes on the properties of commercial glasses and melts. Other base compositions used in this study include wool and E-glasses used for fiber glass, TV front panel glass, and a hard borosilicate glass used for production of glass tubing.

The present study discusses the effect of composition on the properties of the soda-lime-silica glasses, i.e. those based on container and float compositions, from this study. A number of general trends relating various properties to the glass composition and to other properties of the same glasses can be found in these data. Properties considered include density, refractive index, thermal expansion coefficient, glass transformation and dilatometric softening temperatures, transformation range viscosity, and electrical conductivity.

EXPERIMENTAL PROCEDURES

Samples used in this study were produced at Alfred by melting in air in an electrically heated furnace in Pt crucibles. Batch compositions of the container-variation glasses are listed in Table I. Analyzed compositions of these glasses and batch and analyzed compositions of the float-variation glasses are available from the Center for Glass Research at Alfred University. Since details of the experimental measurements can be found in numerous other publications [1-5], only a brief overview of the experimental procedures will be presented here.

TABLE I: Batch composition of container-variation glasses

Sample Number	SiO_2	Li_2O	Na_2O	K_2O	MgO	CaO	Al_2O_3	Others
BASE								
1	81.0	0	11	0	0	7	1	0.0
2	72.9	0	11	2	0	12	1	1.1
3	70.0	1	15	0	0	12	1	1.0
4	72.8	0	15	0	3	7	1	1.2
5	70.7	1	15	2	3	7	1	0.3
6	69.1	1	11	2	3	12	1	0.9
7	71.8	0	15	2	0	7	3	1.2
8	75.3	1	11	2	0	7	3	0.7
9	68.5	1	15	0	0	12	3	0.5
10	73.9	1	11	0	3	7	3	1.1
11	64.7	0	15	2	3	12	3	0.3
12	70.3	0	11	0	3	12	3	0.7
13	73.2	1	15	2	0	7	1	0.8
14	78.8	1	11	0	0	7	1	1.2
15	69.6	0	15	2	0	12	1	0.4
16	75.0	0	11	2	3	7	1	1.0
17	68.2	0	15	0	3	12	1	0.8
18	71.7	1	11	0	3	12	1	0.3
19	74.4	0	15	0	0	7	3	0.6
20	72.8	0	11	0	0	12	3	1.2
21	70.7	1	11	2	0	12	3	0.3
22	73.5	0	11	2	3	7	3	0.5
23	70.6	1	15	0	3	7	3	0.4
24	62.5	1	15	2	3	12	3	1.5

Densities were determined using the Archimedes' Method with kerosene as the immersion fluid. Values are reproducible to ±0.002 gm/cm³. Refractive indices were determined using an Abbe refractometer with resolution of ±0.0001. Thermal expansion curves were measured using a single push-rod vitreous silica dilatometer [1,2]. Average thermal expansion coefficients, which are reproducible to ±0.3 ppm/K, were determined over the range from 100 to 400°C. Glass transformation temperatures (T_g) were determined using the onset of increased expansion method, while dilatometric softening temperatures were determined from the point of maximum expansion on the curve [1,2]. The glass transformation temperature was also determined using a differential scanning calorimeter (DSC) with a heating rate of 20 K/m. The d.c. electrical conductivity was determined using the guard ring method [2,3], with isothermal conductivity reproducible to ±10%. Water contents were determined using infrared spectroscopy and the Beer-Lambert Law, with extinction coefficients for the container-variations of 42.4 and the float-variations of 41.4 L/mol-cm [4]. The accuracy of this method is subject to question due to the assumption of a constant extinction coefficient for all related glasses.

Transformation range viscosities were determined using a beam-bending viscometer designed and constructed at Alfred [5]. This instrument was calibrated using NIST standards 710, 711, and 717. Multiple measurements of commercial samples of a float glass indicate that this instrument yields isokom temperatures reproducible to within ±1 K.

RESULTS

Results of the property measurements are listed in Table II for the container-variation glasses. Results for the transformation range viscosity as well as data for the float-variations compositions they can be obtained by contacting the Center for Glass Research at Alfred University.

TABLE II: Properties of container-variation glasses

Sample Number	Density (kg/m^3)	Refractive Index	TEC (ppm/K)	T_g (°C)	H_2O Conc (wtppm)	Elect. Cond. 200°C-S/cm
BASE	2520	1.5290	9.05	575	232	2.5×10^{-8}
1	2405	1.5019	7.93	573	212	2.2×10^{-8}
2	2503	1.5239	9.47	595	214	1.9×10^{-9}
3	2534	1.5336	11.1	545	151	7.1×10^{-9}
4	2487	1.5266	9.59	561	239	6.2×10^{-8}
5	2494	1.5205	11.1	509	213	4.5×10^{-9}
6	2545	1.5351	10.1	545	280	---
7	2477	1.5150	10.3	558	127	2.8×10^{-8}
8	2444	1.5111	9.23	539	172	2.8×10^{-9}
9	2554	1.5350	11.5	547	183	1.1×10^{-8}
10	2477	1.5201	8.77	553	231	1.2×10^{-8}
11	2559	1.5334	11.5	566	310	3.5×10^{-9}
12	2520	1.5287	8.88	612	204	1.0×10^{-8}
13	2497	1.5170	10.4	528	226	1.0×10^{-8}
14	2430	1.5109	8.79	531	180	8.2×10^{-9}
15	2536	1.5287	11.7	560	186	2.7×10^{-8}
16	2470	1.5130	8.99	567	296	8.3×10^{-9}
17	2556	1.5336	10.6	567	219	2.6×10^{-8}
18	2538	1.5347	10.2	556	301	1.3×10^{-9}
19	2461	1.5106	8.90	568	174	1.5×10^{-7}
20	2517	1.5318	8.69	623	256	1.1×10^{-8}
21	2514	1.5261	10.1	552	290	1.5×10^{-9}
22	2469	1.5138	8.94	575	387	1.2×10^{-8}
23	2502	1.5228	9.82	526	263	1.6×10^{-8}
24	2590	1.5439	12.3	535	182	1.0×10^{-9}

A number of correlations can be made among the properties measured in this study. Many studies [6] have shown that a simple correlation exists, for example, between the refractive index of glasses and their density. A plot of the refractive index of the container and float variation glasses as a function of their density is shown in Figure 1. Each set of data can be fit by a straight line, described by an expression of the form

$$n = Ad + B \qquad (1)$$

where n is the refractive index of the glass, d is the density of the glass, and A and B are constants determined by a best fit of a straight line to the data. Values of A and B obtained for the container and float variation glasses are listed in Table III, along with values obtained by a similar fit to data from Stanworth [6] for ternary soda-lime-silica glasses. Agreement among these sets of data is excellent, indicating that the specific concentrations of the various components of these similar glasses has little effect on the refractive index/density relation for glasses of this type.

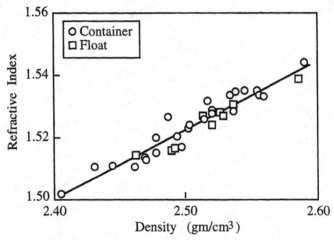

Figure 1: Effect of glass density on the refractive index of container and float variation glasses.

TABLE III: Fitting parameters for Eq. 1

Source	A	B
Container	0.226	0.958
Float	0.220	0.970
Stanworth [6]	0.232	0.935

A linear relation also exists between the transformation range viscosity, as indicated by the 10^{12} poise isokom temperature and the glass transformation temperature, as determined using a differential scanning calorimeter, of these glasses, as is shown in Figure 2. The dashed line on this figure indicates a one-to-one correspondence between these two temperatures, while the solid line represents a best fit to the combined data for all of the glasses. The slope of the best fit line is

Advances in Fusion and Processing of Glass II

within 1% of that of the ideal, one-to-one relation. This line is offset from the ideal line by 12 K, indicating that the viscosity at the glass transformation temperature is approximately 3×10^{12} poise for these glasses when measured using a heating rate of 20 K/m as was the case in this study.

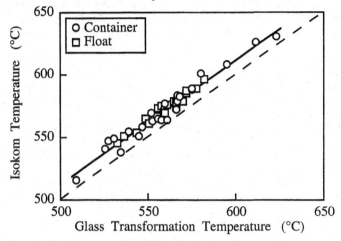

Figure 2: Relation between the 10^{12} poise isokom temperature and the glass transformation temperature of container and float variation glasses.

The viscosity at the glass transformation temperature is a function of the heating rate used in the measurement [7], as is shown in Figure 3, which presents the results for a float-variation glass of a series of DSC measurements of T_g with heating rates varying from 5 to 30 K/m. The log of the viscosity at T_g appears to be a linear function of the heating rate used in the T_g measurement, with values ranging from just over 10^{13} poise for a heating rate of 5 K/m to about 2×10^{12} poise for a heating rate of 30 K/m.

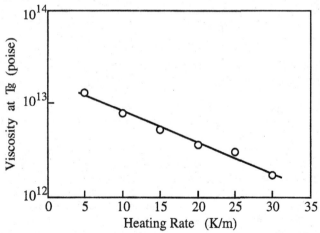

Figure 3: Effect of heating rate on the viscosity at T_g for a soda-lime-silica glass.

Results of this study demonstrate the strong effect of minor concentrations of lithium oxide on the viscosity and T_g of soda-lime-silica glasses. The container-variation glasses studied here can be subdivided into 4 groups as a function of the alkali oxides present in the composition: glasses containing only soda, those containing soda with either lithia or potash, and those containing all three alkali oxides. The relation between T_g and the total alkali oxide concentration in the glass is shown in Figure 4, where open symbols represent lithia-free glasses and solid symbols represent glasses which contain 1 wt% lithia. The T_g of these glasses decreases with increasing concentration of modifier oxides, as expected from the results of many studies [7]. The relative magnitude of the effect of lithia is much larger than the effect of the other two alkali oxides. Even though the glasses only contain 1 wt% lithia, the highest T_g of any glass containing lithia is lower than that of any glass which is lithia-free, i.e. a glass containing only 12 wt% alkali oxide, where 1 wt% of the alkali oxides is lithia, has a lower T_g than a lithia-free glass containing 17 wt% total alkali oxides. This effect is even more striking when one considers that the total alkaline earth oxide contents of these glasses vary from 7 to 15 wt% as well. It appears that the effect of lithia in controlling the T_g and viscosity of these glasses is greater than that of any combination of the other 4 modifier oxides.

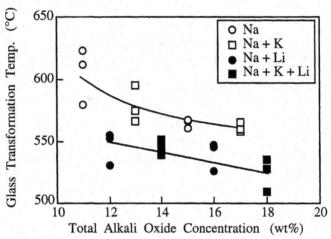

Figure 4: Effect of total alkali oxide concentration on the T_g of container-variation glasses.

The thermal expansion coefficient (TEC) of these glasses is primarily controlled by the silica concentration, as is shown in Figure 5. In general, the TEC of these glasses decreases with increasing silica concentration. The effect of the identity of the modifiers present is somewhat smaller than the effect of the silica concentration. The glasses with TEC greater than the trend line in Figure 5 are high in alkali oxide content relative to the alkaline earth concentration, while those below this line tend to be high in alkaline earth content relative to the alkali oxide concentration. This finding is consistent with results for simple soda-lime-silica glasses, where replacement of soda by lime at constant silica contents reduces the TEC [1].

The conductivities at 200°C of these glasses lie within a range a factor of ≈ 100, as is shown in Figure 6. The lowest conductivities are for glasses containing all 3 alkali and the maximum amounts of CaO and MgO. The highest conductivity is for a glass containing no lithia, potash or MgO, the highest amount of soda, and the least possible amount of CaO. The line is a best-fit straight line to all of the data.

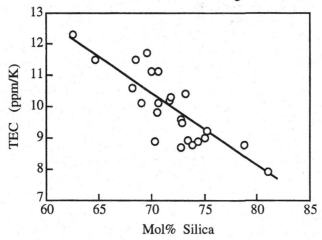

Figure 5: Relation between the thermal expansion coefficient and the silica content of container-variation glasses.

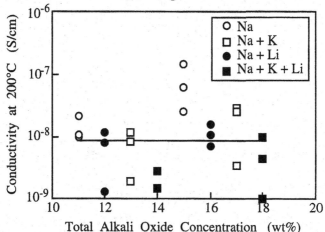

Figure 6: Effect of total alkali oxide concentration on the electrical conductivity of container-variation glasses.

DISCUSSION

Most of the results of this study are consistent with those obtained in studies of simple ternary alkali-alkaline earth-silica glasses [1,6,7]. The density and the refractive index of these glasses are linearly related and are relatively independent of the effect of a small substitution of lithia or potash for soda or of a similar replacement of a small amount of lime by magnesia. The glass transformation temperature of these glasses occurs at a reasonably constant viscosity. It is very important that the heating rate used in a DSC be specified when comparing T_g

values with viscosity, however, since the viscosity at T_g is a strong function of the DSC heating rate [7]. The common practice of indicating that T_g occurs at some constant value, without further information about the DSC measurement, should be discouraged. The primary role of the silica concentration in determining the thermal expansion coefficient of these glasses was expected, since the effect of overall network connectivity largely controls thermal expansion of silicate glasses [1]. Apparently, changes in the total non-bridging oxygen concentration are more important than the identity of the modifier ion used to charge balance these ions.

The most surprising result of this study is the very large effect of lithia on the viscosity and T_g of these glasses. While lithia is expected to decrease viscosity, the average decrease of more than 30 K in T_g induced by only 1 wt% of lithia is impressive. These results support arguments that use of lithia as a flux to lower melting temperatures and thus energy costs should be reconsidered. While it may prove that the reduction in energy costs cannot offset the increase in batch cost of adding lithia, the continued increases in energy costs expected in the future may well make use of lithia increasingly more favorable.

CONCLUSIONS

The behavior of the complex alkali-alkaline earth-silica glasses considered here can be predicted from those of simple ternary systems. The results of this study should prove useful in predicting the effect of minor changes in batch composition in the production of container and float glasses. The surprisingly large effect of lithia on viscosity and glass transformation temperatures of these glasses suggests that lithia may well prove useful in reducing melting temperatures and hence the cost of energy used in producing these glasses.

ACKNOWLEDGMENTS

Portions of this work were funded by the NSF University/Industry Center for Glass Research at Alfred University (CGR) and portions by the US Department of Energy, Office of Industrial Technologies, grant #DE-FG07-96EE41262.

REFERENCES

[1]J. E. Shelby, "Density and Thermal Expansion"; pp. 133-57 in *Introduction to Glass Science and Technology*, by J. E. Shelby, Royal Society of Chemistry, Cambridge, 1997.

[2]B. M. Wright, J. E. Shelby, "Phase Separation and the Mixed Alkali Effect," *Phys. Chem. Glasses*, **41**[4] 192-8 (2000).

[3]J. C. Lapp, J. E. Shelby, "D.C. Conductivity in Sodium and Potassium Galliosilicate Glasses," *Advanced Ceram. Materials*, **1**[2] 174-8 (1986).

[4]M. G. Mesko and J. E. Shelby, "Water Solubility and Diffusion in Melts of Commercial Silicate Glasses," *Glastech. Ber. Glass Sci. Technol.*, **73** [C2] 13-22 (2000).

[5]J. E. Shelby, "Viscosity and Thermal Expansion of Alkali Germanate Glasses," *J. Am. Ceram. Soc.*, **57**[10] 436-9 (1974).

[6]J. E. Stanworth, "Table 13"; p. 61 in *Physical Properties of Glass*, by J. E. Stanworth, Clarendon Press, Oxford, 1950.

[7]J. E. Shelby, "Viscosity of Glassforming Melts"; pp. 107-32 in *Introduction to Glass Science and Technology*, by J. E. Shelby, Royal Society of Chemistry, Cambridge, 1997.

VISCOSITY OF COMMERCIAL GLASSES IN THE SOFTENING RANGE

Alexander Fluegel, Arun K. Varshneya, Thomas P. Seward, and David A. Earl
New York State College of Ceramics
Alfred University
2 Pine Street
Alfred, NY 14802

ABSTRACT
The medium range viscosity (log viscosity / Pa·s = 6 to 9) is of vital interest for forming and annealing in the glass industry. It is essential for the design of melting and annealing furnaces, as well as for the forming processes. In this work we determined the viscosity of 150 glass composition variations centered on the float, container, low-expansion borosilicate, TV panel, wool, and textile fiber glasses by parallel-plate viscometry. The composition-viscosity relationship was calculated through multiple regression.

INTRODUCTION
This study was part of a larger project of the NSF Industry/University Center for Glass Research (CGR) [1] to give the glass industry a database and a method for calculating the properties of technical glass melts within the composition and temperature limits of interest. As part of it, the viscosities of 150 industrial glass variations (including float, container, low-expansion borosilicate, TV panel, wool, and textile fiber glasses) have been determined between log (η / Pa·s) = 1-12 by various groups. In this paper we concentrate on container, low-expansion borosilicate, textile fiber, and TV panel glasses within the range of log (η / Pa·s) = 6-9, as measured by parallel-plate viscometry.

EXPERIMENTAL DESIGN AND PROCEDURE
Experimental design
Member companies of the NSF Industry/University Center for Glass Research (CGR) selected six groups of industrial glasses for the study: float, container, low-expansion borosilicate, TV panel, wool, and textile fiber glasses. Starting from one base composition per glass group, supplied by CGR member companies, twenty-four further composition variations per group were selected using a

Plackett-Burman design based on weight percent. The concentration limits and all oxides of interest were given by CGR member representatives. Twenty-five variations of a glass group with five independent variable components (converted to mol oxide / mol SiO_2) would allow for analysis using a full quadratic model, including all linear effects, 2-component interactions and squared terms. Even though each of the glass groups contained more than five variable components, not more than twenty-five variations per group were designed in order to keep the experimental work within reasonable limits. Therefore, full quadratic models could not be developed, and some of the 2-component interactions were partially aliased. Model coefficients were selected based on their individual significance levels using a stepwise regression procedure.

Glass preparation

Glass batches (1-2 kg) were prepared from analytical-grade chemicals: silica, alkali and alkali-earth carbonates, Al_2O_3, H_3BO_3, Na_2SO_4, Fe_2O_3, Co_2O_3, Cr_2O_3, TiO_2, ZrO_2, NaF, CeO_2, PbO, ZnO, Sb_2O_3, As_2O_3. The batches were ball-milled for two hours and melted for 4-5 hours in a platinum crucible at 1400-1600°C. The melts were poured into a steel mold, cooled, crushed, re-melted for ~30 minutes, poured into a steel mold again, and annealed. All glasses were made either at Alfred University (container, float, and wool glasses) or at Corning Inc (TV panel, low-expansion borosilicate, and textile fiber glasses). Three of the low-expansion borosilicate glasses and two of the textile fiber glasses proved to be unsuitable for further analysis due to phase separation and/or crystallization.

All glasses were analyzed chemically by Corning Laboratory Services (Corning, NY) and Integrex Testing Systems (Granville, OH). Following the chemical analysis, a linear correlation analysis (Pearson's matrix) [2] of the variable component levels showed no correlation between the linear coefficients and partial aliasing regarding 2-component interactions and squared terms.

Parallel-plate viscometer measurements

The principles of parallel-plate viscometry are described by Varshneya [3]. For this study, a Theta model "Rheotronic R" parallel-plate viscometer was used. The temperature reading and the LVDT displacement of the viscometer were calibrated first, using a thermocouple calibration multimeter, and samples of known height respectively. A NIST viscosity standard 710A (soda-lime glass) was used for calibration. The standard deviation of all repeated 710A measurements over the 620-820°C range was 0.4°C. The calibration was confirmed by NIST standard 717A (borosilicate glass) from 600-780°C. During calibration with the NIST standards, no error trend was observed over temperature and time, assuming a constant temperature distribution within the furnace during all measurements. In addition, no temperature trend was observed over the cross-section of the vertical furnace chamber.

Cylindrical samples (diameter 8.5-9 mm, height 4-6 mm) were prepared through core drilling and grinding off both ends until they were parallel to ± 10 μm. The accuracy of the sample dimensions was sufficient, as demonstrated by the repeated 710A calibration runs mentioned above. The samples were placed into the viscometer between two thin electronic-grade alumina substrates. The applied load was 500 g, and the heating rate was 1 K/min. A data logger recorded the temperature and the sample height every 20 seconds, where the expansion of the silica push rod was compensated automatically through a reference rod directly adjacent to it. Additionally, the expansion of the alumina substrates was compensated. Next, the viscosity was calculated within the range of circa dh/dt < 10^{-5} cm/s, assuming no-slip conditions to the contacting alumina substrates, as established by Varshneya [4]:

$$\eta = \frac{2 \cdot \pi \cdot M \cdot g \cdot h^5}{3 \cdot V \cdot dh/dt \cdot (2 \cdot \pi \cdot h^3 + V)} \qquad (1)$$

where η = viscosity in Pa·s; M = applied load; g = gravity acceleration; h = sample height; V = sample volume; dh/dt = deformation or sag rate. Next, the isokom temperatures for log (η / Pa·s) = 6, 7, 8, and 9 were calculated, after fitting the experimental data to the Vogel-Fulcher-Tammann equation. The standard deviation of the temperature errors during repeated viscosity measurements of different sections of the same glasses due to inhomogeneities in the glasses and due to other experimental irregularities was 2°C. This means that the standard error of the viscosity models must be > 2°C, otherwise the model error would be unrealistically low due to "over-fitting."

Multiple regression analysis

Multiple regression analysis was done in the programs "Design Expert" and "Multiple Correlation Analysis" with the analyzed concentrations in mol / mol SiO_2, as independent variables, and the four isokom temperatures in °C as dependent variables. All data sets proved to be suitable for the analysis. Initially, linear fits including all glass components were performed for each isokom temperature. No outliers were found within all data sets analyzed. Then, a component was excluded from the model if there was less than a 90% confidence level in its significance ("Student-t" test parameter << 2), which resulted in a reduction of the model standard error. Finally, all 2-component interactions and non-linear influences were analyzed stepwise based on the model:

$$T_{isokom} = F0 + \sum_{j=1}^{n} \left(F1_j \cdot C_j + F2_j \cdot C_j^2 + \sum_{k=j+1}^{n} (F3_k \cdot C_j \cdot C_k) \right) \qquad (2)$$

where F0-F3 are the model coefficients, with F0 being the intercept, F1 the linear coefficients, F2 the squared terms of the same component, and F3 the coefficients for 2-component interactions. All F2 and F3 were set to zero for the linear models. The n in Eg. (2) is the total number of glass components (excluding silica), j and k are individual numbers of the glass components, and C are the analyzed concentrations in mol / mol SiO_2. In cases where 2-component interactions were partially aliased, the most significant interaction was selected for the model, resulting in the lowest model error.

RESULTS

Table I. Compositions of the investigated glasses in mol%, as analyzed

LO	SiO_2	B_2O_3	Al_2O_3	CaO	Na_2O	K_2O	BaO				
Avg:	77.30	11.33	2.84	1.06	6.08	0.96	0.43				
Min:	66.89	8.01	1.25	0.00	3.07	0.00	0.00				
Max:	86.21	14.28	4.67	2.43	8.95	2.19	0.89				

Co	SiO_2	Al_2O_3	MgO	CaO	Li_2O	Na_2O	K_2O	Fe_2O_3	Cr_2O_3	TiO_2	SO_3
Avg:	72.17	1.24	2.24	10.14	0.87	12.25	0.66	0.08	0.06	0.19	0.10
Min:	62.90	0.60	0.02	7.18	0.00	9.76	0.01	0.01	0.00	0.01	0.01
Max:	81.27	1.94	5.03	13.25	2.13	14.79	1.37	0.18	0.13	0.41	0.22

E	SiO_2	B_2O_3	Al_2O_3	MgO	CaO	Na_2O	K_2O	Fe_2O_3	TiO_2	F⁻	
Avg:	58.88	4.07	8.22	4.00	22.32	1.02	0.16	0.16	0.39	0.78	
Min:	42.62	0.00	5.77	0.69	15.17	0.00	0.00	0.00	0.00	0.00	
Max:	73.60	8.45	9.99	8.79	28.05	2.27	0.37	0.33	0.81	1.93	

TV	SiO_2	Al_2O_3	MgO	CaO	SrO	BaO	Li_2O	Na_2O	K_2O	TiO_2	CeO_2
Avg:	70.04	1.63	1.16	1.93	3.74	3.43	0.55	8.36	5.42	0.26	0.14
Min:	62.16	0.76	0.00	0.00	0.61	0.80	0.00	5.83	4.08	0.06	0.00
Max:	87.10	2.62	2.68	4.26	7.03	6.31	1.29	11.34	6.84	0.46	0.30

	ZrO_2	PbO	ZnO	As_2O_3	Sb_2O_3	F⁻					
Avg:	0.82	0.44	0.65	0.05	0.10	1.27					
Min:	0.00	0.00	0.00	0.00	0.04	0.00					
Max:	1.76	0.96	1.65	0.11	0.17	3.01					

LO - low-expansion borosilicate, Co - container, E - textile fiber, TV - TV panel

Table I shows the average, minimum, and maximum concentrations in mol% of the investigated glasses, as analyzed chemically. "LO" symbolizes the low-expansion borosilicate glasses, "Co" the container glasses, "E" textile fiber glasses, and "TV" the TV panel glasses. The concentration ranges (max-min) of the individual components in mol / mol SiO_2 (SiO_2 average), multiplied by the corresponding linear model coefficients shown below give approximations concerning the relative influences of the components, and the significance/error ratio can be calculated from the standard deviations in Tables II-IV.

Table II. Model coefficients for the low-expansion borosilicate glasses

LO	$F1_{B2O3}$	$F1_{Al2O3}$	$F1_{CaO}$	$F1_{Na2O}$	$F1_{K2O}$	$F1_{BaO}$	F0	$F3_{AlB}$	$F3_{AlNa}$	$F3_{KNa}$	σ
6	-405.1	766.9	-140.2	-1044	-934.3	-400.2	955.2	0	0	0	13.9
7	-374.0	523.5	7.9	-753.3	-693.6	-15.5	863.5	0	0	0	10.9
8	-355.6	356.4	131.2	-537.9	-527.6	151.9	796.8	0	0	0	9.0
9	-344.5	234.1	233.9	-371.2	-405.1	212.7	746.2	0	0	0	8.1
6	-425.1	743.2	0	-1074	-984.4	0	958.3	0	0	0	13.4
7	-374.6	524.5	0	-753.5	-694.0	0	863.6	0	0	0	10.2
8	-346.9	376.5	0	-520.5	-500.1	0	795.5	0	0	0	8.7
9	-331.4	269.4	0	-343.3	-361.1	0	744.2	0	0	0	8.6
6	-207.9	2376	0	-1142	-3208	0	934.0	-6026	-8438	26104	9.2
7	-272.7	1663	0	-710.5	-2232	0	846.4	-3017	-7988	18144	7.0
8	-312.7	1205	0	-399.3	-1527	0	781.3	-1219	-7553	12214	6.6
9	-339.2	890.4	0	-164.5	-986.8	0	731.1	-81.11	-7127	7569	7.4

In Tables II-IV the model coefficient data are presented in three sets of four rows. The first column shows the viscosity levels (log (η / Pa·s) = 6, 7, 8, 9) the coefficients are valid for. The first set of four rows corresponds to the linear model including all components, the second set of four rows excludes insignificant components from the linear model, and the last set includes significant components plus interactions and non-linearities. Beginning with the second column, the linear coefficients F1 for each component and the intercept F0 are displayed. Then follow the coefficients for the non-linear models: $F3_{AlB}$-$(Al_2O_3)*(B_2O_3)$, $F3_{AlNa}$-$(Al_2O_3)*(Na_2O)$, $F3_{KNa}$-$(K_2O)*(Na_2O)$, $F2_{Na}$-$(Na_2O)^2$, $F3_{AlCa}$-$(Al_2O_3)*(CaO)$, $F3_{LiNa}$-$(Li_2O)*(Na_2O)$, $F2_B$-$(B_2O_3)^2$. Finally, the last column gives the model standard error σ for each viscosity level in °C.

All coefficients are valid only within the concentration range stated in Table I. When divided by 100, the coefficients equal the temperature change (+/-) needed by an increase of 0.01 mol oxide (or interaction) / mol SiO_2 in the base glass to maintain the same viscosity.

Table III. Model coefficients for the container (Co) and textile fiber (E) glasses

Co	F1_Al2O3	F1_MgO	F1_CaO	F1_Li2O	F1_Na2O	F1_K2O	F1_Fe2O3	F1_Cr2O3	F1_TiO2	F1_SO3	F0	F2_Na	F3_AlCa	F3_AlNa	F3_LiNa	σ
6	823.2	-32.9	74.5	-1858	-764.5	-756.5	427.3	2390	-239.8	1405	885.5	0	0	0	0	8.6
7	781.3	-11.2	147.2	-1811	-670.1	-714.9	781.8	3214	-115.1	1671	800.2	0	0	0	0	7.5
8	698.8	-8.4	185.6	-1767	-613.8	-723.0	924.4	2921	-75.5	1510	744.9	0	0	0	0	7.3
9	611.7	-12.4	207.9	-1728	-576.8	-745.5	964.0	2243	-67.3	1208	706.1	0	0	0	0	7.8
6	840.5	0	78.0	-1861	-755.0	-745.7	0	0	0	0	885.6	0	0	0	0	8.1
7	810.5	0	152.8	-1803	-652.1	-696.3	0	0	0	0	800.6	0	0	0	0	7.5
8	728.1	0	191.1	-1756	-596.5	-706.1	0	0	0	0	745.3	0	0	0	0	7.2
9	636.6	0	212.3	-1717	-563.4	-733.6	0	0	0	0	706.4	0	0	0	0	7.3
6	2544	0	111.0	-3942	-2762	-705.5	0	0	0	0	1052	5709	-4674	-6249	11894	3.9
7	2921	0	204.7	-3852	-2237	-640.8	0	0	0	0	925.7	4581	-5374	-7946	11731	3.5
8	2615	0	238.8	-3715	-2357	-660.1	0	0	0	0	887.1	5049	-5246	-6793	11205	2.9
9	2098	0	248.9	-3569	-2673	-704.8	0	0	0	0	881.6	5972	-4860	-4714	10572	2.7

E	F1_B2O3	F1_Al2O3	F1_MgO	F1_CaO	F1_Na2O	F1_K2O	F1_Fe2O3	F1_TiO2	F1_F-	F0	F2_B	σ
6	-575.5	486.8	-149.6	-192.7	-628.7	482.7	550.0	252.9	-205.8	963.1	0	14.4
7	-538.5	427.5	-130.9	-141.2	-653.1	481.8	456.9	227.6	-275.4	901.9	0	12.9
8	-513.5	378.4	-119.8	-104.6	-651.4	459.9	434.2	201.7	-328.8	855.4	0	11.9
9	-496.2	337.2	-113.3	-77.6	-636.8	427.2	446.5	177.0	-371.6	818.9	0	11.3
6	-572.5	480.6	-144.2	-191.8	-640.6	0	0	0	0	965.2	0	13.3
7	-533.1	411.8	-123.0	-142.1	-671.8	0	0	0	0	904.4	0	12.3
8	-506.5	355.3	-109.8	-106.8	-675.1	0	0	0	0	858.3	0	11.8
9	-487.9	308.3	-101.6	-80.8	-664.2	0	0	0	0	822.1	0	11.6
6	-1157	303.2	-140.6	-220.6	-806.4	0	0	0	0	1004	4143	11.8
7	-1052	254.4	-119.9	-167.6	-818.9	0	0	0	0	938.5	3675	11.1
8	-977.0	212.7	-107.0	-129.9	-808.5	0	0	0	0	889.2	3332	10.8
9	-921.0	177.0	-99.0	-102.1	-787.0	0	0	0	0	850.6	3067	10.7

All coefficients not shown in Tables II and III are insignificant.

Table IV. Model coefficients for the TV panel glasses

TV	F1$_{Al2O3}$	F1$_{MgO}$	F1$_{CaO}$	F1$_{SrO}$	F1$_{BaO}$	F1$_{Li2O}$	F1$_{Na2O}$	F1$_{K2O}$	F1$_{TiO2}$
6	906.0	70.1	390.4	-29.5	-197.3	-1492	-813.0	-389.4	-3277
7	801.3	88.2	465.4	52.3	-111.3	-1406	-740.5	-351.9	-2551
8	757.1	99.6	504.3	104.0	-61.3	-1366	-680.6	-325.5	2011
9	744.3	106.4	523.7	138.1	-31.6	-1351	-630.1	-305.9	-1594

	F1$_{CeO2}$	F1$_{ZrO2}$	F1$_{PbO}$	F1$_{ZnO}$	F1$_{As2O3}$	F1$_{Sb2O3}$	F1$_{F-}$	F0	σ
6	-1512	818.1	-1237	-424.4	5417	-5274	-802.0	884.9	16.0
7	-954.5	849.0	-1153	-378.4	5998	-3292	-801.9	794.7	14.3
8	-635.0	841.0	-1109	-350.2	6561	-2443	-788.0	728.0	13.1
9	-442.6	815.5	-1089	-334.2	7112	-2171	-768.8	676.6	12.2

TV	F1$_{Al2O3}$	F1$_{CaO}$	F1$_{SrO}$	F1$_{BaO}$	F1$_{Li2O}$	F1$_{Na2O}$	F1$_{K2O}$	F1$_{TiO2}$	F1$_{ZrO2}$	F1$_{PbO}$	F1$_{F-}$	F0	σ
6	961.0	391.6	-31.4	-213.5	-1438	-828.8	-410.4	-3370	860.0	-1198	-777.2	876.9	15.7
7	837.4	469.0	50.3	-124.8	-1354	-751.8	-360.3	-2673	880.4	-1103	-772.0	790.7	13.8
8	784.6	508.9	101.9	-73.9	-1314	-690.5	-327.0	-2153	868.1	-1050	-754.0	726.1	12.9
9	768.7	528.4	135.8	-44.1	-1297	-640.0	-303.7	-1750	841.4	-1022	-731.4	675.9	12.6

All linear coefficients not shown in this Table are zero. Non-linear coefficients were not analyzed.

For refinement of the non-linear models in the Tables II and III, and for establishing a non-linear model for the TV panel glasses, further glass compositions need to be analyzed.

DISCUSSION

The above coefficients can be used to calculate the isokom temperatures in °C (Table V), e.g., for a typical container glass with the composition 74 SiO_2, 1 Al_2O_3, 0.5 MgO, 11 CaO, 13 Na_2O, 0.5 K_2O (mol%) and compared with an earlier model by Lakatos et al. [5]:

Table V. Model comparison, isokom temperatures and ▶ in °C

▶	LAK	FL, all lin.	FL, sign. lin.	FL, sign. non-lin.
6	765.32	768.06	770.88	764.86
7	713.34	710.01	715.00	710.00
8	671.30	669.16	673.98	668.56
9	636.60	638.82	642.63	635.96
▶ of diff.		3.06	2.14	1.46

The second column shows the isokom temperatures in °C calculated by Lakatos' linear model (LAK), next column using our linear model for container glasses including all components (see Table III), the fourth column using our linear model including the significant components only, and finally the last column using our non-linear model. The last row in Table V displays for each model comparison the standard deviation of the differences (LAK-FL). One should bear in mind that the accuracy of viscosity models tends to decrease with increasing variable components and phase separation/crystallization tendency. Further models over extended viscosity ranges and all six glass groups are available at the Center for Glass Research at Alfred University.

The authors thank the NSF Industry/University Center for Glass Research and the US Department of Energy (Grant DE FG 07-96EE41262) for financial support.

REFERENCES

[1] T. P. Seward, "Aspects of the Glass Melt Properties Database Investigations at Alfred University", 61st Conference on Glass Problems, C.H. Drummond, ed., American Ceramic Society, Westerville, OH, (2001) 149-163

[2] S. Dowdy and S. Wearden, "Statistics for Research", John Wiley & Sons, 1983

[3] A. K. Varshneya, "Fundamentals of Inorganic Glasses", Academic Press, Boston 1993

[4] A. K. Varshneya, N. H. Burlingame and W. H. Schultze, "Parallel Plate Viscometry to Study Deformation-Induced Viscosity Changes in Glass"; Glastechn. Ber. 63K (1990), 447-459

[5] T. Lakatos, L.-G. Johansson and B. Simmingsköld, "Viscosity temperature relations in the glass system SiO_2-Al_2O_3-Na_2O-K_2O-CaO-MgO in the composition range of technical glasses"; Glass Technology Vol. 13 No. 3, June 1972, 88-95

Emissions, Recycling and Other Environmental Issues

BORON CHEMISTRY IN FLUE GASES FROM BOROSILICATE GLASS FURNACES

J. Simon
Borax Europe Ltd.
1A Guildford Business Park
Guildford, GU2 8XG, UK

ABSTRACT

Volatile boron components in furnace gases from borosilicate glass furnaces can contribute to particulate stack emissions. The effective removal of boron species from these gases depends on the ability to form particulates through cooling or by chemical reaction. The particulates can then be separated by filtration or electrostatic precipitation and recycled in the process. The predominant volatile boron species are metaboric acid (HBO_2) and sodium or potassium metaborates ($NaBO_2$, KBO_2) depending on the composition of the glass and the furnace gases. Thermodynamic modelling of vapor-phase equilibria provides useful insights into factors that influence boron removal. A specific example is used where dust from an electrostatic precipitator is characterized using a combination of techniques including chemical and thermal analysis. Applying thermodynamics to the overall flue gas composition allows the composition of the dust to be estimated via chemical reactions which have led to the formation of the solid phases. The presence of excess acidity in the furnace gas, typically due to SO_2, leads to the formation of boric acid which has a relatively high vapor pressure, and has a major effect on the removal of boron.

INTRODUCTION

Boric oxide is used in the production of a wide variety of glasses; fiberglass, insulation glass wool, technical borosilicates for many applications including cookware, laboratory, pharmaceutical and lighting, and ceramic frits for tiles, tableware and metals. The function of the borate is normally to reduce melting and forming temperatures while controlling other properties such as thermal expansion or durability.

Boron is volatile at glass-melting temperatures and contributes to the formation of particulates in cooled gas streams (1). The removal of boron from the gas stream is desirable for environmental reasons and where borate values can be recovered for recycle. Since the boron is present with other volatile components, it is necessary to understand the interactions in the gas stream when subjected to cooling and the implications for both emissions and recycling of dust.

Electrostatic precipitator (EP) dust has particularly bad handling properties and is difficult to recycle back to the batch at a consistent rate. Varying dust recycle rates can have an adverse influence on the melting process by changing: redox, foaming and glass composition. The composition of the resultant dust will also vary; this compounds the situation and can affect furnace operation. An examination of EP dust provides an insight into the processes which lead to its formation, why it has poor handling characteristics, and how its recycling influences the melting process.

EXPERIMENTAL

The EP dust used in this study was obtained from an oxy-gas fired furnace producing a soft (high-alkali) borosilicate glass. The gas had been cooled using air dilution, which effectively reduced the concentration of CO_2 and H_2O. The dust was subjected to chemical analysis, XRD, TGA/DTA and microscopy. The composition of the gas stream was calculated, and thermodynamic modelling was used to predict dominant species formed on cooling of the furnace stack gas to various temperatures. The condensed solids are mainly amorphous, and it is not possible to be certain about the compounds that are formed except where identified by XRD. The use of thermodynamic modelling is useful in helping to understand possible chemical reactions and the formation of borate species. XRD showed a large proportion of amorphous material and only confirmed the presence of sodium carbonate and chlorides of sodium and potassium.

Fluorine and chlorine were determined by water extraction followed by ion chromatography. Carbonate was determined directly by a Leco carbon analyzer. All other elements were determined by HCl extraction and analysis of the extract by ICP; results are detailed in Table 1. The acid-insoluble fraction was retained for microscopy.

Table I: Chemical analysis of dust

Acid soluble:	Wt. %
Na_2O	32
K_2O	6
CaO	3
MgO	2
B_2O_3	40
CO_2	11
SO_3	3
Cl	3
F	0.3
Insoluble in HCL	0.3

MICROSCOPY

The dust comprised very fine agglomerated particles of condensed volatiles with a small proportion of entrained batch dust. Microscopic examination was carried out on both dust and insoluble material from acid extraction, see Figure 1. Most of the dust consisted of μm-size, or possibly finer particles aggregated into grains of the order of 10 to 100μm diameter. There were also glassy spheres and rounded grains, probably of quartz, up to 200μm, that make up the acid-insoluble fraction. Photomicrographs show crystalline quartz as bright particles under crossed polarizers, right in Figure 1.

Figure 1: Photomicrographs of EP dust; SEM and optical (center and right)

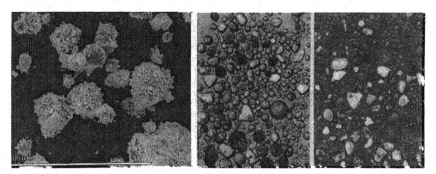

CONDENSATION REACTIONS
Vapor phase

Based on the dust composition it is reasonable to assume the gas stream contains $NaBO_2$, KBO_2, SO_2, HF and HCl in the vapor phase and particulate CaO and MgO. Based on the level of acid insolubles and dolomitic material, entrained particles were estimated to be ~10% of the total. Sulphate is used as the fining agent for this glass and will be evolved in the furnace initially as SO_2. Fluoride and chloride will be present as impurities. The presence of calcium and magnesium is unlikely to be due to volatility and indicates the presence of entrained batch dust which would have been present in the batch as dolomite and limestone. Dolomite is well known to be prone to decrepitation and dust formation. The presence of insoluble siliceous material also indicates mechanical entrainment.

Thermodynamic modelling.

When the gas cools, volatile components will condense and react with themselves and with the products of combustion (CO_2 and H_2O). The objective here was to calculate the speciation using thermodynamic modelling. Using the composition of the dust and the furnace off-gas, it is possible to predict the equilibrium species formed on cooling to various temperatures. Calculations were made by Rio Tinto Technical Services using a free-energy minimization routine in MTDATA (2). This is a software/data package developed by The National Physical Laboratory (NPL), for the calculation of phase equilibria in multicomponent multiphase systems using, as a basis, critically assessed thermodynamic data. Graphs, Figures 2 and 3, show the mass of different solid phases produced at various temperatures. Major phases are shown in Figure 2 and minor phases in Figure 3.

Figure 2: Major equilibrium phases

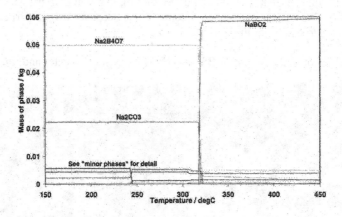

Figure 3: Minor equilibrium phases

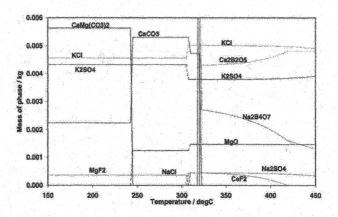

Data show that the deposition of condensed volatiles is essentially complete before being cooled to 450°C. Below 320°C, the dominant or most abundant species are di-sodium tetraborate $Na_2B_4O_7$, and Na_2CO_3, and the minor phases are carbonates of calcium and magnesium, potassium sulphate, potassium chloride, and magnesium fluoride. Sodium metaborate ($NaBO_2$) will be the thermodynamically stable solid phase above 320°C. Below this temperature, CO_2 reacts with sodium metaborate to form di-sodium tetraborate, equation [1], and sodium carbonate; carbonates of calcium and magnesium are also formed.

$$4NaBO_2 + CO_2 = Na_2B_4O_7 + Na_2CO_3 \qquad [1]$$

Thermal Analysis DTA, TGA

Stages in the formation of dust may also be illustrated by thermal analysis: DTA/ TGA. Results (Figure 4) agree well with the thermodynamics. Key stages in the analysis are summarized in Table II.

Table II: DTA/ TGA features

Temperature °C	Weight loss %	Heat flow	Possible cause
20-380	7	-	Loss of water/ CO_2 (?)
380-450	8	Endothermic	Loss of CO_2
850	0	Endothermic	Melting
>1000	>60	Endothermic	Borate and chloride volatility

Figure 4: Thermal analysis results

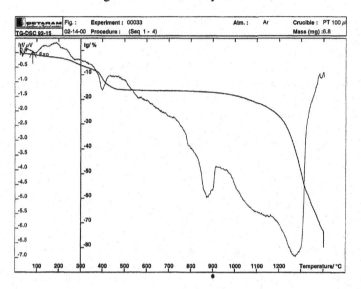

Thermal analysis, heating up the dust, shows distinct properties of the solid which are indicative of the physical and chemical changes on cooling of the gas from furnace to ambient temperature. The high-temperature weight loss (>1000°C) from volatility shows that virtually all the solids have precipitated from the gas phase on cooling to about 1000°C. On heating, there is an endothermic phase change (no associated weight loss) at ~850°C, which is likely associated with formation of a liquid phase, which does not happen during rapid cooling; there is no evidence of melting from the microscopic examination. The weight loss between 300 and 450°C is assumed to be due to the reaction between borate and carbonate to form carbon dioxide. Low-temperature weight loss may be assumed to be mainly loss of absorbed moisture.

DISCUSSION

A formalized dust composition may be estimated from the chemical analysis and the species predicted by thermodynamics as shown in Table III.

Table III: Estimated dust composition

	Weight %
MgF_2	<1
KCl	6
K_2SO_4	5
$CaCO_3$	6
$MgCO_3$	3
Na_2CO_3	17
$Na_2B_4O_7$	40
$NaBO_2$	20

The EP capture efficiency of particulate boron is critically dependent on the nature of excess acidity in the flue gas. Borate particulate can be efficiently scrubbed and captured in the case when the level of acidic $SO_{2(g)}$ is stoichiometrically less than the level of basic $NaBO_{2(g)}$ in the flue gas stream, as described by the reaction in equation [2]. In this case the excess acidity is due only to excess CO_2. The thermodynamic calculations show the stable borate is $Na_2B_4O_7$ in the presence of excess carbon dioxide below 320°C. The normal gas inlet temperature for an electrostatic precipitation is in the range 250 to 350°C. However, a calculation from the chemical analysis shows the ratio of sodium to boron in the EP dust sample is 1.6:1 B:Na, a value intermediate to the calculated ratio of 2.0:1 for pure $Na_2B_4O_7$ and 1.0:1 for pure $NaBO_2$. Thus, it is likely that the reaction of $NaBO_2$ with CO_2 to form sodium tetraborate is kinetically controlled and would go to completion given enough time. Reaction with carbon dioxide can be viewed as the reaction of acidic $CO_{2(g)}$ with basic $NaBO_2$ to form the salts $Na_2B_4O_7$ and Na_2CO_3, equation [1]. It is apparent that the two borates $NaBO_2$ and $Na_2B_4O_7$ do not exist as crystalline compounds in the solid, but as an amorphous solid solution. This reaction may continue after the solids are deposited in the EP and before discharge. However, it is important to note the thermodynamic calculations predict this reaction with carbon dioxide will not proceed beyond the tetraborate composition; i.e., sodium borate cannot be converted directly to boric acid by CO_2 at the temperature within the EP.

When the excess acidity in the flue gas is due to SO_2, formation of volatile HBO_2 and $B(OH)_3$ will reduce the efficiency of boron particulate capture. In this case, SO_2 will react with the $NaBO_2$ to form $Na_2B_4O_7$ and Na_2SO_4 until the $NaBO_2$ is nearly exhausted, equation [2]. Then the excess SO_2 will continue to react with $Na_2B_4O_7$ to form $NaSO_4$ and the more volatile species HBO_2 and $B(OH)_3$, equation [3].

$$8NaBO_2 + 2SO_2 + O_2 = 2Na_2SO_4 + 2Na_2B_4O_7 \qquad [2]$$

$$2Na_2B_4O_7 + 2SO_2 + O_2 + 4H_2O = 8HBO_2 + 2Na_2SO_4 \qquad [3]$$

The formation of boric acid, $B(OH)_3$, and metaboric acid, HBO_2, would reduce the efficiency of removing boron due to their relatively high vapor pressures at high water partial pressure, even at relatively low temperatures. The presence of $B(OH)_3$ and HBO_2 in flue gas having an excessive amount of SO_2 typically requires the use of caustic scrubbing for effective removal of all the boron as sodium borate. Excess alkali is essential for complete removal of boron from the vapor.

The temperature dependence of the reaction between $NaBO_2$ and CO_2 is important in that the transition temperature (320°C) occurs in the temperature range used for electrostatic precipitators, 250 to 350°C. Reaction of $NaBO_2$ in dust with CO_2 will depend on the contact time between the solids and the gas in the EP. This work shows incomplete reaction with CO_2 indicating the potential to continue reaction in the EP that might contribute to formation of caked material, but this has not yet been investigated. Also, the effect of EP temperature on dust properties should be investigated. This work suggests the temperature could be increased without affecting boron removal, since the vapor pressure of sodium metaborate is extremely low at these temperatures.

REFERENCES
1. J.M. Simon, R.A. Smith, Borate Raw Materials, Glass Tech, 2000, **41** (6), 169–73
2. R.G.C. Beerkens, Deposits and Condensation from Flue Gases in Glass Furnaces, PhD Thesis, Technical University of Eindhoven, 1982.
3. Davies, EP Temperature Sensitivity Calculations, Private communication, April 2000.
4. Davies, Boron Thermochemistry, Private communication. Jan. 1999.
5. Roggendorf, H.; Scholze, H.: Kondensations- und Reaktionsvorgange in Abgasen hinter Glasschmelzwannen, Berechnungen und Messungen. Glastech. Ber. 59 (1986), nr. 5., pp. 109-120

Computer Modeling and Process Control

COMPUTER MODELING OF GLASS FURNACE FLOW AND HEAT TRANSFER

Shen-Lin Chang, Brian Golchert
Argonne National Laboratory
9700 S. Cass Avenue
Argonne, IL 60439

Chenn Q. Zhou
Purdue University Calumet
2200 169th Street
Hammond, IN 46323-2094

ABSTRACT

Glass furnaces use combustion heat to melt sand and cullet into liquid glass. Of primary importance to the furnace performance are the quality of the glass product, the utilization of energy, and the emission of pollutants. A computational glass furnace model (GFM) was developed for the glass industry to evaluate the glass furnace performance. GFM simulates the major flow and heat transfer characteristics in both the combustion space and glass melter of the furnace. GFM consists of two computational fluid dynamics codes: one is used to simulate the combustion space flow with spectral radiation heat transfer, and the other to calculate the batch/glass/bubble flow. Recently incorporated features of the model are direct integral solution of spectral radiation transport equations and multiphase flow treatment of the glass flow with the melting of solids and formation of bubbles. The GFM has been validated with experimental data from commercial glass furnaces.

INTRODUCTION

The glass industry has recognized that computational fluid dynamics (CFD) simulation can be an important tool to help improve the performance of glass furnaces. CFD simulations are being used as a tool to analyze flow characteristics of a glass furnace and find ways to improve the energy efficiency and reduce pollutant emissions. In 1998, the Department of Energy's Office of Industrial Technologies funded a consortium of glass companies (initially, Techneglas, Libbey, Inc., Osram-Sylvania, Pilkington-Libbey-Owens Ford, and Owens-Corning. Pilkington withdrew from the consortia, and Visteon later joined the group), two universities (Purdue University and Mississippi State University), and Argonne National Laboratory (ANL) in order to develop a rigorous, coupled glass furnace simulation [1]. Initially, it was decided that the developed simulations would include important coupling issues such as batch melting, foam coverage, and spectral radiation penetration both in the melt and in the combustion space.

It has long been recognized that one needs two separate models in order to build a complete simulation of an industrial glass furnace: one for the combustion space and one for the molten glass. Early on, CFD codes have been used to model the combustion space [2,3] and the glass-melt flow [4,5] of a glass furnace. The early studies found that the flow fields of the combustion space and glass melt region strongly interact with each other. The radiation heat transfer from the combustion space drives the glass melt flow through the melting of the batch and the heating of the liquid glass. Experimentally, it was found that a significant portion of the spectra of the radiation might penetrate into the glass flow [6]. Therefore, a proper spectral coupling of the combustion space and glass melt flows is essential for a realistic modeling of a glass furnace.

Coupled glass furnace simulations are not a new idea [7,8]; in fact, many companies can claim they have coupled models. However, the models developed for the ANL Glass Furnace Model (GFM) have made several significant advances over the last five years. These advances include a new spectral radiation model that conserves energy and is computationally efficient, a true multiphase treatment in the glass melt (explicit transport equations are simultaneously solved for the liquid, solid particles and gas bubbles), three comprehensive validations of the models compared to in-furnace measurements, and, quite recently, a mass coupling to better account for the gas evolution from batch/glass reactions between the glass melt and the combustion space. This paper will present many of the models associated with the GFM as well as some pertinent computational results, with the emphasis being placed on melt model formulation and results.

METHODOLOGY

Three distinct computational regimes are present in any glass furnace simulation: combustion modeling, molten glass modeling, and radiation heat transport modeling. A communication routine was developed to allow the transfer of pertinent information from each one of these regimes to the other (see Figure 1). When the project began, the communication between the computational regimes was entirely energy transfer. As a result of the measurement campaigns and a computational investigation, it has been determined that mass transfer (gas exchange) between the melt and the combustion space is also essential to proper modeling of a glass furnace.

The combustion space is run first, computing the local temperature, pressure and species concentrations throughout the volume. Then, a spectral radiation model is run to calculate the local radiation heat flux, particularly on the glass surface. With this information, the melt model is executed, and the local glass temperature, velocity, and species concentrations are determined. The local glass surface temperature and local gas release rates are then used as inputs into the next combustion calculation.

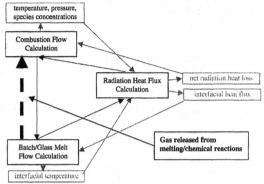

Figure 1. Iteration Routine for a Furnace Computation

COMBUSTION SPACE MODELING

The combustion space is modeled in the GFM using an ANL-developed CFD code, ICOMFLO, which was created especially to simulate combustion [9] on a structured grid. The code has been modified specifically for glass furnace combustion. Combustion flow simulation is very complicated and often plagued by numerical instability problems. ICOMFLO uses a three-step partially, de-coupled computational scheme and divides the combustion species into two groups: major species and subspecies. The three-step scheme includes the computations of (1) combustion hydrodynamics (major species calculation), (2) formation and transport of pollutants (subspecies calculation), and (3) radiation heat transfer (absorption).

Initially, the code computes the major flow properties of the combustion flow in the furnace by assuming only radiation emission in the chamber. In this step, pressure, temperature, density, velocity, and species concentrations are locally computed. A combustion model of the major species is needed in this step to establish an initial temperature field. Next, a kinetic model of the subspecies is used to calculate the formation and transport of the subspecies based on the semi-converged major flow properties computed in the first step. Then, a radiation heat transfer model is used to calculate local net radiation heat flux (the balance of emission and absorption) based on the temperature and pressure calculated in the first step and the species concentrations calculated in the second step. The radiation participating media in a glass furnace include carbon dioxide, water vapor, and soot in the combustion space and glass in the glass melt flow. Since radiative emission and absorption of these media depend strongly on wavelength of the radiation, a spectral radiation heat transfer model is used. In the flow field calculations, the emission of radiation is determined in each iteration. Absorption is only calculated infrequently, since this computational process is very time consuming and the computational results do not change very much once the flow field settles down.

Formulation of Hydrodynamic Flow

In the first step of the calculation, major flow properties are computed with a highly-reduced combustion model. The primary information required from the combustion model for the flow computation is the heat of combustion and mixture molecular weight change due to combustion, which affects mixture density through the equation of state. The species selected for the first-step calculation include nitrogen, fuel, oxygen, and two lumped products (carbon dioxide and water

Conservation Equations: In this step, pressure, temperature, velocity, density, and species concentrations are calculated from the conservation equations of mass, momentum, and enthalpy, the transport equation of species, and the equations of state. The equations of mass, momentum, enthalpy, and species are all elliptic-type partial differential equations. For convenience in numerical formulation, these equations are arranged in a common form, Eq.(1).

$$\sum_{i=1}^{3} \frac{\partial}{\partial x_i}(\rho u_i \xi - \Gamma_\xi \frac{\partial \xi}{\partial x_i}) = S_\xi \tag{1}$$

in which ξ is a general flow property, x_i, $i=1,3$ are coordinates, u_i, $i=1,3$ are velocity components, Γ is effective diffusivity (calculated from both laminar and turbulent viscosities), and S_ξ is the sum of source terms. The general flow property is a scalar, a velocity component, an enthalpy, or a species concentration for the equations of mass, momentum, enthalpy, and species, respectively. Turbulent diffusivity and the source terms are derived from separate phenomenological models.

Lumped, Integral Combustion Model: The computation uses a combustion model based on the integral reacting-flow time-scale-conversion method [10]. The model assumes a simple one-step combustion reaction of the four major species.

$$FUEL + OXIDIZER \xrightarrow{k_1} PRODUCT_1 + PRODUCT_2 \tag{2}$$

Based on the time-integral approach, a separate kinetic calculation was performed to determine the time evolution of the extent of reaction for the flow calculation.

Turbulence Model: The turbulence model is a modified k-ε turbulence model. The model introduces two additional turbulent parameters to determine turbulent viscosity. Turbulent diffusivities for the enthalpy and species equations are calculated from the turbulent viscosity with an appropriate scaling factor.

Formulation of Transport of Subspecies: The combustion in a glass furnace produces many combustion products including CO_2, H_2O, CO, NO, and soot. These products or subspecies are transported by the combustion flow whose major properties have been computed in the previous step. Thus, the transport of these combustion products can be formulated and calculated based on the computed flow properties. The formation of these subspecies can be determined from a set of detailed or reduced kinetics. Soot and gas subspecies models were developed to provide the source or sink terms for the equations.

Soot Model: In most flames, soot is formed in the flame area where fuel is rich. Since soot is the dominant contributor to radiation heat transfer, a soot model was developed to calculate soot concentration in preparation for the following radiation heat transfer calculation. This soot model was based on kinetic models found in the literature [11,12]. The model developed does not explicitly take into account the agglomeration of soot particles, but it does account for local soot formation and oxidization based on the local temperature and species concentrations.

Gas Subspecies Model: The reactions of natural gas combustion can be many. For those who are interested in some of the species involved in the combustion processes, a reduced mechanism is enough for computation. In this study of glass furnace, only seven gaseous species, CH_4, O_2, N_2, CO_2, H_2O, CO, and NO, are considered. A five-step kinetic mechanism for methane burning and NO formation [13] is used for the subspecies calculation. The source terms of the subspecies transport equation can be obtained from the reaction rates found in the literature. Once the solution of the transport equations is converged, the species concentrations are the used in the computation of radiation heat transfer.

Formulation of Spectral Radiation Heat Transfer
     The radiation calculation will provide the local net radiation heat flux for the enthalpy conservation equation of the flow calculation and the radiation heat flux to the boundaries (glass melt flow and the furnace walls). The radiation heat flux can be calculated by solving the radiative transport equation based on local gas pressure, temperature, and concentrations of CO_2, H_2O, and soot calculated from the previous steps.

     By assuming the scattering effect in the combustion flow is negligible, the radiative transport equation is basically the balance of emissive and absorption powers. Local net radiation heat power can be obtained by adding the energy absorbed in a cell from all incoming radiation from all other locations and then subtracting the energy emitted in that cell for every wavelength [14]. In the radiative transport equation, the temperature is known from the flow calculation, but the spectral volumetric absorptivity, κ_λ, is yet to be determined. The spectral volumetric absorptivity of the radiatively participating media will be determined from gas and soot radiation models.

GLASS MELT MODELING
Solid Batch Flow:

In an Eulerian approach, batch particles are divided into a number of size groups, and particles of a size group are treated as a continuum. Conservation equations of solid mass, momentum, and energy are formulated for each size group. The mass, momentum, and energy conservation equations of the batch flow are used to solve for local batch particle number density (throughout the melt, not just on the surface), velocity, and temperature. The conservation equation of solid mass (or number density) for a size group k is derived as,

$$\sum_{i=1}^{3} \frac{\partial}{\partial x_i}(m_{pk}n_k u_{i,p,k}) = S_n \tag{3}$$

in which, m_{pk} is the mass of a size-k particle, n_k is particle number density, x_i, i=1,3 are coordinates, $u_{i,p,k}$ is particle velocity, S_n is the source term. Particles start to melt when their temperature reach the melting point. The melting process shifts particles to a smaller-size group. The shifting produces a sink term for the size group, a source term for the smaller-size group, and a source term for the liquid glass.

The solid momentum and energy equations of a particle size group k can be derived as,

$$\sum_{i=1}^{3} \frac{\partial}{\partial x_i}(m_{pk}n_k u_{i,p,k}\xi - \Gamma_\xi \frac{\partial m_{pk}n_k\xi}{\partial x}) = S_\xi \tag{4}$$

in which, ξ is particle general property, representing a particle velocity component, or particle temperature, Γ is diffusivity, and S is the source term. Interfacial drag and heat transfer models are needed to generate source terms for the momentum and energy equations.

Liquid Glass-Melt Flow:

It is assumed that the glass melt is a Newtonian fluid and the flow is laminar and steady state. For the liquid glass-melt flow, the flow properties needed to determine the state of the flow system and to evaluate its performance are pressure, density, temperature, and velocity components. The properties can be solved from the conservation equations of mass, momentum, and energy, and equations of state. The conservation equations of the liquid flow can be arranged in a common form, Eq. (6).

$$\sum_{i=1}^{3} \frac{\partial}{\partial x_i}(\rho u_i \xi - \Gamma_\xi \frac{\partial \xi}{\partial x_i}) = S_\xi \qquad (5)$$

in which ξ is a general liquid glass flow property, u_i, i=1,3 are velocity components, Γ_ξ is diffusivity, and S_ξ is the sum of source terms. The general flow property represents a scalar, a velocity component, or an enthalpy for the mass, momentum, and energy equations, respectively. Density, viscosity, and thermal conductivity are functions of temperature [15].

Phenomenological Models:

The forces on solid batch include buoyancy force, solid pressure, and interfacial drag. The buoyancy force is due to density difference between solid batch and liquid glass. Batch particles are packed together on the top of the liquid glass. When batch moves, particles exert force to the neighboring particles by contact. The force is called solid pressure. The drag force is the shear stress on the interface surface between the batch and the liquid. The buoyancy force, solid pressure, and drag force are the source terms of the solid momentum equation. The drag force is also a sink term of the liquid momentum equation.

The heat fluxes on the batch flow include radiation heat flux from the combustion space, conduction heat flux from the neighboring batch particles, and interfacial heat flux from the liquid flow. The radiation heat flux is calculated from linked combustion and radiation computer codes. The conduction heat flux is accounted for as the diffusive (second) term in the energy equation. The radiation heat flux is a source term in the solid energy equation in the batch coverage area and is a source term in the liquid energy equation outside the batch coverage area. The film conduction heat flux is a sink term of the liquid energy equation. A melting rate can be derived from the film conduction heat flux and becomes a source term of the liquid mass equation.

The batch-melting model operates on the assumption that when the batch temperature reaches the melting temperature, the temperature remains at the melting temperature, and radiation heat flux is used to melt batch particles. Note that the radiation melting occurs on the topside of the batch, and the film conduction melting occurs on the bottom side of the batch. The batch-melting rate is equal to the ratio of the sum of radiation and film conduction heat fluxes to the heat of melting. The melting rate becomes a source term of the liquid mass equation and a sink term of the solid mass equation. The melting also shifts the batch particles from a larger-size group to a smaller-size group. The shift rate is calculated based on local particle number density. Batch coverage is determined from local particle number density and velocity.

Bubble Flow in Glass Melter

In a glass melter, bubbles are formed on the interface with solid surface (batch particles and wall) and when gas species are produced from chemical reactions. Bubble size grows when reactions of liquid glass produce more gas species or shrink when gas species are dissolved into the liquid glass. Similar to the solid batch and liquid melt, an Eulerian approach is used to formulate the gas bubble flow. Bubbles are divided into size groups. For each size group, bubbles are treated as a continuum. Mass, momentum, and energy are conserved for each bubble continuum.

Bubbles are formed in various sizes and can be divided into a number of groups according to their masses represented by their initial bubble size. The mass of a bubble can be calculated from its radius and density. Figure 2 shows an example of the grouping of bubbles.

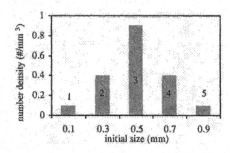

Figure 2: Example of Bubble Size Distributions

Conservation Equations for Bubble Flow: Each bubble group is treated as a fluid continuum. Conservation equations of mass, momentum, and energy are derived in a Cartesian coordinate system. Consider a bubble group k with a mass $m_{b,k}$; the mass conservation equation can be derived by balancing the mass convective and diffusive fluxes with the source terms. The convective mass is the bubble mass transported by velocity $u_{i,k}$ in the x_i direction. The diffusive mass is the bubble mass diffused by density gradient. The diffusion rate is proportional to a diffusion constant Γ. The source terms included the following: (1) bubble generation from batch reactions, (2) bubble increment by the fining reactions of the liquid glass that increase bubble mass, and (3) bubble decrement by the dissolution of bubble gas that decrease bubble mass. Thus, the mass conservation equation of the kth bubble group becomes:

$$m_{b,k} \sum_{i=1}^{3} \left[\frac{\partial}{\partial x_i}(n_k u_{i,k}) + \frac{\partial}{\partial x_i}(\Gamma \frac{\partial n_k}{\partial x_i}) \right] = S_{batch} + S_{fining} + S_{dissolution} \qquad (6)$$

Note that the bubble mass $m_{b,k}$ is taken to the outside of the bracket, because it remains constant in the entire flow.

Balancing the momentum convective fluxes with the source terms derives a momentum equation. The convective momentum is the bubble velocity transported by the bubble mass. The source terms include drag between bubble and liquid glass, and the buoyancy force due to the density difference of the bubble and the liquid glass. An energy equation is derived by balancing the energy convective fluxes with the source terms. The convective energy is the bubble enthalpy transported by the bubble mass. The source terms include interfacial heat transfer between bubble and liquid glass and heat of dissolution.

Phenomenological Models for Bubble Flow Formulation

The source terms used in Eq. (6) arise from the use of a reduced reaction for soda-lime-silicate glass batch.

$$(CaCO_3)_{z_1}(Na_2CO_3)_{z_2}(SiO_2)_{z_3} \longrightarrow (CaSiO_3)_{z_1}(Na_2SiO_3)_{z_2}(SiO_2)_{z_4} + (z_1 + z_2)CO_2 \qquad (7)$$

In the reaction, solid batch $(CaCO_3 \bullet Na_2CO_3 \bullet SiO_2)$ is converted to liquid glass $(CaSiO_3 \bullet Na_2SiO_3)$ and gas species CO_2. The reaction (or melting) rate was determined empirically. An empirical melting rate R_m can be expressed as a function of temperature [16]. Some gas is released to the combustion space and the rest is trapped in the liquid glass. The gas trapped in the liquid can form bubbles of a variety of sizes.

Fining Reaction: Fining agents are added to the glass flow to react with glass species and generate gas. This gas makes bubbles grow larger so they may rise to the liquid surface before exiting the melter. Again, an empirical reaction-rate formula is needed to determine the fining reaction rate and the gas generation rate. Since the refining gas goes into bubbles through the surface, the fraction of refining gas for each bubble group can be derived. The additional fining gas makes bubbles grow in size, and some will shift to a larger-size group. Thus, bubble number density decreases for the additional fining gas to this size group. However, bubble number density increases for the additional fining gas to a smaller-size group. By summing the two terms, the source term of the fining reaction for bubble group k can be determined.

Dissolution Reaction: The gas in a bubble can be dissolved into liquid glass. The fraction of dissolution gas from each bubble group can also be derived. The dissolution makes bubbles shrink in size, and some will shift to a smaller-size group. Bubble number density decreases for the loss of the dissolution gas from this size group. However, bubble number density increases for the loss of the dissolution gas from a larger-size group.

Transport Equations of Bubble Properties:

A gas bubble has many properties, such as pressure p, temperature T, density ρ, molecular weight ω, and M species concentrations ϕ_m. Some bubble properties are determined from state equations, and some need to be solved from transport equations. One of the species concentrations can be determined from the state equations, but the others have to be solved from the transport equations. A transport equation is derived by balancing the convective and diffusive fluxes with the source terms. The transport equation is similar to the mass equation. The source term can be the batch, fining, and dissolution source terms, if the reactions involve the species.

Numerical Scheme

The CFD code uses a control-volume approach to convert the governing equations to algebraic equations on a discretized grid system. The grid system is staggered and consists of four grids: one for each of the three momentum component directions of the liquid phase momentum equations, and a scalar grid for all the other equations. The algebraic equations are solved iteratively with the boundary conditions. There are three major iteration routines in the multiphase flow simulation. The first is for the liquid glass flow calculation, the second is for the solid batch flow calculation, and the third is for the gas bubble calculation. In the calculations of liquid-phase equations, Patankar's SIMPLER computational scheme [17] is used to solve the pressure-linked momentum and continuity equations.

Computational Results

Simulations were developed for a commercial scale glass furnace that has six batch screw feeders (charger end) and one throat. Flow property distributions were computed and are shown from three different views (top, side, and end). The top view is at the glass surface (x-z plane). The side view is the vertical-longitudinal plane (x-y plane). The end view is the vertical-span plane (x-z plane). The actual numbers on the figures are not shown to protect proprietary information.

Figures 3a, 3b, and 3c show the top view of the computed batch melting rate, liquid glass temperature and velocity vectors, and batch particle number density of the glass melt flow on the melter surface, respectively. A uniform radiation heat flux from combustion space was assumed for these calculations. In both Figures 3a and 3b, the red color represents the highest value, and the blue color represents the lowest value. In Figure 3c, the blue color represents the highest particle number density, and the yellow color represents the lowest number density.

Figure 3a shows the batch melting rate distribution. The melting rate distribution will also be correlated to the release of gas into the melt and into the combustion space. In the figure, six batch lines are clearly shown by high melting rates. Figure 3b shows glass temperature and velocity vectors on the melter surface. The temperatures are highest near the exit of the melter and lower near the inlet of the batch. The low-temperature regions indicate the batch coverage areas, which are also displayed very clearly in batch number density distributions, as shown in Figure 3c. On the melter surface, a strong back flow appears that brings molten glass at higher temperatures from the exit region to the batch inlet. This hotter glass flows underneath the batch coverage, providing the additional energy needed for melting the batch other than radiation heat flux on the surface.

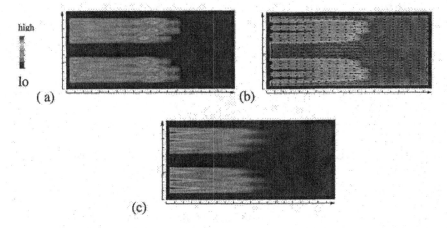

Figure 3: Computed Glass Properties on the Melter Surface: (a) melting rate, (b) temperature/velocity, (c) batch particle number density

Bubble Formation and Transport

Figures 4a and 4b show the gas-bubble number density distributions for bubble sizes of 0.25mm and 1mm in diameter respectively. The liquid velocity vectors are also shown in these figures. For smaller bubble size (see Figure 4a), the bubbles are spread wider and deeper, which follow the liquid flow pattern more closely, because of less momentum and buoyancy force. For bigger bubble size (see Figure 4b), gas bubbles are more concentrated around the batch and near the glass surface, because of larger momentum and buoyancy force.

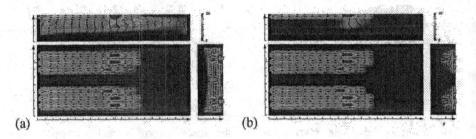

Figure 4:   Distribution of Gas-Bubble Volume Fraction: (a) 0.25mm and(b) 1mm Bubbles Size

SUMMARY

Several of the advanced, multi-phase models for the ANL GFM have been presented. These models allow for more detailed analysis of glass melting furnaces, since batch particles and bubbles may now be tracked throughout the melt. Work is continuing on the GFM in order to include more detailed glass/batch chemistry and a more explicit mass coupling between the melt and the combustion space.

REFERENCES

[1]Golchert, B., S.L. Chang and M. Petrick, "A Consortium Approach to Glass Furnace Modeling," ICAST 99 Proceedings, the 15th International Conference on Advanced Science and Technology at Argonne National Laboratory, 3 April 1999, pp. 152-162.

[2]May, F., and H. Kremer, "Mathematical Modeling of Glass Melting Furnace Design with Regard to NOx Formation," Glastech. Ber. Glass Sci. Technol. 72, No.1 (1999).

[3]Chang, S.L., B. Golchert, and M. Petrick, "Numerical Analysis of CFD-Coupled Radiation Heat Transfer in A Glass Furnace," No. 12084, Proceedings of the 34th National Heat Transfer Conference, Pittsburgh, Penn. (August 20-22, 2000).

[4]Viskanta, R., "Review of Three-Dimensional Mathematical Modeling of Glass Melting," Journal of Non-Crystalline Solids, 177:347-362 (1994).

[5]Kawachi, Shinji, and Yoshio Iwatsubo, "Diagnosis and Treatment of Bubbles in Glass Production Using a Numerical Simulator," Glastech. Ber. Glass Sci. Technology, Vol. 72, No. 7, 207-213 (1999).

[6]Vogel, Werner, Glass Chemistry, 2nd Edition, Springer-Verlag, Berlin, 1994

[7]Schnepper, Carol, Benjamin Jurcik, Christel Champinot, and Jean-Francois Simon, " Coupled Combustion Space-Glass Bath Modeling of a Float Glass Melting Tank Using Full Oxy- Combustion," Proceedings of the Fifth International Conference on the Advances in Fusion and Processing of Glass, Toronto, Canada (July 27-31, 1997).

[8]Hoke, Bryan C., and Robert D. Marchiando, "Using Computational Fluid Dynamics Models to Assess Melter Capacity Changes when Converting to Oxy-Fuel," Proceedings of 18th International Congress on Glass, San Francisco, Cal. (July 1998).

[9]Golchert, B., S.L. Chang, C.Q. Zhou, and M. Petrick, "Validation of the Combustion Space Simulation of a Glass Furnace Simulator," IMECE 2001, Nov. 11-16, 2001, New York, NY.

[10]Chang, S.L., and S.A. Lottes, Numerical Heat Transfer Part A, 24(1):25-43 (1993)

[11]Kennedy, I., "Models of Soot Formation and Oxidation," Prog. Energy Combust. Sci., 23:95-132 (1997).

[12]Fairweather, M., W.P. Jones, and R.P. Lindstedt, "Predictions of Radiative Transfer from a Turbulent Reacting Jet in a Cross-Wind," Combustion and Flame, 89:45-63 (1992).

[13]Nicol, D.G., P.C. Malte, A.J. Hamer, R.J. Roby, and R.C. Steele, "Development of a Five-Step Global Methane Oxidation-NO Formation Mechanism for Lean-Premixed Gas Turbine Combustion," International Gas Turbine & Aeroengine Congress & exhibition, 98-GT-185, Stockholm, Sweden (June 2-June 5, 1998).

[14]Chang, S.L., B. Golchert, C.Q. Zhou, and M. Petrick, "An Investigation of the Effects of Firing Patterns on Heat Transfer and NO_x Formation in a Glass Furnace,"National Heat Transfer Conference, Anaheim, CA, June (2001)

[15]Stanek, S., "Electrick Melting of Glasses", Elsevier Scientific Publing Company, New York (1977)

[16]Kramer, Von F., "Gas Profile Measurements as a Means of Determining gas evolution during Glass Melting," Glasstech Ber., 53 pp.177-188 (1980).

[17]Patankar, S.V., "Numerical Heat Transfer and Fluid Flow", Hemisphere, Washington, D.C. (1980).

SENSOR DEVELOPMENT FOR THE GLASS INDUSTRY

M. Velez, M. Karakus, W. L. Headrick, R. E. Moore
University of Missouri-Rolla
Ceramic Engineering Dept.
Rolla, MO 65409-0330

B. Varghese, R. Zoughi,
University of Missouri-Rolla
Applied Microwave Nondestructive Testing Laboratory (amntl)
Electrical and Computer Engineering Dept.
Rolla, MO 65409-0330

S. Keyvan, H. Fan, R. A. Rossow
Center for Artificial Intelligence in Engineering and Education
Mechanical and Aerospace Engineering Dept.
University of Missouri-Columbia
Columbia, MO 65211

D. G. Robertson   J. M. Almanza
University of Missouri-Rolla   CINVESTAV, Carr. Saltillo-Mty.
Metallurgical Engineering Dept.   Km 13, Apdo. Postal No. 663
Rolla, MO 65409-0330   25000, Saltillo, Coah., Mexico

ABSTRACT
    This work summarizes the different activities undertaken at the University of Missouri-Rolla on developing and testing techniques and instrumentations for the glass industry. Current work includes: (1) on-line measurement of NaOH vapor in the combustion chamber, (2) on-line monitoring of wall refractory thicknesses, and (3) analysis of flame images for burner control and optimization. Limitations and advantages are listed and compared to other current techniques.

INTRODUCTION

The higher demand on quality of glass products, the need to meet future legislation on wastes and emissions, and the need to improve energy efficiency require new sensors for process control. On-line sensors provide a direct measure of some relevant properties such as glass flow, melting rate, viscosity, strength, color, refractory corrosion, emissions, etc., which need to be controlled to optimize the glass melting process. For best applications, a sensor should not change the environment or affect the property being measured, and the sensor should not be degraded by the environment. The advent of non-traditional methods of melting glasses will also require non-traditional on-line sensors under very demanding conditions. This has prompted the US DOE (Dept. of Energy) to invest in "The Industries of the Future" to ensure that R&D resources are allocated to maximize benefits [1]. The status of related sensor technology is presented in Table I. This work summarizes activities at the University of Missouri-Rolla on (1) measurement of alkali species in the combustion chamber, (2) monitoring of wall refractory thicknesses, and (3) analysis of flame images for burner control.

Table I. Status of Sensor Technology in the Glass Industry [based on 2,3].

Issue	Related Topic	Sensor/Principle
Combustion Space	Glass temperature, Gas temperature and velocity	Laser Doppler Velocimetry, radiation flux probe, and gas analysis [4]
Corrosion of superstructure	NaOH vapor concentration	D_{Na} line [5], Gas extraction [6] β-Alumina cell [7], LIFF [8,9]
Corrosion of glass-contact refractories	Acoustic impedance of refractories	New piezoelectrics [10,11] and echo-impact [12,13]
	Dielectric constant of refractories	Sensor-based microwave (radar) techniques [14]
Flame characteristics	Radicals, CO, Soot	Flame spectra analysis [15]
	Digital images	Computer analysis [16]

SODIUM HYDROXIDE MEASUREMENTS

A common practice in monitoring combustion gases in the furnace chamber is the gas sampling and extraction using a stainless-steel water-cooled probe, or a platinum tube, connected to condensation bottles [5]. The dry gases can be monitored at the end of the circuit with a chromatograph. Based on gas flow rates of extraction and chemical analyses, the gas composition inside the chamber can be estimated. Such techniques were used in a simulation pilot-size glass melter [17] to establish a basis of knowledge where comparison can be made to other techniques. Improvement was made by using Na-sensitive aqueous electrodes and the development and testing of a sensor based on Na-β"-alumina electrolytes [7], with commercial materials [18]. This sensor has proven to respond immediately to changes in NaOH vapor concentration in a combustion chamber at very high temperatures (1200 to 1500 °C). Figure 1 shows the schematics of this cell/sensor.

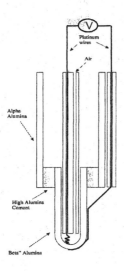

Figure 1. Schematic diagram of the Na-β″ alumina cell design.

REFRACTORY WALL THICKNESS MEASUREMENTS

The refractory walls of glass tank furnaces degrade during their operational lifetime, while premature furnace shutdowns may reduce productivity or even further damage the refractories due to temperature fluctuations. Two direct modes are being tested for refractory wall thickness measurement under various conditions: (1) Frequency-Modulated, Continuous-Wave (FM-CW) Radar Monitoring and (2) Time Domain Reflectometry (TDR). The two techniques require relatively simple microwave hardware, which could allow many measurements to be taken, yielding on-line information.

Frequency-Modulated, Continuous-Wave Radar Monitoring

The system basically requires a signal transmitter or horn, which may also be made from high-temperature materials of appropriate dielectric constant. This technique provides good results with a high degree of accuracy. Measurements have been carried out on AZS samples from room temperature to about 1200°C, and the normalized output frequency spectrum for a 15-cm and a 25-cm thick sample are shown in Figure 2. To determine the thickness from the output spectrum, the dielectric constant for the refractory samples must first be determined. The dielectric constant for the AZS sample was about 4.13 (X-band: 8.2-12.4 GHz) [19]. The measured thicknesses for the two samples are 14.7 cm and 26.5 cm, respectively. Monitoring refractory wall thickness for walls with reasonably high loss tangents can be made possible by incorporating a matched adapter into the design. Since the design is rugged and portable, the system can be easily implemented in the industry.

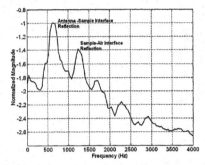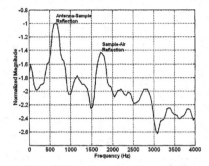

Figure 2. Room-temperature spectrum from the FM-CW radar for 15 and 25 cm sintered AZS (~15% porosity).

Limitations: The reflection at the antenna-wall interface increases the effective system noise level. Higher system noise reduces detection sensitivity. If the sample is too thin, then the reflection from the sample-air interface is overlapped by the noise or cannot be properly deciphered. However, this limitation can be overcome by increasing the bandwidth of the frequency sweep, or by increasing the modulation frequency.

Time-Domain Reflectometry at Microwave Frequencies

This technique has been tested using an open-ended rectangular waveguide using a microwave band-limited pulse launched into the refractory wall, while the reflection of the pulse from the back side of the wall is received and analyzed with a vector network analyzer (VNA). Reflections occur whenever there is an impedance contrast in the medium of propagation. This impedance contrast is provided at the wall-air interface. To determine the refractory wall thickness, the velocity of propagation of the pulse in the medium as well as the time of flight (TOF) must be determined. The TOF is the time taken for the pulse to travel one way through the wall. The TOF can be determined from the output of the VNA. The velocity of propagation can be determined if the dielectric constant for the refractory is determined, which can be determined by calibrating the VNA with a sample of known thickness. The dielectric constant at room temperature for the AZS samples was determined to be approximately 6.28 (S-band: 2.6-3.95 GHz) [20]. The room-temperature outputs after envelop detection for 20-cm and 25-cm thick AZS samples is shown in Figure 3. The VNA output is passed through an envelop detection filter to simplify the analysis of the output data and to determine the peak of the reflected pulse. The measured thicknesses for the two samples are 19.2 cm and 25.1 cm respectively.

Limitations: When the wall is too thin, the reflection from the front end and the reflection from the back end will overlap. This can be overcome if the TOF is greater than the width of the pulse. This can be achieved by working at higher frequency bands, for instance using a G band (3.95-5.85 GHz). Currently, an expensive VNA is used to provide and analyze the TDR pulses. A portable and rugged TDR could be designed for industrial applications.

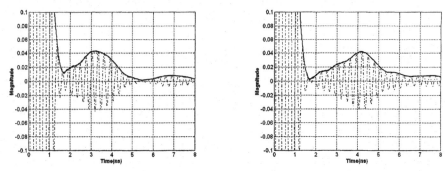

Figure 3. Room-temperature output from the VNA for 20 and 25 cm sintered AZS (~15% porosity).

FLAME ANALYSIS

Combustion flames from natural-gas-fired furnaces for glass melting using oxy/fuel are investigated in this work. The purpose of flame analysis is to understand the correlation between flame imagery and control parameters such as oxygen fuel ratios or heat input in a furnace, as well as flame temperature profiling. The pilot-scale glass furnace utilized is a 0.1-0.5 MBtu/hr furnace that can melt from about 50 kg to 1,000 kg of glass/day [17]. The furnace is controlled through a LabView hardware and software control system. Figure 4 shows the spectrometer probe and the camera set-up at this pilot-scale glass furnace.

Figure 4. Spectrometer Installed at Pilot-Scale Glass Furnace

Various combustion-controlled experimentations were carried out at this furnace. The oxygen/fuel ratio was varied from 1.8 to 2.4, with several combustion-control experimentations in both step and ramp-up fashion. Using a spectrometer, spectral intensity data were collected over the ultraviolet/visible regions. The data were analyzed for specific radical chemiluminescence and the electromagnetic emission spectrum. Direct correlation and dynamic response were observed from the emission band from the hydroxyl flame radical, OH, to burner stoichiometry and flue-gas NOx emissions [21]. The results show a great promise for online combustion monitoring at the burner level for gas-fired glass furnace applications.

Data such as flame images, natural-gas flow rate, oxygen fuel ratio, and spectrum intensity from fiber-optics spectrometer were collected from the pilot-scale natural-gas-fired glass furnace. The result shows direct correlation between furnace operating condition and flame combustion behavior. These data were then incorporated in a LabView program to demonstrate, through simulation, an on-line combustion-monitoring condition. Figure 5 shows a screen shot of the LabView simulator for the control experimentation changing the natural-gas flow rate from 140 scfh to 110 scfh (standard conditions ft^3/h) in a step fashion. The VI and UV spectra are also shown in the lower part of the screen shot of the LabView simulator in Figure 5. The VI spectrum data are utilized to calculate flame temperature using a black-body radiation model. Table II shows the result for the three oxy/fuel ratios.

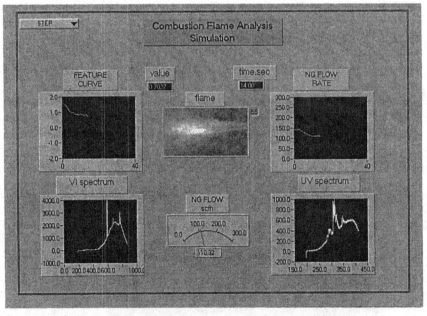

Figure 5. Screen shot of LabView simulation in pilot-scale furnace

Table II. Flame-Temperature Calculations	
Temp (°K)	O/F Ratio
1699.2	2.4
1736.4	2.1
1757.4	1.8

CONCLUDING REMARKS
The University of Missouri-Rolla Refractory Group has worked over the last decade on models and testing related to refractories for the glass industry. The facilities include an oxyfuel simulator furnace, a high-temperature digital camera, gas chromatograph, and spectrometer. Continuing work is being established on the following tasks: (1) developing a stable alumina sensor for detecting NaOH on-line; (2) developing on-line monitoring techniques for scanning refractory walls in glass melters; (3) measuring emissions under special operating conditions; (4) test burners; and (5) applying flame-analysis results in industrial environments.

ACKNOWLEDGEMENTS
We express our appreciation to the NSF Industry-University Center for Glass Research under Grant No. 0128040; DOE Inventions & Innovations Program Grant No. DE-FG36-02G012050; and Grant No. DOE DE-FC07-00CH11032.

REFERENCES
1. DOE OIT portfolio 2002: http://www.oit.doe.gov/glass/portfolio.shtml
2. A.J. Faber, R.C.G. Beerkens, "Technology Development of Sensors and Control for Glassmaking," *The Glass Researcher*, **8**[2] 4-7 (1999).
3. J. Fenstermacher, "Sensors for Glassmakers of the 21st Century," *The Glass Researcher*, **10**[1] 1 (2000).
4. R.L. Cook, R. Arunkumar, "Sensors for a Coupled Combustion Space-Glass Bath Furnace," *The Glass Researcher*, **10**[1] 8-9; 17; 21 (2000).
5. S. G. Buckley, P. M. Walsh, D. H. Hahn, R. J. Gallagher, M. K. Misra, J. T. Brown, S.S.C. Tong, F. Quan, K. Bhatia, K.K. Koram, V.I. Henry, R.D. Moore, "Measurements of Sodium in an Oxygen-Natural Gas Fired Soda-Lime-Silica Melting Furnace," pp. 183-205 in Proc. 60th Conference on Glass Problems, The American Ceramic Society, 2000.
6. S.S. Tong, J.T. Brown, L.H. Kotacska, "Determination of Trace Impurities in a Furnace Atmosphere at Operating Temperature," *Ceram. Eng. Sci. Proc.*, **18**[1] 208-215 (1997).
7. J.M. Almanza, "Development of a High Temperature NaOH Vapor Sensor Based on a Beta-alumina cell," PhD thesis, University of Missouri-Rolla, 2003.
8. S.F. Rice, M.D. Allendorf, M. Velez, J.M. Almanza, T. Burns, "Laser-Based Sensor for Measuring NaOH and KOH in Furnaces," presented at

the 104th Annual Meeting Am. Ceram. Soc., St. Louis, MO, April 28-May 1, 2002.

9. P. Walsh, *The Glass Researcher*, **10**[1] 3 (2000); http://www.oit.doe.gov/glass/factsheets/monitoringalkali.pdf

10. S.M. Pilgrim, "Ultrasonic Transducers Successfully Test Refractory Walls," *The Glass Researcher*, **8**[2] 20 (1999).

11. R.E. Dutton, D.A. Stubbs, "An Ultrasonic Sensor for High Temperature Materials," pp. 295-303 in Sensors and Modeling in Materials Processing: Techniques and Applications, Edited by S. Viswanathan, R. G. Reddy, J. C. Malas, The Minerals, Metals & Materials Society, 1997.

12. M.J. Sansalone, W.B. Street, Impact-Echo: Nondestructive Evaluation of Concrete and Masonry, Ithaca, NY, Bullbrier Press, 1997.

13. B.J. Jaeger, "Refractory Wall Thickness Measurements in High Temperature Environments and in Thermal and Material Property Gradients Using the Impact-Echo Method," Ph. D. Thesis, Cornell University, 2000.

14. R. Zoughi, L.K. Wu and R.K. Moore, "SOURCESCAT: A Very-Fine-Resolution Radar Scatterometer," *Microwave J.*, **28**[11] 183-196 (1985).

15. W. Von Drasek, „Optimization of Oxy-Fuel Combustion with Optical Sensors," *Ceram. Eng. Sci. Proc.*, **19**[1] 29-49 (1998).

16. http://www.oit.doe.gov/sens_cont/factsheets/flame.pdf

17. M. Velez, L. Carroll, C. Carmody, W.L. Headrick, R.E. Moore, "Oxy-fuel Simulator Glass Tank Melter," pp. 47-54 in *Environmental Issues and Waste Management Technologies in the Ceramic and Nuclear Industries VI, Ceramic Transactions, Vol. 119.* Edited by J. P. Bennett and J. D. Smith, The American Ceramic Society, Westerville, OH, 2001.

18. http://www.ionotec.com/, Ionotec Ltd., 14 Berkeley Court Manor Park Runcorn Cheshire WA7 1TQ, England.

19. B. Varghese, R. Zoughi, C. DeConink, M. Velez, and R.E. Moore, "Frequency Modulated Continuous Wave Monitoring of Refractory Walls," to be published in *Ceram. Eng. & Sci. Proc.*, 2003.

20. B. Varghese, R. Zoughi, M. Velez, and R.E. Moore, "Refractory Wall Thickness Measurement Using Time Domain Reflectometry at Microwave Frequencies," pp 221-234 in 39th Symposium on Refractories, American Ceramic Society, St. Louis, MO, April 2003.

21. S. Keyvan, R.A. Rossow, M. Velez, W.L. Headrick, R.E. Moore, C. Romero, "Combustion Control Experimentations at a Pilot Scale Glass Furnace," to be published in *Ceram. Eng. & Sci. Proc.*, 2003.

ADVANCED OPERATION SUPPORT SYSTEM FOR REDOX CONTROL

O.S. Verheijen and O.M.G.C. Op den Camp
TNO Glass Group
P.O. Box 595
5600 AN Eindhoven
The Netherlands

ABSTRACT

TNO Glass Group has developed a Glass Process Simulator (shortly GPS) to supply operators and glass technologists with the maximum amount of relevant information for optimal furnace operation. In addition to furnace temperatures as measured by thermocouples, GPS provides online information on redox, glass color, emissions, and the total flow and temperature distribution in the furnace. Because the simulation model GPS is approximately 10 times faster than real time, GPS provides information on the future state of the furnace as function of changes in process settings, batch and melt properties, and external process disturbances. Combination of GPS with special developed inline redox sensors enables the GPS-users to simulate the effect of redox changes at the entrance of the furnace on both the glass melting process and product quality. Next, with GPS (scenarios for) process setting and redox state adaptations to react fast and accurate on the redox disturbances at the entrance of the glass furnace can be determined. This paper shows how GPS can be used, in combination with information from inline redox sensors, to control the color of container glass.

INTRODUCTION

Nowadays, mainly temperature control is applied in glass-melting furnaces to maintain process performance and product quality. However, as has been observed in practice and will be shown in this paper, also anticipation of unintentional redox changes in glass-melting furnaces is required. As a consequence of changes in the redox state of a glass melt,

- the fining performance of the glass-melting furnace changes, which may result in an increase of the seed count level,
- extensive foam formation may be observed (especially for E-glass melts [1]),
- problems during the forming process may be encountered due to changed glass forming properties, and
- the produced glass may exceed the color specifications (especially for container glass).

During the container glass production, large amounts of foreign mixed recycling cullet are used as raw material. The use of recycling cullet in glass production is profitable with respect to energy consumption, emission reduction, and acceleration of the batch-to-melt conversion process. A disadvantage of the use of foreign mixed recycling cullet is that the composition and oxidation state of the cullet mixture is not known exactly and varies with time (see figure 1). This makes the redox state of the glass-forming batch a difficult controllable disturbance parameter for the glass melting process.

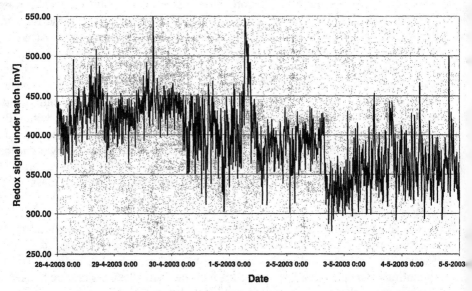

Figure 1   Measured redox state underneath the batch blanket with an inline redox sensor in a container glass furnace.

To avoid the redox state related problems mentioned above or at least to be able to counteract on redox variations in the glass-melting furnace, both a fast simulation model of the glass melting process and (an) inline redox sensor(s) are required. The purpose of the inline redox sensor is to detect (changes of) the redox state of the glass melt batch in the early stage of the glass-melting process (i.e. at least faster than the detected redox change in the produced bottle). A fast simulation model, which calculates temperature distribution, glass melt flow patterns and redox state distribution throughout the glass-melting furnace, is required to:

- predict the effect of detected changes in the redox state of the freshly molten glass forming batch, and
- advise on adaptations of process setting and redox state to counteract on the (undesired) effect of the redox changes.

To support operators and glass technologists in the fast selection of process settings and redox state adaptations to counteract on detected redox changes at the entrance of the furnace, TNO Glass Group has developed a furnace navigation system called GPS, which stands for Glass Process Simulator. This system provides fast online information with respect to for instance glass flow, temperature distribution, emissions, energy consumption, redox state, and glass color.

FAST SIMULATION OF THE GLASS MELTING PROCESS

Together with major glass producers, TNO has developed a detailed model for the simulation of processes in glass melting furnaces, viz. melting of the glass-forming batch, fining and refining of the glass melt to reduce the amount of blisters and dissolved gases in the glass, and homogenization of the glass melt before the glass melt is delivered to the forming section(s). This detailed Glass Tank Model (GTM) is able to simulate any type of furnace with all kind of heating systems such as cross-fired, U-flame regenerative, recuperative, oxy-fuel, gas and/or oil combustion, all electric, or electrical boosting. Moreover, different types of batch chargers (blanket chargers, screw and paddle chargers) and different types of mixing systems such as bubblers and stirrers are available.

Although indispensable for the control of product quality of glass furnaces, GTM does not meet the requirements with respect to computational speed for the continuous simulation of a furnace under changing process conditions. GPS however is a fast dynamic furnace simulator that is based on the detailed TNO-GTM model. A collection of advanced mathematical techniques is used to speed up the model. The resulting GPS, which is fully based on first principles, is a furnace specific product. GPS allows the possibility to test the effect of intended changes of process settings on the glass melting process without risk of production losses. Figure 2 shows the graphical user interface of GPS installed at a major container glass plant in the Netherlands.

GPS simulates both the glass melting process and the combustion process above the glass melt. As a result of the simulation, the distribution of glass melt flows, temperatures, and redox active species are known throughout the furnace. Because the photon conductivity of a glass melt depends on the concentration and type of multivalent ions such as Fe, Cr, S, etc., in GPS the heat conductivity of the glass melt is coupled with the concentration of multivalent ions in the glass melt, which are calculated with the redox model.

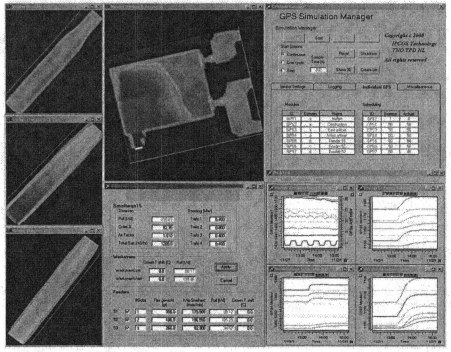

Figure 2   Graphical user interface of GPS installed at a container glass plant in The Netherlands

SIMULATION OF THE REDOX BEHAVIOR OF A GLASS MELT

To describe the transient redox behavior of a glass melt with GPS, the temperature dependent redox reactions occurring in the glass melt are solved simultaneously. For an oxidized green container glass melt, the main important redox reactions are:

- Thermal sulfate decomposition:
  $$SO_4^{2-} (m) \Leftrightarrow SO_2 (g) + 0.5\ O_2 (g) + O^{2-} (m)$$

- Redox equilibria of the polyvalent metal ions Fe and Cr:
  $$Fe^{3+} (m) + 0.5\ O^{2-} (m) \Leftrightarrow Fe^{2+} (m) + 0.25\ O_2 (g)$$
  $$Cr^{6+} (m) + 1.5\ O^{2-} (m) \Leftrightarrow Cr^{3+} (m) + 0.75\ O_2 (g)$$
  $$Cr^{3+} (m) + 0.5\ O^{2-} (m) \Leftrightarrow Cr^{2+} (m) + 0.25\ O_2 (g)$$

- Physical dissolution of the fining gases SO_2 and O_2:
  $$SO_2 (g) \Leftrightarrow SO_2 (m)$$
  $$O_2 (g) \Leftrightarrow O_2 (m)$$

Advances in Fusion and Processing of Glass III

As an example, the temperature dependent concentration based equilibrium constant for the thermal decomposition of sodium sulfate, K_c, is given by

$$K_c(T) = \frac{p_{SO_2} \cdot p_{O_2}^{0.5}}{[SO_4^{2-}]} = e^{\frac{-(\Delta H - T \Delta S)}{RT}} \tag{1}$$

in which p_{SO2} and p_{O2} are the vapor pressures of the fining gases SO_2 and O_2, $[SO_4^{2-}]$ is the concentration of sodium sulfate in the glass melt, T is the absolute temperature, R is the universal gas constant, and ΔH and ΔS are the enthalpy and entropy of the redox reaction, respectively. The glass composition dependent values for the enthalpy and entropy for all redox reactions are determined beforehand by well-defined experiments on lab scale. For the oxidized green container glass, two temperature regimes are distinguished in which different redox reactions predominate. In the low temperature regime, the predominating redox reaction is given by

$$SO_4^{2-} (m) + 2 Fe^{2+} (m) \quad \Leftrightarrow SO_2 (g) + 2 O^{2-} (m) + 2 Fe^{3+} (m)$$

whereas in the high temperature regime, the predominating redox reactions are given by

$$SO_4^{2-} (m) \Leftrightarrow SO_2 (g) + 0.5 O_2 (g) + O^{2-} (m), \text{ and}$$

$$Fe^{3+} (m) + 0.25 O_2 (g) \Leftrightarrow Fe^{2+} (m) + 0.5 O^{2-} (m)$$

The simultaneous solutions of all redox reactions in the glass melt result in concentration distributions of all these species throughout the glass furnace. Figure 3 shows the concentration of Fe^{2+} at the center plane of a container glass furnace.

Figure 3 Calculated Fe^{2+} concentration at the center plane of a container glass furnace

DEMO CASE: THE USE OF GPS FOR COLOR CONTROL
Unfortunately, variations in the redox state of foreign recycling cullet, which is used as raw material for container glass production, is common. As a consequence of the use of foreign cullet with varying redox state, both glass melt temperatures and color of the produced container glass changes. Figures 4 and 5 show the with GPS calculated effect of a step change to more reducing cullet (log

pO$_2$ at 1300°C changes from −4.5 to −4.9) on glass melt temperatures and dominant wavelength of the produced glass.

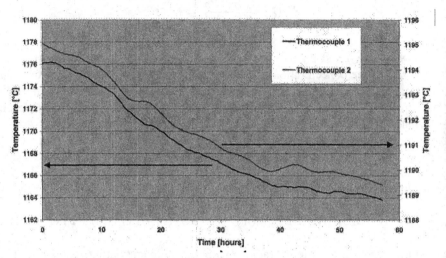

Figure 4   Response of glass melt temperatures on a redox change at the furnace entrance calculated with GPS

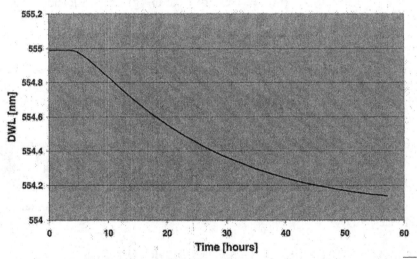

Figure 5   Response of dominant wavelength on a redox change at the furnace entrance calculated with GPS

Applying only temperature control to compensate for the observed temperature and color change, showed that the effect on the dominant wavelength of the glass product was only limited. To meet the color specifications of the container glass, also redox control is required next to temperature control. Anticipation of redox changes in the glass furnace requires information of the redox state of the glass product or preferably information on the redox state of the glass melt in the furnace. The time delay between the detection of a redox change in the glass product and the adaptation of the batch recipe is at least 17 hours. To be able to anticipate in an early stage of the melting process, inline redox sensors are developed [2] for both in the feeder section and the batch charging area (see Figure 6) of the glass-melting furnace.

Figure 6   Inline batch redox sensor positioned in the charging area of a
            container glass furnace.

Figure 7 shows the dominant wavelength of the container glass as function of time in case
- no correction of the redox state of the batch is applied,
- the time at which the redox state of the batch is corrected is based on the time at which the redox change is detected in the produced bottle (i.e. a time delay of 17 hours),
- the time at which the redox state of the batch is corrected is based on the time at which the redox change is detected by the inline feeder redox sensor (i.e. a time delay of 12 hours), and
- the time at which the redox state of the batch is corrected is based on the time at which the redox change is detected by the inline batch redox sensor (i.e. a time delay of 3 hours).

It is clearly seen from the simulations with GPS that with the use of the inline redox sensors, a fast reaction on redox changes in possible with which the color of the produced bottle can be maintained within the specifications.

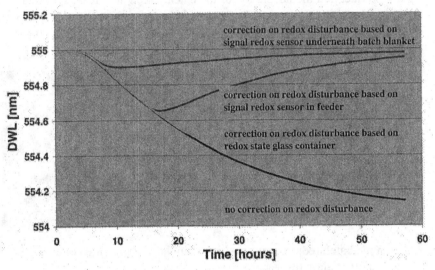

Figure 7   Response of dominant wavelength on a redox change at the furnace entrance calculated with GPS

CONCLUSIONS

In glass-melting furnaces, to avoid melting problems while running the furnace at the lowest possible energy consumption, stable glass melt temperatures are pursued. For a constant product quality (glass color), next to temperature control also redox control is required. This paper showed that the fast Glass Process Simulator (GPS) in combination with information from inline redox sensors enables the GPS-user to react on temperature changes and redox disturbances.

ACKNOWLEDGEMENT

The authors wish to thank Dr. Laimböck from ReadOx & Consultancy B.V. for his contribution to the work discussed in this paper.

REFERENCES

[1]   A.J. Faber, O.S. Verheijen, J.M. Simon, 7th *International Conference Advances in Fusion and Processing of Glass*, Rochester (NY), USA, July 27-30, 2003.

[2]   P.R. Laimböck, R.G.C. Beerkens, J. van der Schaaf, 62nd *Conference on Glass Problems*, Champaign (Urbana), USA, October 16-17, 2001, pp. 27-41.

GLASS FURNACE SIMULATOR – ADVANCED MELTER OPERATION AND TRAINING TOOL

Miroslav Trochta
and Pavel Viktorin
Glass Service, Inc.
Rokytnice 60, 755 01 Vsetin
Czech Republic

Marketa Muysenberg
and Erik Muysenberg
Glass Service B.V.
Watermolen 22, 6229 PM Maastricht
The Netherlands

ABSTRACT

Glass Service, Inc. uses *GS Glass Furnace Model* computational fluid dynamics (CFD) package as a base for a glass furnace *Simulator*. It contains a user-friendly graphical front end, which is a simplified version of furnace operation station that allows modifying furnace settings and display the calculation results in an operator-friendly way. Behind this user interface, coupled glass and combustion solvers and post-processors are running. The *Simulator* itself serves as a "translator"; it translates the user-friendly inputs to not-so-user-friendly parameters, such as boundary conditions for the solvers. In the opposite direction, it reads the "crude" outputs of the solvers and translates them to the indicators which are familiar to operators.

Two example simulators are presented. The first one is a small oxy-fired container furnace. Among other parameters, the user can modify the color of the glass, batch properties (cullet, moisture), way of batch charging (pull and distribution), distribution of fuel among burners, bubbling setup and fuel properties. The outputs include glass and crown temperatures in thermocouple/pyrometer positions, animated glass flow, quality indices and exhaust gas composition, including environmental indicators as NOx.

The second example is a single port taken from a cross-fired float furnace. The user can modify parameters (sizes, angles, turbulence properties, etc.) of three underport burners and can also set up fuel composition, staging, enriched air, etc. Output screens contain crown temperatures, visible flame shape, detailed exhaust gas information, flow patterns and heat fluxes to glass surface.

An example of the Simulator's reaction to changes in furnace operation parameters is presented. The response is, in most cases, faster than that of a real furnace. The simulator has been validated and shows good agreement with reality.

INTRODUCTION

Glass Service, Inc. (GS) has been developing its own CFD package for the last decade. This package is used by GS and other glass companies to analyze and diagnose furnace operation. Although this package features a user-friendly graphical interface, a case setup still requires understanding of basics of CFD and physical phenomena involved. But CFD has a potential to become a useful "interactive" tool, for example, for furnace operators to predict consequences of their modifications of operating parameters. To allow minor changes in input parameters without knowledge of boundary condition types and other CFD solver-specific issues, GS has developed a program interface, called *GS Simulator*. The intention was to make the environment look like control screens on an operator station so that the operator can control the model in the same way the real furnace is operated.

SIMULATOR FUNCTION

GS Simulator serves as an interface between solvers and a user-friendly front end. Its basic tasks are to receive input data from a graphical user interface (GUI) and to "translate" the data into solver files, i.e., to boundary condition parameters, geometry setup, physical properties of glass etc. For example, as the user changes pull rate, corresponding velocity boundary conditions (inlets as well as outlets) have to be modified; if the user changes batch moisture, batch reaction heat and other properties have to be changed; if the user changes the position of a cooler, the geometry of the model must be modified; if cooling rate is changed, volumetric heat sink of a solid object has to be modified.

The Simulator runs Glass Model and Combustion Model. GUI periodically displays all output variables (temperatures, exhaust gases etc) and simulator compares the solvers' convergence status with criteria given by GUI. When all criteria are satisfied, simulator stops the solvers and starts *GS Trace* program(s) that perform particle tracking. When particle tracking is finished, GUI displays the critical trajectory, residence time distribution and melting index distribution. The computation usually takes several hours and the response is faster than response of the real furnace.

CFD SOLVERS

Standard GS solvers (*GS Vitreous* for glass and *GS Combustor* for combustion space) run on the background behind the GUI. They are fully coupled, i.e., they periodically exchange temperatures and heat fluxes on glass and batch surface and in walls.

The only difference from normal modeling study setup is that the grids are slightly coarser. For typical float furnace study, GS would use about one million cells for glass and about 1.5 million cells for combustion space. When we optimize a float furnace model for the Simulator, we start also with this typical high number of cells and then gradually reduce the grid as long as no problems, no errors and no significant differences from the detailed model are observed. The resulting grid contains about 100 000 control volumes for combustion space and about 50 000 control volumes in glass. Next to this, we also adapted and improved the speed of our solvers for this application.

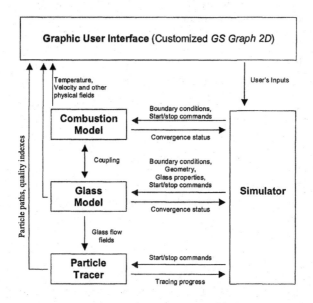

Figure 1. Data flow among solvers, Simulator and Graphic User Interface.

USER INTERFACE

The user interface is derived from *GS Graph 2D*, which means that it inherited all its visualization capabilities. As an extension, it contains control screens to modify operating parameters and other screens that display the results in an operator-friendly way.

Examples of the screens are given below, where we also give more details about particular applications of the *Simulator*.

EXAMPLE SIMULATORS

As the real furnace control screens, hence the *Simulator* GUI screens, usually contain details of furnace design and other sensitive information, we cannot show a simulator of an existing industrial furnace in this paper.

Instead, we have prepared two examples. The first one is an oxy-fired container furnace. The second one is a simulator of a single port of a cross-fired float furnace.

Note that both models are somewhat simplified and real applications of the simulator contain many more parameters and outputs, and allow the user to modify all control variables of interest.

Oxy-fired container melter simulator

This example is a simulator of an oxygen fired container melter. A typical pull rate is 160 MTPD. The user can change the following parameters:

- Pull rate
- Batch charging distribution between left and right doghouse
- Batch moisture
- Batch temperature
- Cullet percentage
- "Glass darkness" – in a real simulator we use iron content instead
- Bubbling rate
- Total gas
- Total oxygen
- Gas and oxygen distribution among the burners
- Gas composition
- Oxygen purity

The input and output screens can be seen on the screenshots below. The screenshots also show two different setups. In the second one (compared to the first one), the fuel distribution among the burners was moved towards the hot spot (less firing above the batch). Such a change increased bottom temperatures and improved the mean melting index.

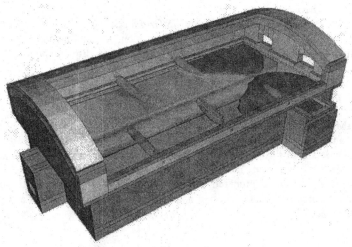

Figure 2. Overview of the furnace simulated

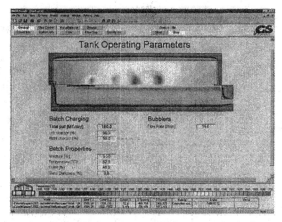

Figure 3. Oxy melter simulator main operating parameters screen

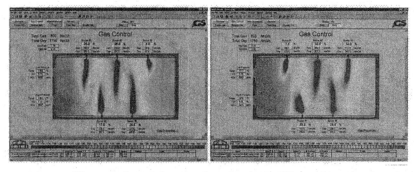

Figure 4. Oxy melter simulator cases: Gas Control screens

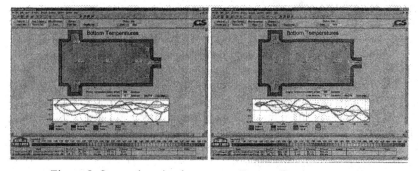

Figure 5. Oxy melter simulator cases: Bottom Temperatures screens

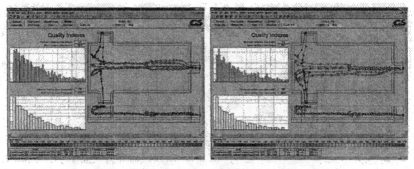

Figure 6. Oxy melter simulator cases: Quality Indexes screens

Cross-fired float furnace port simulator

The other example is not a traditional coupled furnace. This simulator is intended for users who want to get insight into combustion, to understand the responses to changes of firing setup and to find optimum configuration of firing in a single port. What we simulate is a combustion space of one section (port) of an underport air fired, cross-fired float furnace. There are 3 burners under the port. The parameters which can be changed are:

- Primary air flow
- Oxygen enrichment
- Temperature of preheating
- Second stage air flow
- Second stage oxygen flow
- Second stage air/oxygen temperature
- Staging nozzle diameter
- Glass surface temperature
- Fuel flow
- Fuel temperature
- Fuel distribution among the burners
- Burner nozzle diameters
- Vertical, horizontal and spraying angles for each burner
- Swirl frequency for each burner
- Turbulent intensity for each burner
- Fuel composition

The outputs, apart from all physical fields (implicitly given by the GUI) include:

- Combustion ratios, lambda indicators
- Input and output velocities
- Visible flame shape

- Heat balance information – heat in preheated air, heat value of fuel, crown loses, heat flux to glass, advective heat flux to exhaust
- Exhaust gas info – temperature, O_2, combustibles, CO, NO_x

The input and output screens can be seen on the screenshots below. The pictures also show two different cases. In the first one, all the combustion air enters the furnace through the port. In the second one, part of the air is introduced via staging. The total excess oxygen (Lambda total) is the same for both cases and is equal to 1.055. However, in the second case, there are fuel-rich conditions in the flame: the primary lambda ratio is only 0.952. One of the positive effects of this change is reduction of NO_x by approximately 15%.

Figure 7. Port simulator cases: Main Setup screens

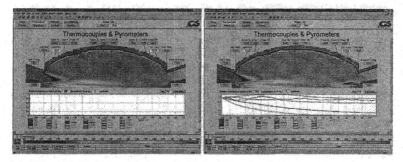

Figure 8. Port simulator cases: Thermocouples & Pyrometers screens

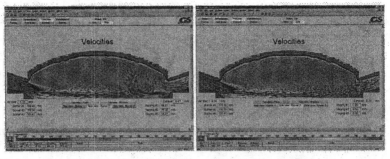

Figure 9. Port simulator cases: Velocities screens

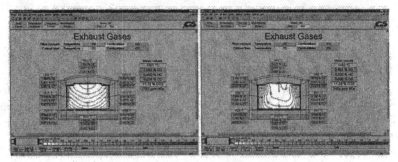

Figure 10. Port simulator cases: Exhaust Gases screens

CONCLUSIONS

GS Simulator's responses seem to be accurate enough to predict consequences of changes in tank operating parameters. The error of simulator is often even beyond accuracy of aging thermocouples. At the same time, it can be used even by people who do not have experience with CFD. The choice of the input parameters and output indicators, screen design and simplicity/accuracy ratio is defined by customer. Thus, this application of CFD can become a useful tool:

- for optimizing furnace parameters (fuel profiles, cooling, etc.) without disturbing the production
- to obtain data that are difficult to measure in real process (residence time, quality indices, flow patterns, temperatures anywhere)
- as a training tool for operators
- to find optimal configuration for making new type of glass or different tonnage

REFERENCES

[1]Schill, Franěk, Trochta, Viktorin, "Integrated Glass Furnace Model". *Proc. of 5th Intl Seminar on Math. Sim. in Glass Melting*, Horní Bečva, Czech Rep., 1999

　　　　　　　　　Advances in Fusion and Processing of Glass III

COMPARISON OF PHYSICAL AND MATHEMATICAL MODELS OF STIRRING IN FOREHEARTHS

Jiří Brada
Glass Service, Inc.
Rokytnice 60, 755 01 Vsetín
Czech Republic

Tomáš Krobot, Jan Kučera
GTM
Špitálská 182,580 03 Hradec Králové
Czech Republic

ABSTRACT
The objective of this work is to compare results of physical and mathematical models of stirring in forehearths and to validate numerical methods used for the simulation and evaluation of stirring action approximation.

Based on similarity criteria, the physical model represents a real forehearth, including dimensions of the forehearth itself and stirrers in the scale 1:4, thermophysical properties of the liquid, stirrers' parameters and forehearth's operational conditions. Flow patterns and the effect of stirring on them are studied so that they can be compared with results of the mathematical model.

The mathematical model solves momentum balance equations and uses a drag force approach to take the action of stirrers into account. The drag force exerted by stirrers on the liquid is used as a source term in momentum balance equations for the flow of the liquid. Drag force calculation uses all crucial parameters of stirrers – their geometry, dimensions and number of revolutions per a unit of time. Although the mathematical model is able to calculate also temperature distribution, isothermal conditions from the physical model were kept to obtain comparable results between physical and mathematical models. The results showed good agreement and applicability of the approach using the drag force calculation.

INTRODUCTION
The homogenization of glass melt may be divided into either a spontaneous one or force one. The spontaneous homogenization always occurs during the glass melting process. It is induced mainly by diffusion, by ascending of bubbles during the glass melt refining and by free convection. In

an ideal case, the glass melt should be entirely homogenous, that is, the sample taken from any place should have the same composition.

This stage, however, is not achievable in practice, because there are processes working against the effect of the spontaneous homogenization, making the glass homogeneity worse. These processes include the corrosion of the refractory material, volatilization of some compounds from the glass melt, imperfect bath mixing, the presence of foreign matters in the batch, the use of foreign cullet (with different composition), etc. The cause of an inhomogeneous glass melt may be an unsuitable technological process or an imperfect glass melting furnace or unsuitable parameters of glass melting aggregate parts.

The most effective method for glass homogenizing is a machine homogenization. There are following applications of mechanical stirrers:
- Glass melt stirring in a glass melting pot (crucible)
- Permanent machine homogenization in working ends, channels or distributors
- Continuous machine glass homogenization in feeders

PHYSICAL MODEL
The most suitable and reliable method of research and development of machine homogenization is physical modeling since the process of machine homogenization may be physically and uniquely defined. As an effect of homogenization, thermal gradient in the stirring zone diminishes; glass melt currents are caused mostly by the effect of stirrers and gathering in throat systems. Therefore, it is possible in most cases to use an isothermal model for studying. If the glass homogenization is simulated in a throat system (e.g. in feeder), the model is operated with closed circulation.

Homogeneity Criteria
Following criteria for evaluation of glass homogenization efficiency are used:
- Critical RPM (revolutions per minute) – is a minimal RPM number, at which no model liquid undergoes the homogenization process.
- Glass homogenization efficiency – the goal of model tests is to find operating conditions to get homogenous liquid behind the stirring system.
- Head loss in the stirrer system - the difference of model liquid level height in front and behind the stirrers.
- Head loss in feeder canal - canal head losses caused the dropping level height along the feeder.

Criteria characterize the property of a physical problem. Galileo and Reynolds numbers ensure the physical similarity of the flow patterns. Reynolds number Re includes the ratio between inertial forces and viscous forces; Galileo number Ga includes the ratio between gravitation forces and viscous forces; Froude number Fr characterizes the rotation movement of stirrers in viscous liquid (the pumping effect); second Reynolds number Re_2 describes flow pattern due to stirrers; the rotation and up/down movement of the gobber are described by the third Reynolds number Re_3 and by the second

Galileo number Ga_2:

$$Re = \frac{Q}{\nu B} = \frac{\nu \eta B}{\rho}, \ Ga = \frac{g B^3}{\nu}, \ Fr = \frac{n^2 d}{g}, \ Re_2 = \frac{n d^2}{\nu}$$

$$Re_3 = \frac{f d_p^2}{\nu}, Ga_2 = \frac{g d_1^3}{\nu}$$

where B is width of canal, Q flow rate, n stirrers' RPM, g acceleration of gravity, d_1 orifice diameter, ν kinematic viscosity, η dynamic viscosity, d_p plunger diameter, f cut/min (gob/min), ρ liquid density, d stirrer diameter, and v stands for flow velocity.

Model Scale
The scale of the model is a compromise between what's desired and what's possible. The full-scale model of the real process would be the best model to study physical phenomena of the process. Usually this is not possible since there are many limitations - dimensions of the equipment, model liquid viscosity, pump output or the monitoring environment and others.

Viscosity of model liquid is calculated from the Galileo number:

$$\left(\frac{B^3 g}{\nu}\right)_{model} = \left(\frac{B^3 g}{\nu}\right)_{real} \Rightarrow \nu_{model} = \nu_{real} \ s^{3/2}, \ s = \frac{B_{model}}{B_{real}}$$

Flow rate is derived using the Reynolds number:

$$\left(\frac{Q}{B\nu}\right)_{model} = \left(\frac{Q}{B\nu}\right)_{real} \Rightarrow Q_{model} = Q_{real} \ s^{5/2}$$

The RPM of stirrers is obtained from the second Reynolds number:

$$\left(\frac{n d^2}{\nu}\right)_{model} = \left(\frac{n d^2}{\nu}\right)_{real} \Rightarrow n_{model} = n_{real} \ s^{-1/2}$$

where s is the scale of model.

MATHEMATICAL MODEL

Governing equations for a steady or time-dependent, three-dimensional flow of a viscous, incompressible liquid are derived from general laws of mass and momentum conservation, which results in the following form:

$$\frac{\partial \mathbf{v}}{\partial t} + \mathrm{div}(\mathbf{v}\,\mathbf{v}) = \mathrm{div}(\nu\nabla\mathbf{v}) - \nabla p + \mathbf{F} \tag{1}$$

$$\mathrm{div}\,\mathbf{v} = 0 \tag{2}$$

where $\mathbf{v} = (u, v, w)$ is the velocity vector, ν is kinematic viscosity of the liquid and p denotes pressure. The momentum balance equation (1) and the continuity equation (2) are solved iteratively using finite volume method called SCGS (Symmetrical Coupled Gauss Seidel, for details see [1] and [2]).

The term $\mathbf{F}$ on the right-hand side is the crucial term for stirring calculation – it is a vector of the drag force, which is exerted by stirrers on the liquid. Under the assumption that a shaft with rectangular blades can approximate stirrer shape, the drag force vector can be written in the form

$$\mathbf{F} = \frac{1}{2}C_D\rho S|\mathbf{v}_r|\mathbf{v}_r \tag{3}$$

where ρ is liquid density, S is the projected area of a blade on the plane perpendicular to the to the flow of the liquid and $\mathbf{v}_r$ stands for the stirrer velocity relative to the velocity of the liquid. The symbol C_D denotes the drag coefficient, calculated as a function of the angle of attack and the Reynolds number:

$$C_D = \frac{C_D^0}{\mathrm{Re}}, \quad \mathrm{Re} = \frac{L|\mathbf{v}_r|}{\nu}$$

where C_D^0 is a function of the angle of attack and L denotes the characteristic dimension of the stirrer (usually parallel to the flow).

PROBLEM SETUP

Physical Model Geometry, Comparison Method

The scale 1:4 was chosen for the physical model to have reasonable kinematic viscosity, pump output and other parameters. The physical model is made from PERSPEX material and it has inner dimensions 0.894 m, 0.228 m, 0.089 m (length, width, depth), spout inner dimensions in its inflow and outflow part are 0.09 m and 0.048 m, respectively; spout height is 0.141 m. The discretization used 204x117x68 = 1 623 024 grid nodes.

There are three sets of stirrers arranged in lines and consisting of 4, 6 and 2 stirrers respectively. Stirrers have the same geometry – a shaft with a spiral-shaped blade on its bottom part. All stirrers are pumping up. The plunger in the spout area has 4 horizontal S-shaped blades and it also rotates.

Tracing of massless points in mathematical model was chosen to compare results of physical and mathematical models. A spherical cluster of massless points in the mathematical model approximates the sphere tracer injected into the model liquid in the physical model (inhomogeneities are modeled as tracers (tracering liquid) having the same composition as model liquid). A typical cluster has a diameter in the range 0.25 mm – 1.5 mm and uses between 100 and 1000 massless points. A single cluster put into the calculated velocity field creates a curve, which represents the movement of the tracer. Whenever the line becomes thicker, it means that the tracer is diffusing and clearing off – it stops to be observable in the physical model, although it is still visible in the mathematical model. This happens especially in the vicinity of stirrers.

Tracers are injected in two plane cuts, whose distance from the axis of the plunger is 585 mm and 240 mm respectively. The two plane cuts differ only in the horizontal distance of points 2 and 4; 50 mm is valid for the first cut and 40 mm holds for the second cut.

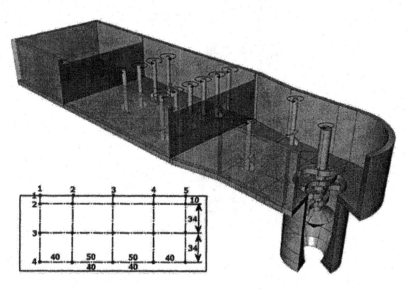

Figure 1: Model geometry with positions of injection points

Glass and Model Liquid Properties, Model Parameters
A solution of glucose syrup and glycerin is used as the model liquid. Its viscosity is related to glass viscosity through the selected scale and the Galileo

number. Based on the selected scale, following properties and parameters were used in the physical model to simulate the real forehearth:

Table I. Comparison of thermophysical properties and parameters

Property [unit] / parameter	Glass melt	Model liquid
Density [kg/cm^3]	2770	1425
Dynamic viscosity [N.s/m^2]	997.2	66.975
Stirrers stirring speed [RPM]	17.5-17.5-12.5	35-35-25
Plunger stirring speed [RPM]	9	18
Pull rate [Metric Tons Per Day]	80	1.29

CALCULATION RESULTS AND COMPARISON
Flow Under the Level

Tracers 1-2 and 5-2 placed 10 mm under the level exhibit a behavior similar to the one of the level tracers - the diffusion of tracers already starts when they are passing the line 1. Tracers are then affected by the stirring effect of the line 2 and go to the lower part of the line 2, where they wear off.

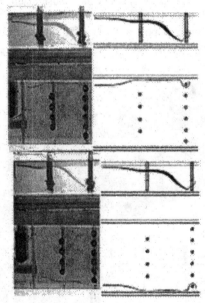

Figure 2: Tracers injected underneath the level

Flow Through Lines of Stirrers

Two tracers were run through the line 1, namely 1-3 and 4-3. Both of them start in the middle of the model liquid height and are affected by the action of stirrers' blades. The general trend in both horizontal and vertical directions calculated by the mathematical model corresponds to the physical model one, although there are differences in the behavior of tracers. The upward action of stirrers predicted by the mathematical model is stronger than in the physical model, which does not create sufficiently strong flow in the horizontal direction ("bundle" or "pack" around stirrers) and tracers get faster directly to the volume with the influence of stirrers' blades. The effect of pumping the model liquid upwards is thus applied on tracers sooner than in the physical model.

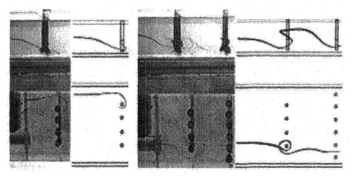

Figure 3: Interaction of tracers with stirrers

Horizontal Tracers

Horizontal tracers are used to visualize flow on or slightly under the model liquid level. They are injected across the whole width of the model to see the influence of stirrers on the flow on the level. Because horizontal tracers are wearing off quite quickly, they were highlighted in the pictures from the physical model. The right-hand part of comparison figure contains also horizontal tracers injected under the level to show the influence of stirring in the line 1.

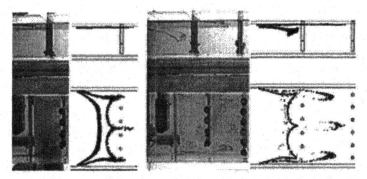

Figure 4: Horizontal tracers under the model liquid level

Vertical Tracers

Two types of vertical tracers were used – several vertical tracers in different position across the width of the model and sequential tracers in the longitudinal axis of the model. Both ways of using vertical tracers show a parabolic velocity profile before the tracers get close to the line 1 and begin to be influenced by it. The parabolic profile in the direction across the width of the model is fully parabolic, while the well-known half-parabolic profile with the peak on the model liquid level is formed in the vertical direction. As they get close enough to the line 1, tracers are accelerated towards the bottom part of its stirrers and are mixed.

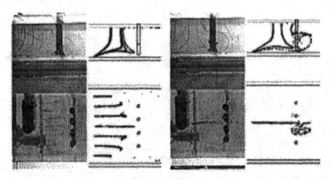

Figure 5: Vertical tracers and their interaction with stirrers

CONCLUSION

The upper third of the glass depth is running into spout without mixing. The pumping action of the stirrers pushes tracer down under the stirrers and consequently pumps it up. The tracer splits into streamlines and continues flowing to the second line of stirrers. The tracer near a stirrer almost wears off, but the color of the tracer is getting darker above the stirrers.

Based on the results from the model test, modifications of stirrer sets were recommended in order to improve homogenization performance of the system.

The mathematical model, although it uses a simplified approach for the geometry and properties of stirrers, is able to calculate flow patterns and predict the basic trends resulting from the action of stirrers. Differences in the areas of stirrers' influence do not affect the general behavior of the whole system in the sense of pumping effects and time needed by tracers to through selected parts of the model.

Future work on the mathematical model is going to be focused on taking detailed blade geometry into account, which will require using the time-transient model, and on simulating up and down movement of the plunger, which will help to improve the agreement between the results from both physical and mathematical model.

REFERENCES

[1] S.P. Vanka, "Block-Implicit Multigrid Solution of Navier-Stokes Equations in Primitive Variables", *Journal of Computational Physics*, **65** [1] 138-158 (1986)

[2] S.V. Patankar, *Numerical Heat Transfer and Fluid Flow*, Hemisphere Publishing Corporation, Washington, 1980

Advances in Fusion and Processing of Glass III

MATHEMATICAL MODELING OF BATCH MELTING IN A GLASS FURNACE

Petr Schill, Miroslav Trochta
Glass Service, Inc.
Rokytnice 60
75501 Vsetin, Czech Republic
Tel.: +420 571 498511, e-mail: schill@gsl.cz

ABSTRACT

The presentation deals with description of the 3-dimensional model of batch melting which is part of the complex software GS-GFM (Glass Service – Glass Furnace Model). The batch model involves batch to glass conversion, gas release and water evaporation based on reaction kinetics via degree of conversion. The degree of conversion is estimated by DTG and DTA measurements. All the batch properties are functions of degree of conversion and of temperature. The gases released from batch (mainly CO_2, H_2O) enter the combustion space. The influence of combustion space quantities (temperature, gas composition, flow, etc.) by batch gases is of significant meaning and is being modeled. The model examples of batch layer melting and of combustion space field modification by batch gases are presented.

THE NEED OF BATCH MODEL
Glass Technology
- Improve knowledge of heat consumption in glass furnace
- Batch melting process influences the flow and temperature distribution not only near batch blanket but inside the whole furnace
- Impact of gases released from batch to the processes in combustion space

Mathematical modeling
- Extremely complex problem because of approximation of micro-phenomena (species chemical reactions, particle dissolution) by macroscopic model means   (limited grid size)
- Various kinds of batch melting models yield different results
- Lack of 3-D batch melting models in common CFD codes

THEORETICAL PRINCIPLES

The conversion from batch into glass is considered as a continuous process described by macroscopic variables of gas degree of conversion α_g and energy degree of conversion α_e. The gas degree of conversion α_g was selected a as representative one. It can be easily determined by transient TG measurements:

$$\alpha_g(t) = R_g(t)/R_{g,TOT} \tag{1}$$

where t is time, R_g is mass of gas released. The formula (1) can be converted to the position variable expression:

$$\alpha_g(x) = R_g(x)/R_{g,TOT} \tag{2}$$

where x is path from input to the actual position.
The local heat sink Q_b [W/m3] is given by kinetic formula

$$Q_b = \frac{d\alpha}{dt}.H_b.\rho \tag{3}$$

$$H_b = (1 - m_c)[(1 - m_w)H_o + m_w H_w] \tag{4}$$

where H_b, H_o, H_w [J/kg] represent the total reaction and conversion heat, resp. reaction heat of pure batch, resp. vaporization heat of water and m_c, m_w are mass fractions of cullet, resp. moister water.

The rate of degree of conversion is expressed by product of two variables

$$\dot{\alpha} = \frac{d\alpha}{dt} = \frac{d\alpha}{dT}.\frac{dT}{dt} = f(T).r \tag{5}$$

where temperature derivative of α, the function $f(T)$ can be determined experimentally from TG measurement and the rate of batch heating r is calculated according to the local batch layer heating from top and from below.

EXPERIMENTS

TG measurements were done for a set of heating rates. The measured values were converted from time- to temperature- dependency. Resulting courses of degree of conversion $\alpha(T)$ and of the temperature derivative function $\alpha`(T) = f(T)$ for white glass-batch are shown on the graph:

446

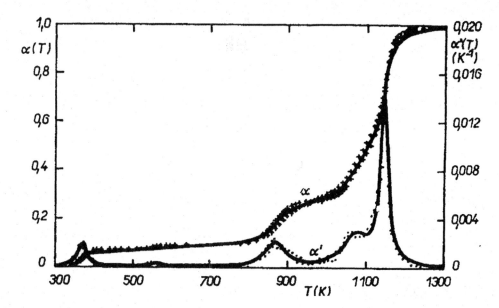

Figure 1. Degree of conversion $\alpha(T)$ and the temperature derivative function $\alpha`(T) = f(T)$ for white glass-batch

MATHEMATICAL MODEL

· The three dimensional batch melting area is modeled by solving the same momentum (Navier-Stokes)-, continuity- and energy- equations as glass-area. The finite difference method with control volumes [2] and Symmetric Coupled Gauss-Seidel (SCGS) [1] procedure are used for numerical solution on non-uniform staggered grid system. Following specific features are applied for batch area:

- The batch layer with free edges is floating on the glass
- The batch area shape (in 3 dimensions) is automatically calculated with respect to the batch existence-stage given by T- and α- limits. Each control volume is marked to distinguish the fully melted and non-melted stage.
- A batch-glass interface is considered and new glass exit velocities are modified by viscosity
- The buoyancy term is neglected
- The batch properties are functions of T and α

The heat sink term in energy equation is given by the above formula (3) for Q_b by using the formulas (4), (5):

$$Q_b = \frac{d\alpha}{dt}.H_b.\rho = f(T).r.\{(1-m_c)[(1-m_w)H_o + m_w H_w\}.\rho \qquad (6)$$

The values of H_o (approx 600 kJ/kg) and of $\alpha(T)$, $f(T)$ have to be measured and the values of H_w, m_c, m_w are given by operation conditions. The function of $f(T)$ is formulated by superposition of particular processes k each of them is expressed by the Lorentz-Cauchy formula:

$$f(T)_k = \frac{f_{km}}{1 + 4(\frac{T - T_{km}}{\Delta T_k})^2}$$

(7)

where the process (peak) characteristics follows:
fkm=max.height
Tk = position
ΔTk=half-width

The mass flow and temperatures of gases released due to chemical reactions and water evaporation are calculated in each control volume and they are transferred to the combustion space model via coupling procedure (which is normally used for energy coupling of glass- and combustion-space). These quantities play important role in combustion model calculation.

MODEL CALCULATION RESULTS
Examples of batch calculation in two glass furnaces are given: one furnace is equipped with two front chargers and the second furnace with one-side batch charging. The figures 2,3,4,5,7,8 show very sharp temperature gradient in horizontal and vertical directions during batch layer melting. This rapid temperature change on small distance brings difficulties by numerical solving the model equations. The CFD code Glass Service – Glass Furnace Model (GS-GFM) involves procedures as SCGS [1] method, exponential approximation for large Pe, Re numbers [2] and quantities-smoothing over grid cells which are able to overcome these difficulties.

Front charging

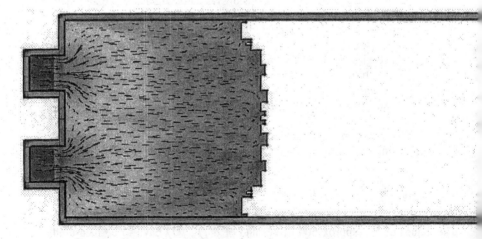

Figure 2. Temperature and flow distribution on batch surface

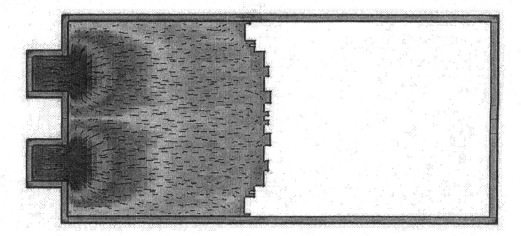

Figure 3. Temperature and flow distribution in horizontal cut 13 mm below batch surface

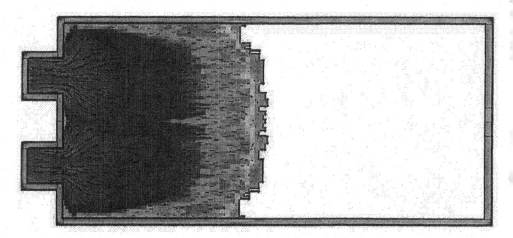

Figure 4. Temperature and flow distribution in horizontal cut 38 mm below batch surface

Figure 5. Temperature and flow distribution in longitudinal vertical cut inside the batch layer

Temperature [°C]
0   75 150 225 300 375 450 525 600 675 750 825 900 975 1050 1125 1200 1275 1350 1425

Figure 6. Temperature scale for Fig.2 – 5.

One-side charging

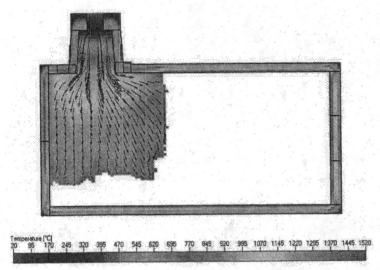

Temperature [°C]
20 95 170 245 320 395 470 545 620 695 770 845 920 995 1070 1145 1220 1295 1370 1445 1520

Figure 7. Temperature and flow distribution on batch surface

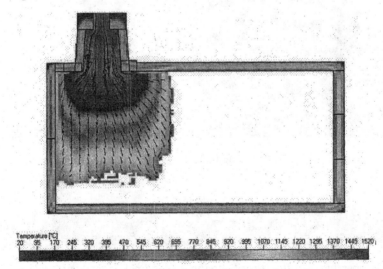

Temperature [°C]
20 95 170 245 320 395 470 545 620 695 770 845 920 995 1070 1145 1220 1295 1370 1445 1520

Figure 8. Temperature and flow distribution in horizontal cut in the middle of the batch layer

CONCLUSIONS

The three-dimensional mathematical model of batch melting is fully incorporated into standard commercial CFD code Glass Service – Glass Furnace Model (GS-GFM). It is using for modeling of industrial furnaces, as for example Container-, Table Ware-, TV-, Float-, Fiber-Tanks. It can work in all kinds of batch feeding as in horizontal front- or side-chargers and in all-electric furnaces with vertical melting. The heat consumed by batch melting is integrated into the total heat balance. The calculation of gases released from batch is important for proper coupling with combustion space model.

REFERENCES

[1] S.P. Vanka, "Block-Implicit Multigrid Solution of Navier-Stokes Equations in Primitive Variables", Journal of Computational Physics, 65 [1] 138-158 (1986)

[2] S.V. Patankar, Numerical Heat Transfer and Fluid Flow, Hemisphere Publishing Corporation, Washington, 1980

Secondary Processing

GLASS SHAPES OUR FUTURE

Jorma Vitkala
Tamglass Ltd. Oy
Vehmaistenkatu 5
33730 Tampere, Finland

ABSTRACT
The use of bent glass is expanding as architectural and automotive designers favour new and softer shapes. For traditional and competitive reasons the automotive industry has been a path-breaker in the adoption of designs that are not only new but also highly cost-efficient. This in turn is a direct reflection from the industrialised mass-production nature of the business. The same options for advanced novelties are open to architectural applications as well, particularly as the glass industry has taken up new and revolutionary manufacturing technologies. The use of the latest findings in convection technology in particular has permitted the creation of flashier and larger glass sizes utilising all the advantages that tempered glass brings. Larger and thinner surfaces that effectively prevent thermal breakage and wind loads are some of the new options. At the same time glass can be used not only as a curtain wall on contemporary buildings, but also made to be part of the rigid structure of the building itself.

INTRODUCTION
Bent glass has been used ever since industrial glass manufacturing started. In the first part of the 20th century bent glass was quite widely used, but this state of affairs changed in the 1960s and 1970s. The use of bent glass was reduced and the reason for that was the strong increase of labour costs. The type of bent glass used was quite labour-intensive as the technologies of those days made use of annealed glass for bent products. This type of glass also fell short of the requirements placed on safety glass. Even though things have changed, annealed glass is still used to some extent particularly in façade structures.

To meet the safety glass requirements laminated glass was put to use over the years and this type of glass was well up to all the requirements of various safety specifications. Laminated, bent glass has been successfully applied for several decades now and it has enabled structures that are architecturally impressive. It is easy to predict that this situation will prevail for a number of years to come.

The use of bent and tempered glass has been relatively modest during the past couple of decades. This has mainly been due to shortcomings in manufacturing which have made the production costs for bent, tempered glass much too high for successful commercial application. In the current millennium, we are already a couple of years into that, new production methods have opened a host of new possibilities for architecture and construction. These new options also raise intriguing challenges for the designers and builders of today and tomorrow's new solutions.

Above all the new technology allows the adoption of all the advantages brought by tempered glass, such as large sizes, ability to take on powerful wind loads, avoidance of thermal breakage etc.

AUTOMOTIVE INDUSTRY THE PATH BREAKER

Automotive industries have been innovative path breakers for design and shapes. In rising to the challenges of new concepts in the tough competitive climate of the automotive industry, they have also been in the forefront of raising new challenges for the glass industry. As a key supplier the glass industry has naturally had a vested interest in trying to meet the new requirements of car designers as effectively and quickly as possible. The requirements put on glazing in contemporary cars have become continuously more demanding and complex. At times meeting these requirements has proven to be difficult and sometimes mistakes have been made, too. But the increasing proportion of glass in cars goes to show, that the inspiration and challenge provided for the glass industry has brought up continuously newer and more cost-effective solutions.

The new technologies that have been built into state-of-the-art glass production lines have contributed to architectural glazing, as well as the glass processing machinery required for its manufacture. Industrial processes in today's competitive climate need to be more economical and more cost-effective than ever even if they, at the same time, produce solutions that are more demanding than ever.

LIMITLESS CHALLENGES

The recent car shows bear ample evidence of the role that glass plays in today's automotive solutions. It is also a remarkable fact that glass is used for other purposes than traditional windshield, side and back lite solutions. The new trends in the car industry put an emphasis on visibility, stylishness and transparency. The last of these three, transparency, is particularly marked and it clearly affected all concept cars especially as far as rooftop glazing was concerned.

Glass has been the subject of expanding dimensions in other aspects of the automotive industry, too. The role of glass as a rigid element in the car body has been utilised for decades. The new solutions and the increased proportion of glazing now pushes forward new solutions for items such as the control of solar energy flow into the car. Breakthrough solutions are around the corner. Electro chromic glass, for the control of energy flows, is already part of today's technologies in the flat glass and construction industries, but new solutions for bent glass are on their way. Solar panels and transparent solar panels on the rooftops of cars are used in the ventilation systems of automobiles at times when the cars are parked as well. The result is improved energy-efficiency and comfort.

The new trends in the car industry put an emphasis on transparency.

The use of glass in automotive applications has seen new functions such as the protection of the structural supports by using ceramic paints and screen-printing techniques. In this way the automotive industry has been able to effectively use the excellent surface characteristics of glass, a material that is reasonably advantageous in comparison with other materials, as has been witnessed by designers and construction teams. The good properties of glass expand the role of the material and the advantages more than offset the constraints related to the weight of the glass. Glass is also a good-as-such surface material with ability to protect and endure at the same time, as it is highly cost-efficient.[1]

The glass functions such as the protection of the structural supports by using ceramic paints and screen-printing techniques

The weight of glass, sometimes seen as a constraint in automotive industry, is, however, not a problem in architecture. That is also why we need to look at glass applications with a maximum of creativity to find new ways in which this fantastic material can be put to effective use.

The excellent surface characteristics of glass expand the role of it as it is highly cost-efficient material.

GPD FOR TECHNOLOGY TRANSFER

The whole idea behind the Glass Processing Days Conference is the sharing of information, efficient transfer of know-how between the different application areas of glass. The leading thought is to provide advantages for the entire industry by information sharing. At this point I would like to repeat the view held by James Carpenter, a very well known architect and innovative user of glass and a highly respected lecturer on international forums. His message is that everything at the GPD and especially the automotive applications, intrigue him precisely because they contain so many new lessons and ideas for all of us.

THE SMOOTH NEW FORM OF ARCHITECTURE

It is quite evident that we are moving towards an increasing use of soft and smooth new shapes in buildings. This development will surely be enhanced by the possibilities for cost-effective production that are brought forward by today's tempering technology. The new technology allows the production of high-quality specialty glass sizes in short series, perhaps only a few products per run. A change of the curvature of the glass no longer becomes mainly a cost item, since the bending characteristics can be set separately for each batch entering the tempering furnaces thanks to new control systems now available.

As a consequence the sizes of bent, tempered glass can also vary and grow.

It is now possible to produce larger uniform glass surfaces than before. A direct advantage is that tempered glass, even large-sized glass, can successfully meet the required mechanical, thermal and wind loads. Today's manufacturing technology permits the production of 3.3 metres x 7.0 metre flat glass units. In Europe the most popular size has become 2.8 metres x 6.0 metres.[2]

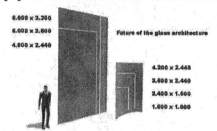

In the case of bent glass the utilization of the benefits from larger glass sizes are only ahead. It is remarkable from a production point of view that short and long series, small and large sizes, can be produced simultaneously on the same machine. The bent glass size may go up to 2.4 metres x 4.4 metres with a minimum curvature of 3.0 metres.

Table 1. Effects of the strength against wind load of various types of glass with various fixing conditions

Glass type		Support condition	Maximum size[(1)]
6 mm	Float Glass[(2)]		4.6 m²
6.4 mm	annealed laminated glass[(3)]	4 edges framed	4.6 m²
6 mm	toughened glass[(4)]		6.8 m²
10 mm	Float Glass[(2)]	Top and bottom edges framed, vertical edges free or butt jointed to adjacent panes	1.47 m height
10.8 mm	annealed laminated glass[(3)]		1.47 m height
10 mm	toughened glass[(4)]		2.72 m height[(5)]
12 mm	Float Glass[(2)]	Bolted at 4 points near the corner	0 m²[(6)]
12.8 mm	annealed laminated glass[(3)]		0 m²[(6)]
12 mm	toughened glass[(4)]		6.6 m²[(7)]
Note 1.	Based on current UK design stresses for a vertically glazed rectangular pane of aspect ratio 1.5 (i.e. height/width = 1.5)		
Note 2.	To EN 572-1		
Note 3.	To EN 12543		
Note 4.	To EN 12150		
Note 5.	The building user may find that the glass appears too flexible if the span exceeds 1.87 m.		
Note 6.	Holes in annealed glass have no usable strength other than to hold the glass in place under its own weight (as in some mirror or furniture applications). They cannot be designed against applied loads.		
Note 7.	The building user may find that the glass appears too flexible if the area exceeds 4.1 m².		

This information assumes the glass is subject only to a wind load 1000 N/m² and to no other loads or design criteria.[3]

NEW CONVECTION TEMPERING TECHNOLOGY

Today's tempering technology supports the production of tempered glass in a more efficient and better quality dimension than ever. It also opens up entirely new ways of combining a traditional radiation furnace with the best features of a convection process.

A radiation furnace has so far been the most reliable and cost-effective tempering furnace around. This has been quite natural, since the high emissivity of glass supports radiation heating. Convection technology, which became one of the best keys of the 1990s to high-quality production, has been significantly developed over the years. The development of numerical modelling has been in a key position in the development of today's convection technology. As shown in the adjacent picture, the development of convection technology received a boost by the introduction of the first generation of Low-E glass on the market in the early 90s.

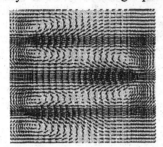

Results from numerical modelling - convective heat transfer

It became necessary to look for new options in heating technology for glass with an emissivity of less than 0.05. The demand for large sizes acted in the same direction.[4]

The latest development in this field is the pro-convection™ furnace. This furnace embodies all the know-how that has been accumulated during a long development process. It revolutionizes the process of convection in a new generation method. No longer are expensive and maintenance-intensive blowers required in the furnace. Convection is now generated by compressed air pressure that also allows the profiling and focusing of convection to the area where it is needed. In flat glass the aim is always to achieve as even a temperature as possible. This means that profiling and convection is required in the centre of the glass sheet. In the case of bent glass the profiling and focusing may concern other sectors of the glass, and heat needs to be focused to a given spot for the best overall quality result of the process.

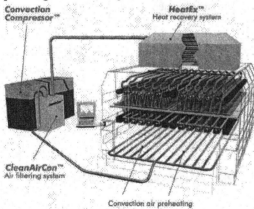

Convection Compressor™

Heatfx™ Heat recovery system

CleanAirCon™ Air filtering system

Convection air preheating

Filtering and pre-heating of the air flow and heat exchange provide a better overall energy economy than would be possible even in a gas furnace.

FUTURE TRENDS

We all wish we could see as clearly into the future as we can see through glass. History and past trends help us in creating sustainable visions for what might become important for the glass business in the future. Personally I have been enriched through my role as Chairman of the Glass Processing Days Conference.

This position has given me an opportunity to follow the development of the glass business very closely and in continuous close co-operation with leading professionals of the business.

Against this background I have listed some new thoughts that I think will influence glass processing and the future of the glass industry.

Glass for Sustainable Development

Sustainable development is an issue that raises new challenges for the glass industry particularly in view of the concern we all share over the future of our environment and the possibilities for re-cycling of different materials. We are also concerned about the adequacy of energy sources, their availability and price. All these factors influence the future of the glass business that is eager to promote a sustainable global development.

The availability and price of energy will affect the development of the glass business most of all. This can be put against the background that around 50% of the energy economy of the world deals with the issues of heating and cooling, illumination and the ventilation of buildings. The other half of the economy circles around transportation systems and industry as well as other users.[5]

Knowing this it is pretty obvious where the biggest savings can be made. Automotive industry, and through that, traffic and industry, are not the major consumers of energy as some people sometimes imagine.

The Effect of Legislation

Glass in buildings is where the major challenges of the future are. Changing legislation and regulations to follow their times or in a manner where they boost a desirable development is a key driver in the future. To put it very plainly: It is vital to follow the path of development and come up with new regulations whenever needed.

The energy regulations introduced in Germany in 1995 and their effect on the use of low emissivity glass used for energy efficient windows are an excellent case-in-point. Today practically all insulated glass units in Germany make use of low emissivity glass.

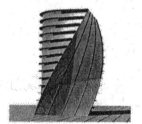  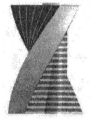

The use of daylighting and efficiency of energy use in heating, cooling and ventilation boost today's demand for coated glass. In the short run the use of different coatings will continue to grow. The coatings developed for different climate regions already improve energy efficiency considerably.

The sizes of architectural glasses have grown strongly and their shapes have changed and become very advanced in recent years. This is partly due to the idea of making better use of daylighting, but also to the general trend towards more transparency and overall openness in the use of glass in architecture. As you can see from the attached pictures, a new trend for the shapes of buildings is about to be born, a trend that favours versatility and an abundance of curved surfaces. This places new requirements on glass producers, processors, and glaziers.

 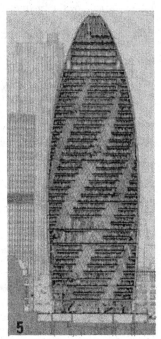

The popularity of double-glazed facades should not be overlooked either. This structure makes major contributions towards better energy efficiency and better control of natural ventilation and energy flows.

SUMMARY

The advent of new glass products increases the options for designers and also boosts the opportunities for glass processors to look for better value-added products.

The use of coated glass on buildings and automobiles will increase in the future, and the sizes of individual and combined glasses will grow. New architecture and designs place demands on new shapes for building and automobile glazing. There is a constant need to respond to changing and increasing requirements. These will not only require control of energy flows but also new properties, such as self-cleaning and new options for communications.

Glass in a communications role in both buildings and automotive applications opens up new and highly interesting application areas. If we add the element of speech identification and the engagement of different sensors on the market and under development, we can vision a window with which one can communicate. A deeper and longer relationship probably should not be recommend, even if glass is one of the world's most fascinating and demanding materials.

Implementing the new visions for the glass industry requires lots of joint action and the development of new processes.

There is a need for the glass industry itself to change as it faces growing competition from other materials. Enhancing the cause of the glass industry is in our joint interest. In doing this we can only wish that our future will turn out to be as transparent, safe and clear as safety glass itself.

REFERENCES

[1]Jorma Vitkala, Exciting automotive glazing trends at the Frankfurt Motorshow. A report by Jorma Vitkala, Chairman, Glass Processing Days, www.glassfiles.com, 2001

[2]Jorma Vitkala, Tempering Quality and the Case of Large Glass Sizes, GPD Proceedings 2001

[3]John Colvin, Hansen Glass Processing Ltd.

[4]Jorma Vitkala, Numerical Modeling - Market or Research Driven, Proceedings of Modelling of Glass Forming and Tempering 2002

[5]Paolo Scaroni, Coatings and e-Commerce - Glass Industry Growth Drivers for the 21st Century, GPD Proceedings 2001

SPONTANEOUS CRACKING OF THERMALLY TOUGHENED SAFETY GLASS: ACTUAL STATE OF DISCUSSION IN RESEARCH AND PRACTICE

Andreas Kasper
Glasstr. 1
D-52134
Herzogenrath, Germany

AUTHOR'S FOREWORD

Good comprehension of the present paper demands pre-information about the reasons of spontaneous breakage of thermally toughened glass. The reader finds several literature sources included in the text, many of them published by the author of the present paper. They are meant to deliver further background information about this complicated phenomenon. In the case of doubts or wish for further information or discussion, please mail to:Andreas.Kasper@Saint-Gobain.com. All kinds of comments and suggestions are welcome.

INTRODUCTION

Toughened glass on buildings occasionally suffers from so-called spontaneous cracking. Often glass breakage is caused by inadequate assembly, accidental damage or vandalism. But in some cases a special kind of glass inclusions, so-called nickel sulfide stones (NiSS), have been proved to be at the origin of the breakage.

NiSS as glass defects have been described in many publications, beginning in the early nineteen-sixties with BALLANTYNE [1]. Since then, the nickel sulfide problem, i.e. the spontaneous cracking of thermally toughened safety glass sheets, is known as a quality and reputation problem of tempered glass [examples: 2]. Glass producers have tried to solve the problem by primary measures, but as they were not completely successful, the phenomenon still remains a very actual problem in view of the increasing number of all glass facades of many modern buildings.

Consequently, in Europe, TC129, a working group of CEN, has developed a new standard: EN 14179-1 "Heat soaked thermally toughened soda lime silicate safety glass - Part 1: Definition and description". Its draft was published in November 2000, and the standard is expected to be finalized by the end of 2003. As its name says, it proposes to the customers a new safe glass product, i.e. a tempered glass, treated with a sophisticated, proven heat soaking procedure to prevent spontaneous cracking. Its spirit is that during Heat Soak (HS) the glass temperature is kept at (290±10)°C during two hours. After this, all kinds of toughened glass sheets, independently from their size and thickness, are assumed to be safe on a building.

The HS process: What is it good for?

It seems that NiSS are inevitably introduced into the glass in a very small number during the glass melting process. Below 320°C they undergo an allotropic transformation, accompanied by a volume increase of 4%. Because of this specific mineralogical property NiSS can cause spontaneous failure of toughened glass. A summary of the "Fundamentals of Spontaneous Breakage Mechanism Caused by Nickel Sulfide" can be found in [3].

HS processing was invented in order to prevent thermally toughened glass from spontaneous cracking on buildings caused by NiSS. By heating up the glass, it destroys potentially defective glass sheets. Through temperature increase, the conversion of the NiSS from their α to β form is highly accelerated, as compared to room temperature. Defective sheets break therefore in the oven instead on a building's façade.

Not only NiSS, but also other defects causing "weak points" in the glass, such as big refractory stones or bubbles (> 500 μm), may singularly (< 10% of the breaks observed) cause breakage during HS process.

An estimation of the safety of HS processed glass, explaining the mathematical and statistical basis, was published recently [4]. It comes out that under application of worst case conditions for every step of the estimation, the residuary break risk is less than 1 break in 400 tons of carefully processed glass.

➢ An important annotation seems to be necessary here. In very seldom cases spontaneous breaks occur on a building although the glass is said to have been HS processed. Unfortunately it is impossible to objectively prove the application of HS, e.g. by chemical or physical analysis of the glass. If, additionally, in such a case the break rate is <u>much</u> higher than that estimated in [4], <u>and</u> the breaks begin after an incubation time of 1 to 2 years (see [3]), then, from the author's (scientific) point of view, it must be suspected that a part of the lot were not at all processed, possibly because of lack of time and urgency of delivery. If the reason for this was that producers actually had a problem with frequent breaks in the HS process, their action was double wrong: just the "bad glass lot" they were processing at that moment is predefined to cause trouble if not properly processed. On the other hand, as far as we know today, if a proper HS is applied, also "bad glass" does not cause spontaneous breakage on a building.

The HS process: Which conditions to apply?

Many different HST (Heat Soak Test) conditions were proposed and used in the past. The most important among them is certainly German DIN 18516, prescribing a furnace temperature (whatever this may be) of 280 ... 300°C, and 8 h of holding time at constant temperature. In no other country an official technical specification exists.

The DIN Standard was more or less carried over by many countries, but often modified in temperature (250 ... 360°C) and holding time (1 ... 12 h).

It was therefore necessary, in the context of CEN harmonization, to develop a new European Standard: EN14179-1, called "Heat Soak Tested Thermally Toughened Safety Glass", defining on this way a new product. On the one hand, it is established to homogenize the HST procedure in all European countries, and, what is much more important, on the other hand, in order to implement HS process conditions which guarantee a safe product with a very low level of residual break rate.

Sturdy industrial experiences are at the basis of the CEN-TC129/WG2 work concerning this standard [5]*: During 5 years, 1258 breaks in HST were recorded by microphone in a HS oven. More than 75% thereof occurred during the heating stage of the process. After 2 hours of holding time, a 98.5% break level was reached. This was estimated to be the safety level of the processed glass as well.

On the other hand, the producer in question reports that in 3,850,000 m² (= 67,000 t) of glass processed in his factory since 1969, no break due to NiS was observed after putting the glass in place on a building. This means that the practical safety of the HST processed glass is still much higher than the estimation from HST breakage (less than 1 break in 400 tons, see above).

HS processing was developed and taken very seriously at this producer of thermally toughened glass. In the time before 1995, a 12 hour's cycle was applied; since then, after technical improvement and modernization of his equipment, conditions are similar to those prescribed in EN14179-1.

Because of his convincing arguments, the working conditions of these basic industrial knowledge and experiences were copied into the standard. It designs a classical HS process, usually consisting of the 3 well-known main phases:
➤ To heat up the glass from room temperature to process temperature;
➤ To hold this temperature for a minimum time;
➤ To cool down again to room temperature.
CEN-TC129 / WG2 judged it to be sufficient to officially control the holding phase, and to basically entrust the other phases over to the producer's own safekeeping. In order to anticipate misinterpretations of the Standard's prescriptions, key parameters were constituted:

---

* In [5] also a detailed description of the breakage mechanism caused by critical NiS inclusions can be found.

- ➤ Spacers shall be put between the glass plates to guarantee a minimum distance. Heating air must be able to penetrate easily between the sheets.
- ➤ The glass has to be heated up to a minimum temperature of 280°C. This means that a limit glass temperature has to be reached in every point of the glass stack.
- ➤ A maximum temperature of 300°C shall not be stepped over during holding time.
- ➤ The start of the holding time is clearly defined by the moment when the last temperature recording point on the glass surface reaches the minimum process temperature.
- ➤ The minimum holding time is 2 hours.

The HS process: Control by Calibration

Generally, in the temperature range and glass arrangement in question, glass temperatures are difficult to measure. It is necessary to glue thermocouples on the surface of some sheets, in order to bring them into intense contact to the glass. Glue or special adhesive tapes, after a temperature treatment at up to 300°C, are difficult to remove without traces.

Therefore a direct measurement of the glass surface temperature is not applicative for every-day control of the HS process. That is the reason why EN14179-1 prescribes a calibration routine with fixed temperature measuring points on the glass, and with use of the spacers between the glasses later applied in practice. Glass temperatures are later on referred to the air temperature recorded simultaneously.

During the calibration runs with different loads of the oven, thermocouples are glued on at least 10 points of the glass surfaces, on different glass sheets of the glass stacks. Details can be found in [6].

None of these thermocouples shall be in direct contact with the heating atmosphere: then the atmosphere temperature is measured instead of the glass temperature. For example, one real mistake consists of gluing thermocouples on the edges of the glasses. One may think that in this way it is ensured that on no point of the glass over-temperature can be reached. This is certainly true, but it omits the technical necessity that the heating atmosphere must be hotter than the glass, otherwise there would not be a temperature gradient to cause a heat flux.

Following practical experience, to finally reach 280°C on every point of several tons of glass in an adequate time (!), requires a heating atmosphere temperature of about 320°C. Thermocouples on the glass edges will now measure the atmosphere temperature instead of the glass temperature because they are immediately exposed to the intensive incoming air flow, but the glass, in reality, does not reach this temperature. Therefore, the distance between glass edges and thermocouples is laid down by the standard to be at minimum 25 mm.

But even if the glass edges were over-heated to 320°C, another technical reason shows clearly that this is of no importance. The edges of thermally toughened glass are always surrounded by a compressive stress zone. NiS inclusions, in that zone, are definitively harmless because only if a tensile stress surrounds the stone it is able to cause a break. At the edge, its state, whether α or β, does not matter anyway.

The HS process: Heating rate

Some laboratory results ([6], [7]) indicate an influence of the heating rate on the efficiency of the HS process. This item is actually subject to research; no concrete results have been published up to date.

The HS process: Minimum Temperature

With decreasing temperature, the rate of the phase change of nickel sulfide decreases exponentially. We call this a kinetically controlled reaction.

Following ARRHENIUS' rule (which is in accordance with the results of our DSC measurements with NiS, see e.g. our papers [7]), the transformation rate divides by a factor of about 2 with a temperature decrease of about 20 K. This means that a HS, processed

➢ at (270 ± 10)°C, would need 4 hours,
➢ at (250 ± 10)°C, 8 hours,
➢ an so on in steps of 20 K.

to maintain the same effectiveness on the glass.

We have said above that the HS process prescribed by EN14179-1 bases on a "positive technical example": in practice no breaks occurred on buildings when the conditions were so. This means that the processed glass will be safe, but the argument is not necessarily reversible. It cannot be excluded that also other, weaker conditions produce a safe glass as well. For example in [8], some colleagues from Japan propose the application of a lower temperature (250°C) and a shorter time (15 min). Unfortunately we, in Europe, are not in the state to be allowed to experience this; it would be much too risky because the proof can solely be obtained by the presence or absence of breaks on the buildings, with a minimum time horizon of several years. If the application of a different condition would not lead to immediate success, this would be deadly for the future use of tempered glass in façades.

Consequently, attention has to be paid not to under-run the minimum temperature of 280°C during the entire holding time.

The HS process: Maximum Temperature

300°C was fixed as the maximum temperature in the holding phase because of two arguments.

1. De-tempering, at this temperature, is more or less negligible (<10% loss of superficial compressive stress), but has to be respected in borderline cases. Special attention has to be paid to glasses with very high tempering degree

in use because the de-tempering rate is proportional to the present stress (see below). Eventually, after HS processing, the glass has to be toughened for a second time.

2. The equilibrium temperature between α-NiS and β-NiS is known to be 379°C, but this is not always a sharp step. Following DSC analysis, the temperature of α-NiS back-formation on heating scarcely depends on the heating rate, but strongly on the composition of the inclusion. For example, pure NiS, when heated up slowly, shows a wide gap between the end of the α to β transformation and the beginning of the back-formation of the α phase. In contrast, $NiS_{1.02}$, heated up with a higher rate, already shows the beginning of the back-formation at about 320°C. (More information about structure and composition of NiSS: see [9]).

Therefore, to be on the safer side, EN14179-1 says that 300°C shall not be stepped over during the HS process, except eventually for a short time, at the end of the heating phase (max. 320°C), but not at all during the holding phase.

The HS process: Tempering degree

The break rate in HST depends on the tempering degree of the glass. Trials showed that there is a remarkable influence. At a higher tempering degree, an increasing number of breaks are recorded. Among the NiSS present anyway, an increasing number becomes critical. This is because a stone, due to its size and chemical composition, can only force a limited stress to the glass. Only if the sum of the stress caused by the inclusion, and the intrinsic stress of the glass, overstep the critical threshold stress, will a break occur (further information: see [10]).

Thicker glass often shows a higher tempering degree than thinner panes. With the same glass quality, their break rate in HST is therefore observed to be relatively higher.

In every case the safety level of the panes after the HS process is sufficient: the same high proportion of critical inclusions is eliminated, and the differences resulting from the variations in glass thickness are included in the basic calculation and statistics.

The HS process: Dimensions of Panes

Glass dimensions for cladding building applications vary over a wide range. Looking at the possible thickness' (6 to 15 mm) and dimensions (1 to 10 m²), a factor of more than 20 can be between the smallest and the largest panes of a HS oven's load.

In every-day practice, it is although excluded to apply different HS procedures for different glass sizes and building facade dimensions. A standard HS process must guarantee a sufficient safety for all of them.

All other influencing factors (tempering degree, HS conditions, ...) kept constant, the mass of a pane determines its break probability in the HS process. This is because the inclusions are evenly distributed all over the glass' volume. If the volume of a single glass pane increases, the risk to find an inclusion in this glass sheet increases proportionally to its volume.

On the other hand, to cover the surface of a given building, a higher number of smaller panes, or a lower number of bigger panes is needed, and the risk relativizes. The differences between the different dimensions decrease, and are solely due to their different thickness. Comparing the thicknesses mentioned above, the residual risk only varies with a factor of 2.5. This is also covered by the basic calculation and statistics.

CONCLUSIONS AND SUMMARY

If the prescriptions of EN14179-1 are fulfilled, "Heat Soak Tested Thermally Toughened Safety Glass" is safe, and fit for its use on a building's façade.

CEN - TC129 / WG2 has elaborated this standard on the basis of sturdy industrial experiences. Defective panes, carrying critical NiS inclusions inside, are destroyed with a high reliability of meanly 98.5%. Referring this figure to the mean break rate of glass panes in the HS process, a residual break rate of less than 1 break in 400 t of tested glass is estimated with worst-case suppositions.

As compared to this estimation from breaks in HS, practical safety on buildings is still very much higher. Data collection since 1969, including 3.8 km² of carefully processed glass from one supplier, did not show one break on a building.

Factors eventually influencing the safety of the HS processed glass are:
➢ The heating rate, but its influence is not exactly known yet.
➢ Minimum temperature must not fall under 280°C. ARRHENIUS' rule says that even with only 20 K less, for the same effect on the glass, the holding time would double.
➢ Maximum temperature must not exceed 300°C during holding time. Too strong de-tempering and an unknown higher break risk would be the consequences. It is though permitted to over-step the limit for a short time during the heating-up phase of the HS process.
➢ With increasing tempering degree, the break rate in HS process increases. This is because the breakage risk is enlarged to, e.g., smaller NiSS, so the percentage of critical stones increases.
➢ Increasing mass and volume of a single pane increases its individual break risk. On the other hand, for a given facade, only the glass thickness plays a role. A maximum factor of about 2.5 would the lie between the safety of thin (6 mm) or thick (15 mm) glass, respectively.

Adequate safety is obtained in every case because all of the imponderability listed above is included in the statistics forming the basis of EN14179-1.

REFERENCES

1   BALLANTYNE, E.R.: Fracture of toughened glass wall cladding, ICI house, Melbourne. Commonwealth Scientific And Industrial Research Organisation (CSIRO), Div. Build. Res. Melbourne (Australia). Report 06 1-5. 1961.

2   [a] WAGNER, R.: Inclusions de sulfure de nickel dans le verre (in French), Glastech. Ber 50 (1977) Nr.11, pp.296-300

[b] Flat glass panels 'explode' in Britain, American Glass Review, Nov. 1993, p.8 (author unknown)

[c] BARRY J.C.: A study of nickel sulphide stones in tempered glass, Ultramicroscopy 52, 1993, pp. 297-305

[d] POPOOLA O.O., COOPER J.J., KRIVEN W.M.: Microstructural Investigation of fracture-initiating Nickel Sulphide Inclusions in Glass, Ceram. Eng. Sci. Proc. 14 (1993) 3-4, pp.184-194,

[e] WALDRON, B.: NiS: is there a problem?, Glass, Nov. 1993, p.439-442

[f] FORD, T.J.: Spontaneous Glass Breakage, Glass Magazine, May 1998, pp. 92-95

[g] Nickel sulphide breakage, Glass Digest, march 1992, p.12 (author unknown)

[h] PAUL, U., AULICH, U.: Nach Glasregen am Lafayette: Baustadtrat stellt Ultimatum, Berliner Zeitung, Nr. 280 v. 01.12.1998, p.1

[i] BRUNGS, M.P., SUGENG, X.Y.: Some solutions to the nickel sulphide problem in toughened glass, Glass Tech., Vol. 36 (1995), n°4, pp.107-110

[j] WILLMOTT, T.: Nickel sulphide inclusions: Proving the 'myth' can be a reality, Glass and Glazing, Oct. 1996, pp.24-26

[k] MERKER, L.: Zum Verhalten des Nickelsulfids im Glas, Glastechn. Ber. 47 (1974) 6, pp.116-121

[l] EVANS, A.G, The role of inclusions in the fracture of ceramic materials, J. Mater. Science, 9(1974) pp.1145-1154

[m] HSIAO, C.C.: spontaneous fracture of tempered glass, Fracture 1977, Vol. 3, ICF4, Waterlow, Canada, June 19-24, 1977, pp.985-992

3   KASPER, A.: Fundamentals of spontaneous breakage mechanism caused by NiS Inclusions. Proceedings of the Glass Processing Days 2003 (ISBN 952-51-5910-2), pp.696-698, also available online at http://www.glassfiles.com/library/lib_article.html?id=751&subcat_id=48

4   KASPER, A., SERRUYS, F.: Estimation of the Safety of Toughened Glass after a Heat Soak Test, July 2002. Online at http://www.glassonweb.com/publications/safetyoftoughenedglass/index.php

5   KASPER, A.: New measurements of NiS transformation kinetics to better understand the HST breakage data. Proceedings of the Glass Processing Days 2001, ISBN 952-91-3526-2. Tampere, Finland, June 2001, pp.87-90. Conference speech, session 10. Online at http://www.glassfiles.com/library/lib_article.html?id=313

6   ELLUIN, JC., KASPER, A., SERRUYS, F.: The Heat Soak Test: it is not the standard but the calibration where it is all about. Proceedings of the

Glass Processing Days 2001, ISBN 952-91-3526-2. Tampere, Finland, June 2001, pp.711-714. Online at http://www.glassfiles.com/library/lib_article.html?id=480

7 [a] KASPER, A., BORDEAUX, F.: Nickel sulphide: New Results to Optimise the Heat Soak Test for Tempered Building Glasses. Glastech. Ber. Glass Sci. Technol. 73(2000)No.5 pp.130-142

[b] KASPER, A.: Nickel sulphide: Supplementary statistical data of the heat soak test. Glastech. Ber. Glass Sci. Technol. 73(2000)no.11 pp.356-360

8 SAKAI, C., KIKUTA, M.: Adapted Heat treatment for phase transformation of NiS inclusions in the heat strengthened and tempered glass. Proceedings of the 6th Glass Processing Days (ISBN 952-91-0885-0) (1999) pp.76-78; online at http://www.glassfiles.com/library/

9 KASPER, A., MOSCHEK, S., STADELMANN, H., ZEIHE, R.: Composition and Structure of NiS Inclusions in Float Glass, and their Impact on the Heat Soak Process. Proceedings of the Glass Processing Days 2003 (ISBN 952-51-5910-2), pp.692-695; on-line: http://www.glassfiles.com/library/lib_article.html?id=750&subcat_id=48

10 SWAIN, M.V.: A fracture mechanics description of the microcracking about NiS inclusions in glass, J. Non-Crystal. Solids 38 & 39 (1980) pp.451-460

SWAIN, M.V.: Nickel sulphide inclusions in glass: An example of microcracking induced by a volumetric expanding phase, J. Mat. Science 16 (1981) pp.151-158

KEYWORD AND AUTHOR INDEX

9 781574 981568